STRATEGIC CORPORATE

MANAGEMENT FOR

ENGINEERING

PAUL S. CHINOWSKY
School of Civil and Environmental Engineering
Georgia Institute of Technology

with
James E. Meredith
School of Civil and Environmental Engineering
Georgia Institute of Technology

New York   Oxford
OXFORD UNIVERSITY PRESS
2000

Oxford University Press

Oxford   New York
Athens   Auckland   Bangkok   Bogotá   Buenos Aires   Calcutta
Cape Town   Chennai   Dar es Salaam   Delhi   Florence   Hong Kong   Istanbul
Karachi   Kuala Lumpur   Madrid   Melbourne   Mexico City   Mumbai
Nairobi   Paris   São Paulo   Singapore   Taipei   Tokyo   Toronto   Warsaw

and associated companies in
Berlin   Ibadan

Published by Oxford University Press, Inc.
198 Madison Avenue, New York, New York 10016
http://www.oup-usa.org

Oxford is a registered trademark of Oxford University Press

Library of Congress Cataloging-in-Publication Data

Chinowsky, Paul.
    Strategic corporate management for engineering / Paul S.
Chinowsky, with James E. Meredith.
        p. cm.
    Includes bibliographical references.
    ISBN 0-19-512467-7
    1. Engineering firms—Management. 2. Strategic planning. I.
Meredith, James E. II. Title.
    TA190 .C49 2000
    624′.068—dc21
                                                99-16338
                                                CIP

Printing (last digit): 9 8 7 6 5 4 3 2 1

Printed in the United States of America
on acid-free paper

To Melissa for all your support,
and to Sydney for the future

CONTENTS

III EXTERNAL STRATEGIC MANAGEMENT ISSUES   243

PREFACE

Where Do You Want Your Company to Be in 5 Years?
Do You Know How to Get There?

The traditional philosophy of management in civil engineering, both in academia and in industry, places great emphasis on the ability to plan and execute projects. Preparing individuals to conduct these planning and execution tasks is seen as the primary role for universities and industry training programs. We see evidence of this philosophy in the actions of our industry's leading researchers, scholars, and executives. Consider the following observations compiled from literature, interviews with executives, professionals and their clients, and the personal experiences of the authors: (1) In the civil engineering and construction management literature, an overwhelming number of paper topics focus on the technical and managerial issues that can affect project budgets and schedules; (2) in graduate-level construction programs, the majority of core courses are designed around project management techniques and responsibilities, and very few offer more than one course in nontraditional areas such as company management or strategic industry analysis; (3) the research reports funded by the leading research organizations indicate that their support is primarily focused on improving the cost effectiveness of projects; and (4) in civil engineering organizations, aspiring executives are generally assigned to follow a familiar project-oriented career path—where employees are provided extensive education and experience in either field engineering positions or in entry-level office positions that focus on enhancing project-management skills.

These observations highlight that today's civil engineering industry, through sharing of research, teaching, and practice, has evolved on a project management model. Professors, researchers, and practitioners use the project management cornerstone as the industry's standard of performance. They use its terminology and theoretical contents to discuss company priorities, employee goals, and student competencies. In contrast, a similar emphasis on corporate management is noticeably lacking. Specifically, the analysis needed to solve diverse problems companies face as they struggle to create competitive and profitable organizations requires a distinct set of knowledge, understanding, and skills. For example, consider the deliberations involved when a company must decide whether to enter a new market. Among other issues, corporate personnel must assess how viable the move will be, determine what competencies the organization must develop, create action plans to achieve those competencies, and decide what investments in knowledge and new technology the company should make. Traditional project management concepts and principles must be augmented with strategic management concepts to assess completely the opportunities and risks associated with these types of decisions.

▓ The Project Management Tradition: Creating a Knowledge Gap

The focus on a project-management approach to civil engineering management requires us to introduce a new element to the classic management traditions within the industry. The reasons for a project management focus can be recited by rote by many long-time industry professionals:

"Civil engineering is extremely cyclical."

"Project success is a key to company success."

"Building trends are difficult to predict."

The evolution of these reasons is based in solid fact. The post–World War II building boom-and-bust cycles have seen companies go from riches to bankruptcy and back again more times than some wish to admit. Similarly, the need to manage projects with tight schedules and budgets has made and broken many company reputations. Finally, the failure to predict building trends correctly has led to as many company failures as it has led to spectacular successes. Given these factors, it is understandable how the anxiety and trepidation associated with trying to avoid becoming one of the many quiet, or one of the few spectacular, failures can start to permeate the industry and reinforce itself into the basis of civil engineering management practice.

In addition to these longstanding factors, several new factors have emerged over the past decade, which are reinforcing the project-management tradition. Reductions in profit margins are forcing companies to tighten cost controls to levels previously unforeseen. Competition from local, regional, national, and even international companies threatens every company in the industry. Changing educational demographics has resulted in a greater number of university-trained civil engineering students entering the workforce than ever before. The combination of these and other factors is forcing companies to examine projects to a greater level than previously seen. The need to conserve every dollar and every man-hour has become an end goal as many industry managers strive to stay afloat one project at a time.

Given the impact of these issues on the civil engineering industry, it is not surprising that civil engineering management education has closely evolved along a similar path. Specifically, the evolution of project management as the overriding focus of university programs has both followed and reinforced industry traditions as each succeeding generation of industry managers graduates and begins the traditional career path. By responding to traditional industry requirements for specific educational skills, university programs have slowly emerged as a mirror image of the industry itself, instilling in students that the planning and execution of a project is the key to professional success. At both undergraduate and graduate levels, courses focusing on the concepts of critical-path-method (CPM) management and project cost controls dominate the curricula of every civil engineering management program. Students have far less exposure to areas related to managing engineering and construction organizations, such as creating competitive corporate strategies, forecasting the impact of new technologies and enhancing client relationships.

When faced with a longstanding tradition, many people would ask why this tradition should be challenged. If the tradition requires adjustment, then it naturally should occur over time to correct itself. Unfortunately, the rate of this evolution is not occurring at a pace that is commensurate with rapidly evolving economic and management challenges. In contrast, the combination of industry and academic focus on project-level management is establishing a knowledge gap within civil engineering management. Civil engineering professionals are entering the industry with only a narrow, project-level base of knowledge sufficient for responsibilities associated with traditional entry-level and project-level positions. However, as these professionals ascend the career ladder, they reach senior and executive-level positions requiring appropriate levels of responsibilities. Rather than facing these responsibilities with a commensurate knowledge base, they are forced to address corporate-level management issues from a project-level perspective. This incompatibility between responsibility level and knowledge level represents the management knowledge gap in civil engineering.

Strategic Management: Breaking the Tradition

Identifying a management knowledge gap represents a first step in transitioning a company to new management focus areas. However, without clear and compelling reasons to move further, this first step may also be a last step. For civil engineering companies, emerging internal and external forces of change are rapidly providing these clear and compelling reasons. The need to move beyond a project-to-project focus to one that charts a long-term strategic plan cannot be ignored much longer. Specifically, it is time for civil engineering firms to ask themselves if they can truly answer the two questions posed at the beginning of this introduction, "Where do you want your company to be in 5 years?" and "Do you know how to get there?" Issues including the need to manage organizational change, forecast emerging technologies, and understand new employee issues require students and aspiring executives to understand a new set of business demands and challenges. Industrial sectors, including manufacturing, high technology, healthcare, and financial services, are accustomed to these demands and continue to produce a large supply of professionals who are knowledgeable of their issues. In contrast, the civil engineering industry has yet to take a similar approach and thus is ill suited to meet evolving corporate demands.

The impact of this knowledge deficiency is potentially very damaging to the industry. First, clients are becoming more sophisticated about modern business issues. Therefore, in their interfaces with industry professionals—whether in the field or in the board room—they will assess how well their problems are being understood. Service will be awarded to those who: (1) identify and understand their needs, and (2) can help them to achieve their business goals. Second, just as clients are becoming more sophisticated, now unforeseen competitors are as well. Clients are finding that competent firms from a diverse range of industries can actually meet their needs. For example, some of the country's largest consulting firms have carved niches in the construction industry, providing services to corporate clients. Unless the civil engineering industry begins to produce a larger supply of professionals who are knowledgeable of changing business demands, the industry will continue to face increasing competition from outside professions.

Given these threats, civil engineering firms must begin to adopt an additional management outlook focusing on the strategic placement and advancement of the civil engineering corporation. Although this new focus is unique to the civil engineering industry, it is solidly founded on both theory and practice. For several decades, management theorists have been introducing manufacturing, service, and technology companies to the concepts of strategic thinking and planning. Through a combination of numeric methods and personnel strategies, these industries have moved beyond the short-term planning traps. Subsequently, these industries focus upon strategic management issues that serve to answer the fundamental questions posed at the beginning of this introduction.

■ Bridging the Gap: A New Approach

The identification of the management knowledge gap as an industry issue is the first step in creating a solution for civil engineering managers. Building upon this step is the focus of this book. To provide managers and students with a single source from which to bridge the management knowledge gap, a step-by-step approach is presented to put into place the strategy required to augment the project management tradition with a new strategic management focus. Through a series of questions, each posed at the beginning of a chapter, the reader will have the opportunity systematically to analyze strategic management knowledge at both a personal and a company level. The questions will challenge long-held beliefs by advocating a broadening of the project management focus to include the adoption of a focus that encompasses broad and far-reaching company missions, visions, and goals. The questions will challenge managers to go beyond current strategic planning exercises and develop actual implementation details. However, in keeping with the focus on engineering and the need to incorporate strategic management principles into a corporate context, each section will expand upon the questions to provide specific methods, checklists, and examples of how to translate theoretical management concepts into tangible company plans. When completed, the answers to the questions will form the basis of a strategic management plan that can be customized for specific market niches, business focal points, or market opportunities.

The strategic questions approach developed for this book addresses a combination of perspectives. From an educator's perspective, the introduction of strategic management concepts for civil engineers requires a modular approach, amenable to the quarter or semester format on which higher learning institutions are organized. The questions provide a basis for these modules by dividing strategic management concepts into individual components. From a manager's perspective, the concepts must be presented in a manner that enables the reader to produce a tangible product. This requirement is met through the sequential nature of the questions, leading a manager through the strategic management process from beginning to end in an organized structure. Finally, from a consultant's perspective, the information must provide new opportunities for market expansion and innovative approaches to solving traditional industry issues. The questions respond to this need by providing a structured methodology from which a consultant can serve as an outside advisor to civil engineering companies. In summary, the question methodology is at once flexible for the reader while also thorough in its capability to lead a manager through the strategic management process.

To assist the reader in understanding the concepts and questions presented in each chapter, a series of cases are presented that highlight the application of a specific concept within the civil engineering industry. The use of cases brings the concepts introduced in each chapter out of the realm of theory into the factual world of implementation. Based on interviews with civil engineering companies from all facets of the industry, the case studies have been selected to illustrate how a concept can be used as a basis for success within a specific company. Each case study is a direct result of company executives answering the same questions as those presented in the book. Therefore, the results presented in the cases provide the reader with a set of examples that are attainable through the appropriate application of the question methodology technique. *A note to the reader:* The names of individuals in the case studies have been changed, but the company names and locations are authentic.

■ The Chapter Layout

The strategic question methodology is presented in the book over the course of ten chapters divided into three sections. As the following descriptions illustrate, the sections provide a step-by-step introduction to the strategic management process, internal strategic issues, and finally, the challenge of external forces on the strategic management process.

Part I: Preparing for the Strategic Management Task

Part I introduces two critical components of strategic management: the knowledge set required to perform strategic management tasks successfully and the strategic planning process that serves as the implementation mechanism for strategic management. Chapter 1 sets the foundation for the book by providing a historical perspective on the development of strategy and strategic management in the business environment. Chapter 2 builds upon this introduction by introducing the strategic management questions and providing the basic elements required for strategic management.

Part II: Strategic Planning: Internal Corporate Issues

In approaching the strategic management process, a fundamental division can be made between issues that are internal to a specific organization and issues that are industry focused. Part II introduces the reader to the internal company issues that must be addressed to implement a strategic management plan successfully. Chapter 3 introduces the foundational elements of visions, missions, and objectives. Chapter 4 builds upon the vision statements by introducing the reader to the concept of core competencies and the processes that can be used to determine the core business of a civil engineering firm. Chapter 5 builds upon these core competencies by focusing on the knowledge resources required to transfer a core competency into market opportunities effectively. Similarly, Chapter 6 introduces the importance of facilitating these resources by implementing a corporate education plan that formally turns a static civil engineering firm into a dynamic learning organization. Finally, Chapter 7 concludes the internal issues by introducing specific economic and fiscal concepts highlighting the interrelationships between the current national and global economy and the identification of strategic market opportunities.

Part III: Strategic Planning: External Corporate Issues

The completion of the strategic questions presented in Part II will provide the reader with a strong foundation on which to strengthen the strategic management capabilities of the civil engineering company. To build upon this foundation, and successfully strengthen the company's position in the civil engineering marketplace, Part III focuses upon the external industry issues that every company must face. Chapter 8 opens this section by providing several methods to assist managers in the analysis of new and existing market opportunities. By challenging long-held beliefs about clients, market forces, and risk, the chapter provides managers with the tools to evaluate current market objectives and project selection guidelines. Given this new perspective, Chapter 9 concludes the strategic question process by preparing the civil engineering firm to expand its market share by taking a competitive battle approach to the civil engineering marketplace. Finally, Chapter 10 provides the steps required to put the new corporate strategic plan successfully into action.

■ A Note to the Corporate Reader

The emergence of strategic management in the civil engineering industry has been slow in developing due to strong project management traditions. However, these traditions are now being broken at an increasingly rapid pace. The roles of economics, clients, technology, and employees all need to be reevaluated as companies move away from a project focus to a long-term strategic focus. The questions in this book are intended to assist managers in the process of moving their company through the strategic management process. However, the strategic management process is only half the battle. To move a company successfully to the next level of strategic planning, it is critical to stop and focus upon the second question introduced at the beginning of this section, "Do you know how to get where you want to be in 5 years?" While it is admirable to write a strategic plan, the plan does absolutely no good unless it is implemented through thoughtful and decisive action. Thus, while the question methodology provides a basis for establishing a strategic plan, it is the reader's responsibility to transfer this plan from a paper exercise to a successful corporate action plan.

■ A Note to the Student Reader

The selection of courses that diverge from traditional project management is always a tough decision. The pressure to take courses that reinforce entry-level skills can be far greater than most people want to fight. The students that took the initial strategic management courses that served as the basis for this book faced the very same pressures. Their enthusiasm and dedication to expanding the initial offerings into formal curriculum offerings demonstrated the need to bring this knowledge to a greater student audience. By opening this book you are following the initial steps of these students and taking a first step toward breaking from the project-management tradition and moving forward toward the obtainment of advanced management skills. This book will provide you with an introduction to the concepts, methodologies, and theories behind the application of strategic management to the civil engineering industry. However, this introduction only builds a foundation; it will be up to you to build upon this foundation by bringing strategic management concepts into the industry and putting the concepts into action.

ACKNOWLEDGMENTS

The completion of any large effort always includes individuals who go beyond those names listed as the authors of the task. The same is true of this text. The final version of this text represents the culmination of an extended teaching and research effort. Originally conceived as an elective course for civil engineering students, the information in this book reflects an ever-increasing interest by the authors in the state of management in the civil engineering industry. Developed over several years of interviews and studies, the pursuit of this interest encompassed numerous individuals in every aspect of our professional and private lives. Although it would take pages to thank every individual, we would like to take the opportunity to thank several groups of people for their support over the course of this endeavor.

First, a special acknowledgement must be conveyed to the School of Civil and Environmental Engineering at Georgia Tech. The idea for a course in strategic management was not a mainstream idea in 1995. However, the encouragement to pursue this idea demonstrated a commitment to the long-term vision of civil engineering education that is needed to ensure the continued development of industry leaders. In particular, several individuals need to be singled out for their encouragement and support. Michael Meyer, Chair of the School of Civil and Environmental Engineering, must be acknowledged for his support of this project from its inception. Very simply, without Dr. Meyer's support, this project would not have been completed. His leadership and vision is reflected throughout this text. Additionally, several faculty members served as sounding boards throughout this project. Randy Guensler, Simon Washington, Karen Dixon, Nelson Baker, Jorge Vanegas, and Keith Molenaar must be acknowledged for their patience in participating in endless discussions on content, development, and direction. Their encouragement during the periods when it was doubted whether this project could be finished is appreciated more than they can imagine.

The second group of individuals who must be acknowledged by the authors is our graduate student assistants. Many hours in the library and on-line are reflected throughout this book. Once again, this book could not have been completed without their efforts. In particular, a special note of thanks must be extended to Kathy Elliott. Spending endless hours tracking down references and photocopying materials, Kathy demonstrated the effort and dedication that every faculty member hopes for in a graduate student. Many other student efforts are reflected in this book, including the thoughts of every student that has taken the Engineering Organizations course by the authors. An acknowledgment is extended to each of these students.

The third group of individuals who deserve a special acknowledgment is the industry leaders who participated in the many hours of interviews that led to the cases and background materials in this book. Many of these individuals took valuable time from their schedules to submit to interviews, follow-up questions, and inquiries of their business operations. Their interest in this topic and their willingness to participate were keys to the final success of this project. Several special acknowledgments must be extended to Clint Mays at Metric Constructors, Paul Little at Turner Construction, and Terry Kazmerzak at Parsons-Brinkerhoff, who provided enthusiasm and vision that was essential to push this idea from a concept to a finished product.

Finally, the greatest acknowledgment must be extended to our families. The focus that is required to finish a book is greater than any author can imagine prior to commencing a project. However, during the course of the project, it is often the family and not the author that keeps this focus in tact. When the author is ready to face any object other than the word processor, it is the family support and encouragement that keeps the project moving forward. When an author believes he cannot write another page, it is the family that provides the encouragement to keep going forward. And it is the family that listens to the minor victories that occur throughout the writing of a book, including the completion of a figure, the completion of a chapter, or even the completion of a difficult paragraph. Although the final version of this text carries the names of only two authors, it is in fact the effort of two extended families that is represented by every word in this book. To every member of these families and to every member previously acknowledged and to every person we may have overlooked, the authors say thank you; your support was and continues to be invaluable.

PART I

PREPARING FOR STRATEGIC MANAGEMENT

STRATEGIC MANAGEMENT 1

A Background

Strategy—the word evokes images of executives meeting in top-floor boardrooms setting company policies, or military commanders meeting in operations centers planning foreign campaigns. For civil engineers, strategic planning often builds on these preconceptions and is viewed as a task undertaken by a group of executives, separate from the organization members. Civil engineers are often made aware of this process only after an executive memo is received announcing the beginning of the annual review process. The memo details the annual request for each individual to set personal goals for the coming year as part of the corporate planning evaluation. Outside facilitators are often solicited to conduct daylong strategy sessions where goals are developed through departmental brainstorming and opportunity analysis segments. In each case, individuals are urged to reach new levels of motivation in an effort to push the company or agency to greater levels of success and recognition. Unfortunately, after building momentum through these sessions, organizations often return to business as usual shortly after the process has ended.

Are civil engineering organizations destined to repeat this annual process with the same results? Will employees continue to believe that the annual review process is an exercise with very little direct benefit to them? Can the effectiveness of the annual planning process be increased? The answers to these questions are not a definitive yes or no. Rather, answering these questions requires a new perspective on the traditional concepts of strategy, strategic planning, and the role of business practices in the civil engineering industry. The traditional role of project management as the central focus of day-to-day operations must be expanded to include strategic issues both inside and outside the organization. Opportunities originating from a strong economy, increased attention to civil infrastructure, and revolutionary technical advancements are building a foundation for sustained growth and increased returns for civil engineering organizations. This book provides organizations with a new viewpoint from which to pursue these opportunities aggressively and break the cycle of planning exercises that fail to achieve desired results.

The evidence for this positive outlook can be gathered from every aspect of daily life. The demographics of metropolitan centers are changing as revitalization converts

3

abandoned warehouses into upscale loft apartments. Cities such as Baltimore and Cleveland demonstrate that areas that were thought to be lost to the passage of time can be revitalized through infrastructure rehabilitation efforts. The bridges built in New England during the last part of the nineteenth century are receiving significant Federal and local attention as the structural stability of the spans is questioned. Environmental concerns such as water quality, wetlands preservation, soil contamination, and air-quality assurance are receiving international attention through the establishment of sustainability as a central engineering concept. These are only a few examples of emerging trends that offer opportunities to organization leaders. The organizations that plan for these trends have the possibility to establish the structures that will transform these trends into organizational success.

However, the selection of a strategic management methodology is not a trivial decision. Any visit to a bookstore will reveal shelves of books that claim to have the solution that will change a mediocre organization into a profit machine. The opportunities to enroll in one-day seminars that provide checklists to transform chaotic practices into industry-leading methodologies are increasing at a rate that appears exponential. With this confusing array of possibilities, organization leaders must decide whether a transition to strategic management principles is appropriate, and which approach provides the greatest opportunity for long-term success. In contrast to the claims of many of these options, there is no magic formula or checklist. The key to long-term success is the building of a strong management foundation that supports the development of strategic visions and implementation plans. These visions and plans are unique for every organization. The needs of a 10-person firm are fundamentally different from those of a 10,000-person organization. However, the concepts that form the strategic management foundation for each of these organizations are the same. Developed over centuries of military, industrial, and business conflict, the fundamentals of strategic management have been refined to address the long-term organization viability. The objective of the civil engineering organization is to learn from this history and mold these fundamentals into a framework that is appropriate for the organization.

■ The Strategic Management History

The history of strategy and strategic management covers a broad timeline from ancient Greece to the twenty-first century. Organizations, practitioners, and researchers from every sector of the professional world have focused on strategy as a primary topic at some point. This breadth of interest results in two distinct advantages for civil engineering organizations: First, an extended history exists from which civil engineering organizations can study and learn. While being the first organization to attempt any particular activity can bring notoriety and recognition, it also brings surprises as unexpected events occur from unexpected sources. In contrast, learning from established history provides a perspective on potential events, influences, and outcomes. Although these perspectives do not eliminate the chance for unexpected results, knowledge provides the greatest opportunity to prepare for as many contingencies as possible. With this guiding principle, civil engineering organizations have the opportunity to study the extensive strategic management history and apply relevant lessons, or disregard concepts that fall outside the operating interests of the industry.

The second advantage of this history is the ability to craft a strategic management model out of the successful components of existing models. Rather than moving through the traditional trial-and-error process, the modification of existing models provides an opportunity to analyze historic results for positive and negative trends. Given this analysis, methodologies can be adopted that focus exclusively on the needs of the civil engineering organization. While this is a general thought that can be applied to any field, it is of particular relevance to strategic management. Since history is destined to repeat itself if the lessons of history are not heeded, management mistakes will be repeated if knowledge of business successes and failures is not understood. With the financial consequences that could result from such mistakes, civil engineering organizations need to learn this history.

The Timeline

The evolution of strategic management concepts spans the course of centuries. From origins in military campaigns to modern multinational alliances, strategic concepts have emerged, transformed, and translated into various incarnations. Figure 1-1 illustrates this history in a series of time segments that reflect the fundamental changes occurring in organization management throughout specific points in history. The following sections interpret this timeline to provide a historic perspective on the influences that have led to the strategic management approach introduced in this book. The timeline provides civil engineers with the foundation to evaluate strategic management recommendations currently being offered by outside consultants, continuing education courses, business books, and internal business development groups. Each section contains a brief discussion of the primary drivers that influenced the time period as well as the lessons that emerged that are still relevant for the civil engineering manager.

Military Origins

Building a foundation for strategic management requires a broader perspective than the boundaries of the business domain. The concepts of strategy did not originate with

	Early History	Middle Ages	18th Century	Industrial Revolution	Post-War Era	Modern Era	21st Century
	336 B.C.	1066	1770	1880	1950	1980	2000
Focus	Military Conquest, Empires, Expansion of Civilization	European Expansion, Colonization, Religious Crusades	Independence of Colonies, Craft Industries, Democracy	Mechanization, Assembly Lines, Efficiency, Industrial Expansion	Corporate Expansion, Growth, Brand-Name Recognition	Globalization, Competitiveness, Communications, New Markets	Autonomy, Virtual Teams, Global Economies

Figure 1-1 The strategic timeline illustrates that strategic management concepts have been developing for over two millennia.

the names that are synonymous today with global business success. Long before the Carnegies, Mellons, Rockefellers, and Vanderbilts succeeded in the worlds of banking, transportation, and oil, military leaders introduced the world to the concepts of strategy. From tribal wars to global conflicts, the human race has continuously engaged in hostile conflict. Whether for reasons of territorial conquest, political ambition, perceived insults, or religious belief, the tribes, cities, and nations of the world have been engaged in conflict since the beginning of recorded history. With this history, numerous situations existed where combatants were faced with overwhelming numeric disadvantages, topographical difficulties, or logistic challenges. Responding to these challenges led to strategic and operational innovations that have evolved to influence today's military powers. However, these innovations are not limited to the world of military conflict and conquest. Rather, underlying concepts such as numerical advantage, opposition intelligence, elements of surprise, and supply-chain protection are equally applicable to the development of successful strategic management policies in the business domain (Quinn 1980).

The environment that spawned this strategic planning evolution was harsh. Returning to the Greek Empire in 336 BC, life was almost exclusively devoted to military service (Mitchell 1931). From the age of 20 until physical limitations no longer made it possible, men were devoted to military service. Women were required to address all domestic requirements of society, while slaves were employed for all labor-intensive tasks. With this focus on military service and conquest, the stage was set for large-scale expeditions and military campaigns. Emerging as a central figure on this stage was Alexander, King of Macedonia (Hart 1954). Focusing on the consolidation and expansion of the Greek Empire, Alexander consolidated Greece, conquered Persia, and moved into India before dying in Babylon at the age of 33. With this geographically distributed empire, Alexander was forced to battle numerically superior forces on foreign territory in each campaign. In this environment of apparently overwhelming odds, Alexander illustrated the ability of well-trained organizations to defeat superior opponents through strategic planning and execution. Illustrated in one of the famous strategic victories of the Ancient World, Alexander defeated the Persians at the Battle of Issus (Figs. 1-2a and 1-2b). Outnumbered almost 20 to 1, Alexander turned a numeric disadvantage into victory by focusing on the boundaries of the battlefield. Understanding that the enemy was trapped between the mountains and the sea, Alexander was able to use the Persians' numerical advantage against them. With the physical boundaries limiting the Persians' ability to move freely and take advantage of their superiority, Alexander split the Persian force and attacked its less fortified flanks. Using the natural topography to keep the Persians locked in place, he left the Persians no place to retreat or rally. In this situation, the better-trained Greeks defeated the poorly trained and led Persians. With this battle, Alexander passed down through the centuries a truism that superior numbers do not guarantee success. Rather, it is the combatant that has the ability to use both internal and external resources to his advantage in a calculated maneuver that ultimately emerges victorious.

Moving forward over a thousand years of history, past the Roman Empire and the Dark Ages, strategy evolution focuses on the year 1066 and the Battle of Hast-

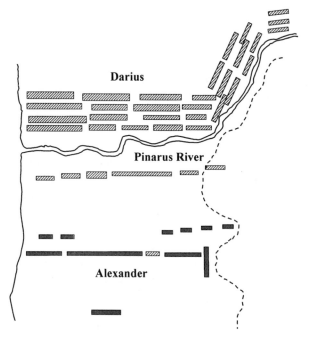

Figure 1-2a Alexander's starting position at the Battle of Issus (Mitchell 1931).

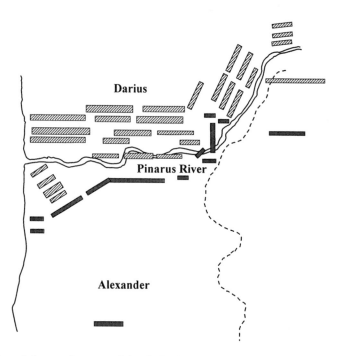

Figure 1-2b Alexander's strategic move to defeat the Persians by splitting the force between the natural boundaries of the battlefield (Mitchell 1931).

ings. Known to every Briton as the end of Saxon rule and the introduction of the Normans, the Battle of Hastings changed history and demonstrated the power of deception, planning, and surprise (Hart 1954) (Fig. 1-3). William, Duke of Normandy, prepared to invade England in 1066 from the French coast. However, anticipating the resistance of the Saxon King Harold, William diverted the attention of the Saxon defenders by establishing an alliance with Norway. Through this alliance, the Norwegian army invaded England on the northwestern shore, 200 miles from William's primary landing at the English Channel several days later. Forced to defend the northern invasion, the English army marched 200 miles from London in eight days and defeated the Norwegian force. However, shortly after the battle, the English army was forced to return the 200 miles and move another 50 miles to the south to defend against William. After almost 500 miles of hard marching, the English army was ill prepared to meet the Norman invaders. Compounding the fatigue factor was the use of mounted knights by the Normans. Capable of rapid attack and defense, the mounted force confused the Saxons by executing feints and charges with the element of surprise. Moving beyond the historical implication of Hastings, the battle handed down to succeeding generations the importance of surprise and flexibility. By forcing the defenders to react to diversions, the Normans used their flexibility to defeat the Saxons by focusing on specific locations and letting the defenders attempt to cover a distributed area with minimal resources.

The Middle Ages provide little in terms of strategic advancement beyond that displayed previously by commanders such as Alexander and William. However, as the world moved into the modern age, military campaigns become more sophisticated with the introduction of modern weapons and the formalization of cavalry as a rapid fighting force. In this time frame Napoleon Bonaparte, considered to be one of the

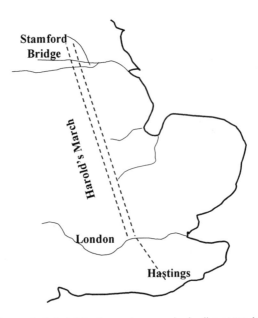

Figure 1-3 William of Normandy defeated the Saxons by a surprise landing at two locations (Mitchell 1931).

greatest strategists of all times, contributes to the world of strategy by refining and formalizing military strategy. During the period 1796 until 1815, Napoleon became the greatest force in Europe by demonstrating the power of offensive campaigns executed with well-trained forces able to operate with rapid movement (Hart 1954). As demonstrated early in his command during the Northern Italy Campaign against the Austrians, Napoleon balanced disadvantages in position with advantages in speed and strategy. Understanding his enemy, Napoleon was clearly aware that the Austrians expected a traditional battle on a set battlefield. Using this expectation against them, Napoleon overwhelmed the Austrians by splitting his force and moving half of his troops through the mountains under the cover of night to attack from the rear. Not expecting such a move, the Austrians were forced into retreat from a two-pronged attack on their position. Such a move was characteristic of Napoleon. Defying conventional wisdom, Napoleon advocated rapid, coordinated attack as the means to defeat equal or greater strength forces. Where possible, superior numbers of Napoleon's forces could be used to defeat defenders in focused attacks on critical areas, thus disorienting the enemy and accelerating victory. In either case, Napoleon's strength lay in his knowledge of the enemy and the use of tactics that were unexpected by opposing leaders. Publishing his Maxims of War in 1830, Napoleon formalized these thoughts into a document that has since been studied by historians, military leaders, and business executives for insights into the use of offensive strategies to keep opponents off balance and unprepared for future attacks (Mitchell 1931).

Lessons Learned As highlighted previously, the examination of military leaders as innovators of strategic management provides today's civil engineering manager with several underlying lessons. Beyond the concepts of planning, surprise, and flexibility that were introduced by the great leaders, the primary importance of the military history lies in the perspective that each of the leaders brought to their campaigns. Although each of the leaders had significant persuasive powers, the reason for their successes resides in an understanding of their opponents (Sun Tzu 1963). The gathering of intelligence about enemy positions, local topography, and the personalities of opposing leaders was paramount to the success of the campaigns. Although each of the leaders employed unique strategies to achieve victory, the development of the strategy would not have been possible without the underlying knowledge of the opponent. Placed in the anticipated context of twenty-first-century civil engineering organizations, this lesson can be applied to the competitive factors surrounding the industry. Understanding the maneuvers that competitors are likely to employ to either block entrance into an identified market or obtain market share of existing clients provides the opportunity to formulate strategic responses. Although these maneuvers do not entail the physical battle of traditional warfare, the mental aspect of the competition is similar. Placing an organization in a position to defeat competitive advances is no less important today than it was to Alexander 2000 years ago.

Eighteenth-Century Craft Society

Moving from the military arena to the business domain, the strategic timeline focuses on the late eighteenth century. In 1770 the United States of America is still a revolutionary dream and Great Britain has an empire that spans the globe. The industrial revolution is a century away and the apprenticeship process for learning trades has changed little in concept since the days of the first civilizations. Although political, legal, and military leaders are challenging traditional concepts, the business community looks very much the same as its early predecessors. Specifically, the eighteenth century remains an era of the crafts. In this time period, long-term strategy focuses on the continuation of the craft through the training of apprentices over extended periods of time (Douglas 1921; Drucker 1993). The focus on creating individual deliverables, whether it is a piece of furniture, an article of clothing, or a simple pen, requires the master craftsman to spend countless hours teaching apprentices the nuances of the individual trades. Although this focus on individual crafts is the centerpiece of eighteenth century business, it is not isolated from today's multinational corporations. The craft industries of the eighteenth century were faced with the issues of material procurement, employee development, and product control that continue to exist in a distantly related form (Pfeiffer 1986).

In the context of the eighteenth century, these issues had a slightly different perspective. First, the issue of material procurement was different in terms of scale and diversity. For most eighteenth-century businesses, diversity was not an option. The craft and apprenticeship process required individual businesses to focus on a single area to ensure that the expertise existed to complete work contracted by buyers (Safley and Rosenband 1993). The time expended to master a single craft precluded any individual business from diversifying into multiple operating units. The result of this single product focus was a material procurement strategy that emphasized the acquisition of a minimal number of raw material types. For example, the furniture craftsman was concerned wholly with the obtainment of the hardwoods required to craft a projected number of pieces over a projected period of time. It is this projected time period that differentiates the eighteenth-century craftsman from today's business organization. In contrast to the majority of materials that can be acquired today with dialing a telephone or sending a fax, the eighteenth-century craftsman did not have the guarantees of delivery that exist for today's industrialized society. With an absence of an extended road network, air travel, or rail service, the eighteenth-century craftsman was forced to work with unpredictable delivery dates and material availability. In many cases, the raw materials originated from overseas, which resulted in projected delivery times of 18 or 24 months. With this uncertainty, the craftsman was forced to procure as much material as possible at the time the order was placed (Smail 1997). The concepts of just-in-time delivery and lean production were unfathomable at the time. Rather, the craftsman was forced to control the use of materials once they were obtained to ensure that they would last until the next shipment could be obtained. Therefore, material control rather than procurement became the overriding feature of a successful craft operation.

Employee development did not fall far behind material procurement as a concern for the eighteenth-century craft operation. As introduced previously, the time period remained the focus of the apprenticeship system. The absence of sophisticated machinery placed the responsibility on the craftsman to ensure quality of product through each individual creation. Training an apprentice who could meet these quality demands was a process that required years of dedication and learning. Rather than attending extended years of higher education, the apprentice contracted with a craftsman to learn the craft through dedicated service and work (Douglas 1921). With a dependence on hand labor, the master craftsman undertook considerable risk when selecting an apprentice. The failure to select an apprentice with the dedication to obtain a craftsman status could result in the teacher spending years of work training an individual who never achieves the predicted level. Given the small size of individual businesses, failures such as these could not be afforded by the individual craftsman. Although the year is 1770, these concerns can easily be translated to the twenty-first century. The current emphasis on craft training and the acquisition of appropriate skills by entering civil engineers resembles the difficult decisions for the eighteenth-century craftsman. The long-term viability of an organization, whether it exists today or 250 years ago remains focused on the quality of its people. Successfully preparing these individuals to ascend through the ranks of the organization is a lesson that remains constant throughout the evolution of organizations. Whether it is apprentices from the eighteenth century or today's entering engineers, the training and education process continues throughout the career path of the individual. As soon as an organization reduces this emphasis, the organization makes a conscious decision to proceed down a road to obsolescence.

The third area of concern for eighteenth-century craftsmen, product control, reads like a page out of today's business press (Szostak 1989). With a dependence on hand-made products, ensuring consistency was a principal concern for many small industries. Without the automation control of today's factories, the master craftsman spent considerable time overseeing the work of less experienced employees. Adding to the importance of this control procedure was the adverse result that existed if the appropriate quality was not achieved. Similar to the situation faced by today's civil engineering professional, the craftsman was faced with the prospect of recreating the deliverable if the quality level was not equal to the shop standard. Given the fact that each piece was an individual effort, the lost time and expense could be significant. Preventing these losses became a prime objective of the craft operation. Early efforts in the textile industry reflect this focus with innovations appearing in the areas of weaving and cloth manufacturing (Cookson 1997). Additional advances in the areas of military weaponry and furniture production provide insights into the desire of eighteenth-century craftsmen to reduce losses emanating from poor-quality products (Farrell 1993). These factors closely resemble the discussions that emerge in today's civil engineering organizations. The search for effective quality control methodologies is not a phenomenon that appeared spontaneously in the latter half of the twentieth century. Rather, the concern for quality product delivery is a historical lesson that exists as a continuous lesson for today's organization leaders.

Lessons Learned With a focus on individual crafts and long apprenticeship periods, the question of relevance to the eighteenth-century operating environment must be addressed. As illustrated, this relevance emerges from the foundational issues faced by craft shops of this time and today's civil engineering organizations. The pre-Industrial Revolution period witnessed the formalization of the craft system and the emergence of the master craftsman as a respected individual within the community. With obligations to employees, customers, and suppliers, the craftsman could not rely solely on technical skills honed through years of apprenticeship and practice. Rather, rudimentary business knowledge was required to guarantee that basic business responsibilities were met. This broadening of knowledge serves as the foundation for this book. The engineer is the modern era's craftsman. Substituting university training for lengthy apprenticeships, the engineer hones knowledge and skills to generate individual solutions for unique problems. However, as with the eighteenth-century craftsman, the engineer cannot survive on technical knowledge alone. Rather, responsibilities from employees, customers, and suppliers compel the engineer to broaden personal knowledge to include the basics of business. In contrast to the early craftsman, today's engineer has the opportunity to go beyond this basic knowledge to encompass areas such as strategic management and long-term planning. The pace of change has accelerated exponentially since the days of the early craftsmen. Keeping pace with that change requires knowledge that extends beyond the classic craft domain. The individuals who are able to respond to this need will be in the greatest position to advance beyond the apprenticeship to achieve the level of master craftsman.

Industrialization and the Early Twentieth Century

The third step in strategic management development accelerates forward 100 years to the industrial revolution. Many argue that this era resulted in the single greatest impact on management and strategic thinking. And it is difficult to argue against the Industrial Revolution as a defining event in the development of strategic management. The Industrial Revolution fundamentally changed the definition of workers, work, and business organizations (Unyimadu 1989). The introduction of industrial capabilities began the decline of the craft-based society and the entry into the world of assembly lines and mass production. Apprenticeships and the need for highly trained craftsmen subsided in favor of a demand for machine operators (Hekman and Strong 1981). The focus on providing services to a limited area disappeared almost overnight as the automobile and the transcontinental railroad opened previously unimagined markets (Szostak 1996). Concurrently, the opportunity to challenge existing manufacturers became available as distribution methods broke physical boundaries between similar business entities (Garrison and Souleyrette 1996). Each of these items combined to establish a new focus on the optimization and efficiency of organizations as well as the opportunity for expansion and market dominance. The introduction of mechanization, public finance, and multiple divisions introduced new methods to achieve these goals and transformed the twentieth century business environment.

The first of these areas, mechanization, changed the world of business more than any other single advancement in the history of civilization up to that point. Mechanization challenged industries from every market sector to reevaluate their employees, products, operating procedures, and markets. First, organizations were challenged to reevaluate the role of the employee in the fabric of the organization. Whereas craft industries viewed apprentices as investments in the future, mechanization transferred the emphasis of success from workers to machinery. The organization with the ability to acquire a mechanized alternative to the human-based production cycle achieved productivity and cost control advancements that were not possible by craft-based production (Hammond and Hammond 1937). Second, organizations were challenged to reevaluate their products. The introduction of industrial-age products such as the automobile and the train challenged the existence of traditional products such as saddles and stagecoaches. The relevance to an industrial-based society was a question that faced every organization and its management team. Third, the craft-based methodologies for completing projects were challenged, as the master craftsman could no longer function as a hands-on teacher and manager. The focus of mechanization was on efficiency and production, not on individual crafting of products (Volti 1996). Henry Ford's introduction of the assembly line and standardized parts permanently altered the operating environment for the traditional craftsman (Winter 1996). The development of large manufacturing plants and the associated expansion of workers established the need for the master craftsman to move from the factory floor to the manager's office. Mechanization required individuals to supervise the operations of the facility rather than participate in the production of the final product. Finally, mechanization challenged craft industries to reevaluate their perception of markets. Prior to the Industrial Revolution, an individual town or neighborhood was served by a limited number of individual craft operations based on the needs of that locale. Mechanization and the associated change in distribution opportunities changed this social dynamic.

The focus on mechanizing craft-based industries dominated Industrial Revolution business discussions. The earliest business schools and business publications focused on topics such as inventory control, unionization, and worker development ("Harvard" 1997). Each of these topics emphasized the acquisition, management, and operation of industrialized plants. Successfully achieving this mechanization required one fundamental element, financing. Although some of the early industrialists such as Andrew Carnegie and Cornelius Vanderbilt achieved great success through individual efforts, the vast majority of Industrial Age organizations required assets beyond those in their personal possession to achieve mechanization. Given this need and increased interest by the general public in business operations, a common interest was established in the financing of industrial organizations. The culmination of this interest occurred in the expansion of the New York Stock Exchange (NYSE) from its beginnings in 1792 to a powerful source for economic investment (Meeker 1922). With the expansion of the NYSE, the general public was offered the opportunity to invest in the industrialization of the United States without having to personally undertake the business operation. This opportunity to invest was eagerly accepted throughout the turn of the century, and activity significantly increased during the 1920s. The message that rapidly emerged from this investment activity was that a demand existed for investment opportunities in industries that could promise a strong vision toward the future (Sobel 1965). Although excessive speculation led to the eventual crash of this market, the fundamental focus on forward-looking organizations remained intact. As long as the management team could provide a strong argument for growth, the general public would risk financial assets to capitalize the growth effort.

The financing of the industrialization effort combined with the mechanization of the shop operation created a new organization position, the professional manager. Professional managers expanded their focus beyond product operations to the exploration of new business opportunities. With new communications capabilities and new product distribution channels, opportunities existed to expand the operations and focus of the organization. With this mindset, the early twentieth century witnessed the beginning of diversification and multidivisional policies. Originated within early organizations such as DuPont, diversification and the opening of multiple divisions represented a strategic advancement for the business community ("Harvard" 1997). No longer were organizations tied to the restrictions of a single product line or market. Through the launching of a new product, or the acquisition of an existing organization, multidivision organizations could strategically attack multiple markets at a single time. Through these acquisitions, oil and chemical companies could move into transportation, banking entities could move into the production of end products, and any organization could legitimately make a case to extend the reach of their business beyond traditional boundaries. Once this concept was proven in the marketplace, the role of the master craftsman was permanently altered. No longer could management afford to have organization leaders working directly on product. The extension of the organization to multiple divisions required constant oversight and the coordination of multiple management levels. Middle managers were required to oversee operations within each division, while senior managers assumed the tasks of coordination and strategic direction. With this shift, the separation of management into a distinct educational, professional, and societal class became a permanent element of the industrialized society (Pfeiffer 1986).

Lessons Learned The industrialization of the economic powers around the world established a new focus in the world of management. No longer was management a secondary focus of the shop owner. Management became an independent topic requiring knowledge of operations, finance, markets, and planning. The complexity of operating an organization escalated beyond that which any traditional shop owner could conceive. The introduction of business schools established a new class of managers that previously would have been viewed as ancillary to the basic purpose of the organization. However, the realization that opportunities existed outside the traditional domain of the operation created a need for skilled professionals with knowledge of both product and management. Although many organizations learned through difficult times that leveraging public financing beyond the capability of the organization could lead to financial disaster, the Industrial Revolution clearly demonstrated that the general public would take a risk on an organization if it demonstrated the potential for growth and profitability. That lesson remains true today. Whether a private company or a public agency, the general public is willing to invest time and money, either directly through investment or indirectly through taxes, to assist in the growth of that organization if it believes that the organization is being managed thoughtfully and forcefully.

This message of appropriate management leads to a second lesson emanating from the practices of Industrial Revolution operations, the focus on operational effectiveness. As demonstrated by Frederick Taylor, a pioneer in the field of industrial engineering, an optimum operation can be achieved if the proper analysis is applied (Nelson 1980). Formalized in the domain of the industrial engineer, optimization and operations research dominated the business world for decades. In fact, many organizations today focus heavily on operational effectiveness as the key to success and profitability. However, the Industrial Revolution demonstrated that limits exist on the benefits that can be gained through operational effectiveness. The task-based perspective of operational effectiveness creates a negative atmosphere for creating strategic visions (Markides 1997). Although operational effectiveness can optimize the operation of any organization, a balance is required between effectiveness and relevance. Managers can become so enamored with the possibility of increasing operational effectiveness that perspective is lost concerning strategic goals. It must be remembered that the goal of the organization remains on markets, customers, and viability. Although a change in operating procedures may save a minimal amount for a given procedure, a far greater benefit may be achieved through the identification of new markets that provide new consumers for the organization's service. This strategic focus emphasizes an external perspective on the operating environment rather than an internal focus on operations. Taylor demonstrated that operational effectiveness is important, but history has taught the manager that a balance between operational effectiveness and external vision is a superior approach to achieving strategic success.

Postwar Management

The Industrial Revolution changed the mindset of managers, but it was the fourth step in the strategic timeline, the postwar years, that redefined strategic management. In the period from the 1950s through the 1970s, industry and academic leaders redefined strategic management. In this time period, large corporations such as General Electric invested heavily in strategic management departments that focused exclusively on developing numeric models to assist in strategic decisions (Davis 1987). The names of organization leaders such as Thomas J. Watson at IBM became today's business legends as they initiated visionary practices for identifying, pursuing, and winning new markets. These leaders built on postwar expansion and optimism as the United States became one of the dominant powers in the Western Hemisphere. However, to build on this enthusiasm, while at the same time protecting the assets of millions of shareholders, organization leaders required new tools that could assess the potential of these new markets. This knowledge was provided through the formalization of business as an education focus. Business schools began to funnel into the marketplace a new generation of managers who were trained in the analysis of numbers and markets as facts rather than personal areas of expertise ("Strategic" 1996). Representing the final break from the craft-based industries, these managers developed the large-scale organization into the epitome of industrialized society. Rapidly, the concepts of decentralization, strategic models, numeric analysis, and management education became synonymous with strategic management concepts.

The first of these areas, decentralization, marked a turning point in strategic management philosophy. The craft tradition fostered a strong sense of ownership by the master craftsmen that owned a given shop. Often handed down over several generations, the individual shop represented the primary asset of an individual family. Consequently, the master craftsman believed that direct oversight of product delivery was central to successful operation. Although this philosophy changed slightly with the Industrial Revolution, the direct oversight of operations remained a central tenet in the philosophy of many managers. Many of these managers either worked their way up through an organization during an entire career, or inherited the business from earlier generations. In this manner, a sense of ownership remained within the organization. The postwar years witnessed the first significant divergence from this philosophy. With the increase in multiple divisions, middle- and senior-level managers could no longer retain direct oversight over every segment of the diversified operations. Therefore, the concepts of decentralized management entered the management domain (Stewart 1990). As illustrated in Figure 1-4, the decentralization concept provided extended decision-making authority to division-level managers. Each manager received extended authority for planning, operations, and product delivery. Each di-

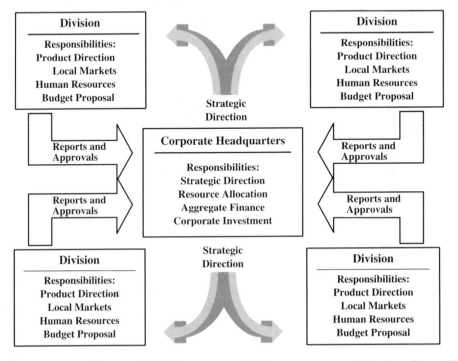

Figure 1-4 Decentralization provided division managers with the autonomy to react to local conditions without waiting for corporate authorization.

vision became a separate profit center that could be evaluated in terms of performance during each planning period. Given this autonomy, divisional managers retained a minimal connection to the corporate parent through reporting and expenditure approval requirements. As demonstrated by organizations such as 3M and General Electric, decentralization provided the opportunity for large organizations to operate as smaller entities (Markides 1997). With the ability to respond to market changes and customer demands, independent operating units could make decisions that otherwise would have been delayed due to interaction with the corporate headquarters. Through this methodology, postwar managers attempted to find a balance between the need for deep-pocket backing and traditional, shop-level customer response.

The move to decentralization provided division managers with autonomy to pursue individual market opportunities. Concurrently, it also created the problem for corporate managers to determine which divisions had the potential to be viable ventures. This need for organization-level decision making ushered in the era of the strategic model developers (Porter 1980). At the core of this development effort was the initiation of portfolio analysis. In this practice, each independent unit is evaluated based on its potential both to add value to the overall organization and to diversify organization pursuits. Popularized by The Boston Consulting Group, portfolio analysis was quickly adopted as a basis for several, influential planning models, including those by Arthur D. Little and General Electric (Davis 1987). Figure 1-5 illustrates the basic matrix concept that emerged from this model development effort. The matrix classifies divisions into four distinct regions based on their potential for growth and market penetration. Using vocabulary that has become infused into the operating dictionary of organizations, the matrix separates operating

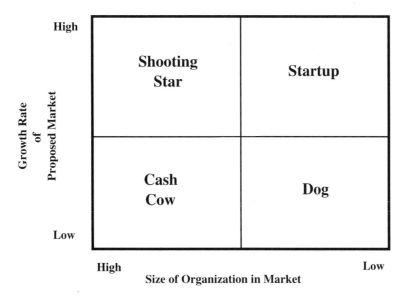

Figure 1-5 The portfolio analysis matrix separated operating units into simple revenue-producing concepts (after Davis 1987).

units into cash cows, shooting stars, dogs, and startups. These four classifications are defined as follows:

1. *Cash Cows.* The lower left region of the matrix defines operating units that function as profitable revenue streams with little risk. Through a combination of dominant size and low growth rates, the organization can depend on consistent returns with little risk from competitors.

2. *Shooting Stars.* The second income-producing category, shooting star divisions, are characterized by rapid growth in areas where the organization has a dominant presence. The combination of these attributes creates an environment where the organization can build quick profit margins while reducing risk from competition.

3. *Dogs.* Divisions that fall into this category are the least desirable for an organization. Characterized by low growth rates and strong threats from competitors, dogs are the greatest threat to organization profits. A drain on corporate resources, divisions in this category should be removed from the portfolio as quickly as possible.

4. *Startups.* The final region incorporates divisions that have high risks, but the potential for large returns. Entering into markets where strong competition exists, but strong growth is occurring, startups provide the opportunity to develop new revenue streams based on calculated risks.

The portfolio analysis method has emerged from this fundamental basis as complex models that incorporate numerous variables including customers, inflation, and market presence. In each case, the models focus on strategic growth and the risk associated with both existing and new markets. Expanded to address the concerns of organizations of all sizes since the introduction of the personal computer, numeric planning and analysis models have become the bellwethers of strategic policy for many individuals. As illustrated in Figure 1-6, the basic portfolio analysis concept was extended through numeric analysis to generate statistical, probability, and decision analysis reports (Collis and Montgomery 1991). Moving as far as possible from the intuitive approach espoused by early managers, the focus on statistical analysis moved strategic management into the realm of formal methodology and practice. Breaking from the tradition of intimate knowledge of product and process, formal planning models enabled managers to make decisions based solely on statistical facts. Postwar management thus moved from a focus on expansion through knowledge of competition and markets to a focus on strategic expansion based on models and statistical analysis.

The transition to a numeric-based method for managing organizations had a secondary effect on new manager education and training. Although loyalty based on years of service remained high on the list of many organizations, new knowledge and education were now required to enter the corporate office. Providing this knowledge became the purview of graduate business education. This formalized education re-

Bubbles Represent Sales for Each Business Unit

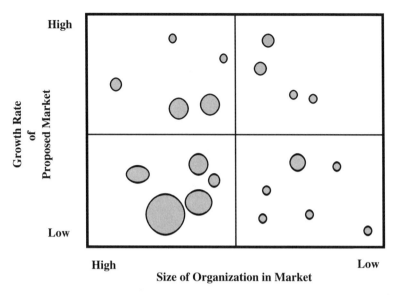

Figure 1-6 The basic portfolio analysis technique was expanded to incorporate statistical and numerical modeling for business unit evaluation.

quirement was reflected in an unprecedented increase in graduate management programs (Reisman 1994). During the postwar time frame, students interested in organization operations, strategy, finance, and technology found an outlet for their interest in graduate business programs. These students were hired by employers who coveted their analytic ability to frame intuitive management decisions in the context of statistical models. From this perspective, the professional manager was elevated from an administrative position to a strategic analyst, capable of bringing case histories introduced in the classroom into the analysis of strategic direction. Whether positive or negative, this emphasis on formal business education permanently changed the complexion of management. Management was no longer a position earned through service and loyalty, but rather emerged as a focal area with its own education and training formalities.

Lessons Learned The transition from experience-based management to scientific management changed the landscape of strategic management. Given analytic tools to analyze the viability of growth and expansion options, organization leaders were relieved of knowing intimate details of products and customers. Rather, by analyzing the numbers, the managers had the opportunity to make scientific judgments on the merits of operations. Additionally, by classifying operating units within a portfolio analysis matrix, organization leaders were provided with a scientific method for evaluating middle-level managers. From an engineering perspective, this scientific approach should have set a foundation that supported the desire for more objective management evaluation criteria. However, the lesson learned from this era was that scientific analysis alone does not provide an adequate perspective for strategic decisions. Many of the organizations that pioneered this scientific approach to strategic decision making, including General Electric, have retreated from the numbers approach with the realization that strategic management requires a significant human element (Dearden 1987). Although statistics and numbers appear persuasive on paper, the predictive ability of statistics is only as good as the stability of the past on which they are based. If the world is changing faster than statistics are able reasonably to predict, then the statistics are of minimal use for accurately forecasting the viability of an organization endeavor. As telecommunications and economics interconnect the world, the experience element is once again emerging as a vital strategic planning component. Therefore, the lesson handed down from the postwar era is that the reliance on scientific analysis can obscure the judgement that is gained over years of experience. Discounting this experience in favor of statistics, without understanding what the numbers represent, is a dangerous game that can result in many more dogs than shooting stars.

Modern Management Foundation

The transition from the postwar years to the modern management era encompasses a gray dividing period. An equivalent progression such as the Industrial Revolution or an event such as World War II did not occur during this fifth stop on the strategic timeline. However, the strategic management domain witnessed fundamental change during this time period due to several events including the increased availability of personal computers, the increased focus on the relationship between strategy and competition, and the globalization of the business environment. In 1979, the business world was introduced to Michael Porter's analysis on the relationship between strategy and competition (Porter 1979). Shortly after, in 1982, the business world was introduced to the personal computer. Additional events at the time included the introduction of digital communications, the emergence of the Japanese economic power, and the reemergence of energy as a global issue. Individually and combined, these events challenged conventional thoughts in strategic management (Drucker 1988; Kotter and Schlesinger 1979). Beliefs including the invincibility of the Western economic powers, the ability to dominate markets based on past performance and the physical boundaries of communications were each crumbling. However, at the same time, the economic recovery of the United States was introducing one of the strongest bull runs on the New York Stock Exchange. Establishing strategic policies that addressed these communication, globalization, and economic factors became the focus as the modern age of strategic management began.

The first of these issues, globalization, is a side effect of the communications advances appearing over the past twenty years. It is no longer possible for organizations to work in isolation from world events (Levitt 1983; Bradley, Hausman, and Nolan 1993; Barnevik 1994). The globalization of economies, corporations, and trade alliances has rendered the isolationist obsolete at the beginning of the twenty-first century. In today's business climate, a crisis in any part of the world results in a global domino effect in financial, political, and commercial circles. For example, an economic crisis in Asia resulting from poor investments and loans from global lenders results in a domino effect as follows:

1. Local manufacturing capacities are reduced as capital investment becomes harder to obtain.
2. Reduced production results in lower demand for oil from Middle Eastern countries to produce energy for factories.
3. Lower demand for oil results in higher gasoline prices for European and North American countries.
4. Higher gas prices result in reduced demand for automobiles by the general public.
5. Reduced automobile demand results in large auto manufacturers reducing orders to the thousands of suppliers in the supply and distribution chains.
6. Reduced orders result in an overall drop in the supply of consumer cash, since reduced demands result in workforce reduction.
7. Finally, this drop in cash flow then affects every consumable-based business in the economy through reduced purchases.

Although this is only one example, similar examples can be developed in any business sector. The interconnectedness of countries has created an environment where an organization in any business sector must be aware of events occurring throughout the world. However, from an alternative perspective, this globalization is creating opportunities in places that were considered untouchable markets only a few decades ago. Regions such as Latin America, Southeast Asia, and Eastern Europe represent economic opportunities for every type of business. For example, in the civil engineering domain, demands for infrastructure improvements are greater than any single organization could possibly address. From this perspective, the village of the master craftsman that contained physical boundaries imposed by access and distance has evolved into a global village with the only boundaries being those that are self-imposed by the organization.

Globalization has resulted in a secondary focus for strategic planners, an expanded focus on competition and its relationship to strategy. As stated by Michael Porter (1996), "Competitive strategy concerns how to create competitive advantage in each of the businesses in which a company competes." Seemingly obvious on the surface, this statement addresses many organization concerns. First, organizations must address whether their current members have the capabilities to compete in a global economy. This implies that individual divisions must be aware of competitors in all regions that they may consider entering. Second, organizations must examine their technical capabilities in terms of communications, information transfer, and broad foundational knowledge. If a division is going to compete in a region where it has no expertise in either local customs or operating conditions, then a reevaluation may be necessary. Finally, organizations must examine their commitment to divisions that propose expanding either geographically or in services delivered. The challenges in today's business environment preclude many of the experimental or tentative ventures that organizations previously undertook. As illustrated in Figure 1-7, challenges from customers, suppliers, existing competitors, new market entrants, and potential substitutes expand the scope of competitive challenges that today's organization must address. The global network of investors will instantly penalize an organization if it fails to address each of these competitive forces. The first indication of weakness could result in financial ruin for many organizations. However, the organization that successfully analyzes the relationship between its resources and competitive strategy has the opportunity to leverage global support by demonstrating vision and strength in its strategic plan and related commitments to success.

A primary opportunity being seized by organizations to demonstrate this vision and strategy is through the use of advanced communications (Drucker 1988). The introduction of concepts such as virtual offices, distributed computing, knowledge repositories, and global networking eliminates many of the boundaries previously faced by organizations with limited resources. Today, organizations of all sizes have equal opportunities to create an electronic presence to potential clients and markets. The use of advanced advertising and multimedia is no longer restricted to multinational conglomerates. Rather, organizations of all sizes can invest in relatively inexpensive tools that provide the foundation for attention-grabbing presentations. Additionally, com-

Figure 1-7 Civil engineering organizations are facing threats from a diverse set of external forces, including both traditional and nontraditional areas.

munications have broken the physical boundaries associated with traditional offices. Today, a virtual organization may be developed for any project. In this virtual organization, experts from around the world are connected via electronic networks into single, cooperative team. To the client, the team is a collection of the best experts required for a given project. For an organization, the team represents a one-time commitment without the overhead associated with permanent employees or regional offices. Electronic alliances such as these represent the first step in the emergence of advanced communication alternatives. Similarly, first steps in the advancement of alternative project delivery systems such as design–build, design–build–operate, and privatization are exploring the opportunities provided by enhanced communication technologies. No longer is an organization confronted with the either–or situation of accepting a project and committing to new employee recruitment or declining a new opportunity. Rather, electronic connections provide the opportunity to enter new market niches by leveraging the combined resources of many organizations.

Lessons Learned Although it is difficult to evaluate a time period while existing within its bounds, several trends are emerging from the modern strategic management era. First, flexibility is emerging as a strong component for corporate success. The ability to respond rapidly to customer requirements, market opportunities, and changing economic conditions is essential for the successful organization. Second, the ability to form alliances and partnerships as hedges against risk, litigation, and financial loss is becoming increasingly necessary as world events result in unexpected changes to project plans. Third, investment in technology and human resources that reflect the move to a knowledge economy is separating traditional and progressive organizations. Organizations that choose to rely on name recognition and tradition will have a difficult time providing equivalent services as those provided by knowledge-focused organizations. Finally, a focus on external influences is becoming increasingly important as the diversity of challenge increases. Strategic management is no longer limited to the capacity of organizations to procure projects. Rather, strategic management is expanding to combine traditional civil engineering strengths with responses to newly emerging drivers and motivators.

Twenty-First-Century Strategy

The year is 2010, and it is the last stop on the strategic management timeline. The communications age has arrived, and with it the ability to communicate information to any project site around the world. The global economy influences decisions from material procurement to human resources. Is this the future? It is not the intention of this book to predict the future of business operations in the civil engineering field. However, since the focus of strategic management is the development of roadmaps for future goals, it is appropriate to examine some of the indicators that appear to be over the horizon.

1. *Virtual organizations.* What emerged as a unique strategic asset in the 1990s will become standard practice in the twenty-first century. As projects grow in size and risk, the need to assemble international coalitions will correspondingly increase. Organizations that establish the foundation for virtual organizations will have the opportunity to formalize the operating procedures for managing these global alliances.

2. *Flexibility.* Rapid response will become a mantra for civil engineering organizations in the twenty-first century. The development of new materials, equipment, and the opening of new geographic markets will require organizations to respond rapidly to evolving client requirements. Streamlining traditional procedures to allow project-level managers to make strategic decisions will enhance the required flexibility.

3. *Employee autonomy.* The increased emphasis on the knowledge economy will result in employees who introduce unique solutions to traditional problems. Issues such as wetland reclamation, hazardous waste mitigation, and soil stabilization may receive unique solutions as employees apply new techniques modified from related industries. Encouraging these ideas will require organizations to allow employees to break from long-held traditions and operating procedures that emphasize hierarchy and approval procedures.

4. *Knowledge assets.* Will the knowledge economy evolve into a new orientation? Absolutely. The only absolute certainty that emerges from history is that change and evolution are inevitable. However, the organizations that prepare for change by building strong foundations in current technologies and human assets are best prepared to evolve during the next period of history. Given this basis, the emphasis on acquiring knowledge assets will increase during the next era. Strategically, the organization that acquires, utilizes, and promotes these assets will have the greatest opportunity to leverage competitive advantage.

There are no guarantees that the previous elements will be the driving forces in twenty-first-century strategic management. However, if history provides any lesson for the future, it is that preparing for change is the best insurance against obsolescence. With this in mind, the history of strategic management provides the foundation for the twenty-first-century civil engineering organization. By analyzing the lessons of history, organizations have a starting point from which to focus on the elements of future organization success. Specifically, an assemblage of the proper strategic knowledge, together with an analysis of organization capabilities, provides an organization with the greatest chance to compete in rapidly changing business conditions.

DISCUSSION QUESTIONS

1. Early military strategists such as Alexander the Great and William of Normandy focused significant attention on knowing their enemies. For these leaders, their enemies were easily defined for military purposes. In today's civil engineering business domain, how does an organization define enemies, and what information should organization leaders obtain on these enemies before entering battle?

2. The eighteenth-century craft society placed emphasis on knowing many parts of a craft business. In today's civil engineering profession, specialization of knowledge is highly regarded. To expand to a strategic business perspective, does specialization of knowledge need to be reduced, and if so, to what degree will future organizations reflect the eighteenth-century craft business?

3. Industrialization introduced management as an independent profession to address the broad concerns of manufacturing operations such as those by Ford and Rockefeller. Do civil engineering organizations require such management experts, or is the business limited to the point where civil engineering experts can sufficiently address organization concerns?

4. The portfolio analysis technique developed after WWII introduced new terms in the business vocabulary. In today's civil engineering market, where are the opportunities for cash cows, shooting stars, and startups?

5. Globalization is redefining every industry around the world. How should civil engineering organizations prepare to enter the global marketplace and what are the unique challenges for a global organization?

6. The virtual organization represents one of the greatest emerging strategic changes for civil engineering organizations. What are the unique challenges that must be addressed in a virtual organization as opposed to a traditional office situation?

2 THE STRATEGIC MANAGEMENT FOUNDATION

Embarking upon this strategic planning exercise is not a trivial decision. If every member of the organization does not clearly understand the goals of the process, then commitment to a complete examination of the organization's operations, processes, and future cannot be guaranteed. Although the pressures of organizational performance can often obscure the broader social, economic, and professional context in which long-term plans are developed, it is these broad contextual areas that make long-term planning an essential issue for civil engineering organizations. As introduced in the following sections, rapidly changing social and technological issues are creating a professional environment that will look very different in the coming decades than that experienced in today's organizations. Specifically, three catalysts are converging to motivate civil engineering organizations to undertake long-term planning. First, the emergence of broad societal and professional issues are affecting core engineering concerns, including the acquisition of employees, the development of markets, and the use of information. Second, the project management tradition that has served as the centerpiece of management education and training is being challenged as to its capability to address long-term issues. Finally, traditional assumptions of civil engineering knowledge requirements are being challenged as the role of business and societal issues emerge as greater influences on the civil engineering industry.

This chapter provides the foundation for organization to address these changes within a strategic management focus. The foundation is constructed from three components: context, definition, and methodology. The first of these components, context, provides a background for the current emphasis on strategic management by outlining the business and society forces impacting the civil engineering industry. Building upon this component, the second component synthesizes these forces into a baseline set of terminology and definitions. Finally, the core component of this book, the question–answer approach to strategic management analysis, is introduced and layered onto the foundation. The construction of this foundation sets the first challenge for organization leaders—to *examine the current organization focus to determine if the knowledge and desire is resident for adopting a new strategic management focus.*

■ The Context of Emerging Issues

Technology, communication, and economic advances are fundamentally changing the global perspectives of time, distance, and spatial boundaries (Chinowsky 1997). Two decades ago organizations could identify themselves as local, regional, national, or international in scope and expect that these definitions were clearly defined. However, with the rapid emergence of technological innovations, these boundaries have been blurred to the point where any organization could theoretically join a design or construction project in any location (Ross 1997). Concurrently, the concepts of company loyalty, traditional competitors, and employee development are changing at a pace that has not previously been encountered in post-Industrial times. As summarized in Table 2-1, these changes are affecting every facet of civil engineering organizations. It is the emergence of these issues that requires a broader management perspective than that provided by project management. Although the strategic questions posed in this book address these issues at length, the following sections provide a brief summary of the catalysts that emphasize the need for long-term planning by civil engineering organizations.

Knowledge Workers

The first emerging issue that will change the face of the civil engineering organization is the evolution of a third generation of worker development. As outlined

TABLE 2-1
Example Impact of Technology on Civil Engineering Operations.

Category	Traditional	Technology-based
Design visualization	Hand drawings and sketches	Three-dimensional models and animation
Project controls	Ledger-based accounting and financial analysis	Computer-based document and financial management
Marketing	Static brochures and promotional literature	Web-based and CD-ROM-based computer animations and continuous updates
Team communications	Teleconferences and express mail	Videoconferencing, e-mail, and fax machines
Data collection	Manual surveys, lab-based sample analysis, analog measurements	Automated samples, field-based analysis, digital connections, virtual test centers

in the historical background, the workforce was dominated for centuries by a craft focus. These workers spent their entire careers perfecting manual skills to reach the level of master craftsman. Following the industrialization of the economy and the expansion of support functions such as engineering, finance, and law, the focus of worker competencies transitioned from craft-based skills to the acquisition of professional knowledge (Grant 1996). In this environment, workers replaced the craft apprenticeship with professional development. In this second phase of worker development, knowledge of industries was formalized into professional education requirements. Rather than learning a craft through long periods of apprenticeship, the new worker combined university education with professional experience to achieve professional recognition. Through this combination, the worker achieved a strong emphasis in the narrowly defined career field. Epitomized in the engineering domain, workers progressed through formal engineering education to entry-level positions that started a series of professional steps that continued to build upon previous experience. After a period of time, the worker had the professional experience required successfully to address problems in the selected field. Now, a third generation of knowledge workers is emerging in all industries, including the civil engineering industry. In contrast to the narrow professional focus of previous professionals, the new knowledge workers are entering the engineering industry with a broader educational perspective that includes influences from economics, marketing, management, and computing, as well as the traditional engineering focus. The new knowledge worker is breaking from the accepted boundaries of professional problem solving to incorporate technological and communications advances to bring new perspectives to traditional and emerging civil engineering projects.

Originally seen in automobile and aerospace manufacturing, this transformation focuses on the day-to-day tasks completed by staff personnel. In manufacturing, operations traditionally required workers with machinery skills developed over long periods of employment in an organization. Large numbers of skilled workers were required to operate the assembly lines that produced quantities of identical or nearly identical components. Knowledge of diverse assembly and engineering procedures was not as valuable as the skill required to keep the assembly line moving at each station. In contrast, today's manufacturing facilities are characterized by highly automated machinery featuring robotics, automated vision systems, and artificial intelligence components. In these facilities, concepts such as just-in-time delivery and component customization have become the norm as technology provides greater opportunities for customer response. Operation of these advanced manufacturing facilities requires workers with advanced knowledge in manufacturing engineering concepts. Each worker requires the knowledge to operate an automated segment of the facility that formerly was manned by multiple teams of skilled operators. With the rapid development of automated machinery, ensuring optimum manufacturing performance requires constant knowledge updates. In this manner, the manufacturing economy is transferring from a skill-based economy to one that values knowledge as the key to operational effectiveness (Drucker 1993).

The next generation of civil engineering professional is similarly witnessing the emergence of knowledge-based tasks as a central focus of organization operations. With this transformation, the educational and professional experience required of tomorrow's civil engineering professional will be fundamentally different from that of twenty or thirty years ago. The industry will increasingly place a premium on workers with expanded professional perspectives as issues that were previously considered secondary to the core engineering business move to the forefront. Expertise in areas such as technology, automation, economics, and market development will be as valuable as project-based knowledge such as computer-aided drafting, scheduling, or material calculations (Chinowsky and Guensler 1998). The ability to access information from sources such as the Internet, government and corporate databases, and private agencies will become a critical attribute as information exchange and acquisition becomes a fundamental component of the business operation. With this focus on broader knowledge as a complement to traditional engineering knowledge, human resource approaches must be altered. The assumption that an engineer will spend an entire career moving through a uniform series of professional positions will be invalid. Knowledge workers will have the flexibility to move between corporate environments while transferring broad industry and business knowledge. The challenge for management is to establish new policies and procedures for attracting these knowledge workers and leveraging their broad professional knowledge.

New Markets and New Competition

The emergence of knowledge workers illustrates one component of human resource issues that are challenging civil engineering organizations. A second challenge for the industry is the area of emerging markets and competition. Historically, the civil engineering industry has been able to divide markets in one of two ways: (1) either as public or private clients, or (2) through their classifications as heavy, industrial, commercial, or residential clients. In the former classification scheme, business practices are characterized based on the type of client that contracts for services. Companies focusing on public sector projects rely on traditional marketing schemes that establish qualifications for short-list bid negotiations with public agencies. Similarly, companies working in the private sector focus extensively on perfecting bid and negotiation processes to the point where the project proposal is elevated to an industrial art form (Clough and Sears 1991).

The division of the client base into the traditional civil engineering classifications of heavy, industrial, commercial, and residential provides greater focus than the public versus private division, but retains the marketing and competitor identification advantages. Specifically, the traditional work categories provide the opportunity for organizations to entrench themselves into a narrowly defined competitive market. Through this entrenchment, the industry elevates the leaders, challengers, and followers into a well-organized division of companies that battle for projects in an intense, but ordered, field of competition (Gould 1997). Since competition from outside organizations is a secondary concern, security is defined through closely held market areas. Through these well-defined competition boundaries, each organization could traditionally anticipate fair profits of 8–10% on the majority of projects with the understanding that competitors were assuming a similar profit margin. Thus understanding the organization's place in this hierarchy provides a defined arena for establishing goals, objectives, and strategies.

Unfortunately, this stability is slowly changing. The days of 8–10% profit margins being an unquestioned standard are fading. Competitive pressures within the sectors and hierarchical levels are forcing profits below 5% in many projects. The drop in profits is forcing organizations to examine opportunities in sectors previously considered secondary markets. The balance of competition and markets is creating an arena where the number of eligible competitors is becoming greater than the number of available projects. With the supply-and-demand balance tipping to the owner's side of the balance sheet, the evidence is mounting that organizations must examine alternatives to their traditional markets. Boundaries accepted as the limits of organization focus can no longer constrain the organization from exploring alternative income opportunities (Goodman 1998). The entire life cycle of a civil engineering project represents opportunities for organization services. For example, development activities, including location analysis, project development, and preliminary design analysis, represent potential expansion at the front end of the project life cycle. By entering a project in the early development stages, an organization has the opportunity to compete in an arena where organization size is far less significant a factor. Small companies have the potential to provide sophisticated project definition services as well as organizations two or three times their size, since the work is focused on broad information gathering, analysis, and interpretation, rather than the development of drawings or the construction of facilities (Harrigan and Neel 1996). Additional opportunities exist throughout the project life cycle. However, the desire to find these opportunities is required before they can begin to return profit and benefits.

The development of nontraditional markets as a source of new revenue streams should be a powerful catalyst for organizations. If opportunity alone does not provide this catalyst, then the emergence of nontraditional competitors should prompt decision makers to stop and take note (Yates 1994). The fact that the engineering-construction field represents approximately 6% of the Gross Domestic Product has been an economic carrot for outside industries for decades. Previously, the hurdle blocking these players from entering the market was the stratification of companies and the focus on industry sectors. Without a costly investment in human resources that understood the core engineering and construction topics, it was difficult for outside competitors to break industry boundaries. However, the emergence of the project life cycle as an emerging market is breaking this economic boundary. Specifically, consulting companies that have built a foundation on providing information analysis services are taking advantage of this opening to actively develop competitive positions. For example, multinational consulting firms such as Arthur Andersen and Ernst &

Young build upon their strengths as information analysts to establish project management divisions as entry points into the engineering-construction domain. By combining new engineering personnel with established business-consulting units, these nontraditional competitors, while making small steps, have the resources to make full-scale assaults on the emerging life-cycle markets. Civil engineering organizations have the option to either ignore this movement as an inconsequential attempt to break into an area that relies on experience and tradition, or question the move in terms of whether these competitors are identifying client services that the civil engineering field has overlooked.

Information Revolution

The expansion of traditional markets brings the discussion of emerging catalysts to a focus on the information revolution. While developments in human resources and markets demand that civil engineering organizations respond to changing circumstances in the employee and customer marketplace, the Information Revolution is a catalyst that impacts all aspects of the civil engineering profession. In contrast to the previous catalysts, the Information Revolution is not a first-time phenomenon. Rather, the industry has witnessed several waves of information technology impact on business operations and processes (Table 2-2) (Paulson 1995). The introduction of the

TABLE 2-2
Technology Evolution From Mainframes To Microcomputers.

Years	Computer type	Computing environment	Application types	Cost
First Information Revolution: early 1960s– early 1970s	Mainframe	Remote computer centers with input through batch run punchcards.	Numeric-based structural alculations, statistical analysis, and coordinate geometry analysis	$100,000–1,000,000
Second Information Revolution: late 1960s– early 1980s	Minicomputers and Standalone CAD systems	Dedicated graphics stations and remote computing centers for individual divisions.	Computer-aided drafting, structural analysis, financial management, and administrative control	$8,000–25,000 for a base minicomputer. Significantly more for graphics workstations.
Third Information Revolution: mid 1980s– late 1990s	PersonalComputers and Workstations	Distributed personal computers and workstations for every employee.	Three-dimensional modeling, animation, structural structural project planning, and project controls	$2,500–7,000 per computer

first mainframes into large organizations during the 1960s provided the opportunity to analyze structural designs and construction operations in a matter of hours rather than the several days that were previously required for hand calculations. Advances by companies such as Parsons, Brinkerhoff, Quade and Douglas together with Sargent and Lundy demonstrated the opportunity that computing presented to the civil engineering profession (Spindel 1971; ACM et al. 1968). Although minicomputers and early workstations provided a greater number of companies with the opportunity to conduct the same engineering and construction analysis, this Information Revolution did not significantly move the industry forward in terms of information management. Rather, these computing technologies provided the opportunity for companies to automate skill-based tasks such as drafting and accounting, but failed to alter fundamentally the work processes associated with civil engineering operations. It took the introduction of the personal computer in early 1982 to turn decentralized computing from a concept to a reality.

The introduction of the personal computer enabled companies reasonably to obtain the computing resources previously reserved for heavily capitalized organizations (Teicholz 1985). As seen through the perspective of two decades of personal computer development, this second Information Revolution equalized access to computer technology. The equalization prompted the introduction of the business and design tools that are considered essential components of today's professional office, including spreadsheets, desktop publishing environments, three-dimensional modeling, computer-based scheduling, and estimating. Although this Information Revolution succeeded in equalizing access to computing technology, fundamental changes to job responsibilities were far from evident within core operations. Rather, personal computers provided a new tool for employees efficiently and effectively to complete tasks historically associated with their positions.

Today, the civil engineering profession prepares to enter a third Information Revolution. While the previous two Information Revolutions focused on the basic introduction of computing technologies, the third Information Revolution focuses on the specific introduction of computing technologies that enable individuals to access information repositories and evolving communication pathways (Fruchter 1996). This access has profound implications for the civil engineering industry in several areas, including intraoffice communications, client relations, and site management. At the core of this transformation is the evolution from traditional hierarchical information transfer to the concept of hyperarchical information access and transfer (Evans and Wurster 1997). Information hyperarchies are breaking traditional hierarchies by allowing direct access between all individuals in an organization. Hyperarchical information concepts are breaking communication barriers by allowing any individual electronically to communicate with any other individual, regardless of title, rank, or office. Hyperarchical information access allows any individual, whether he works out of his garage or out of an urban skyscraper, to access information repositories anywhere in the world. Advancements in communication technologies are expanding these hyperarchical concepts by bringing project participants together through video, audio, and virtual reality environments. Additionally, these transformations in information access are setting the foundation for a challenge to civil engineering organization managers. Traditional communication, organization, marketing, human resource, and operational processes will no longer work within a hyperarchical information concept.

Economic Opportunities

The emergence of worker, market, and information catalysts will inevitably impact the manner that organizations conduct business. However, in the process of conducting interviews for this book, a common response given by industry executives was often, "Yes, these issues are beginning to impact aspects of our business operations, but . . . ," followed by some variation of " . . . in the end, economics are the driving factor in our industry." This general statement was then followed by specific reasons that economics constrain the civil engineering industry, including, "The civil engineering industry is constrained by the difficulty in predicting building industry trends," "The industry is constrained by the ultimate focus on budget and schedule," "Large capital requirements for bonding restrict investment opportunities," and "The cyclical nature of the economy prevents long-term planning and investments." Although each of these statements is based in truth, we challenge our readers to abandon these sentiments for just a moment.

A fundamental component of long-term planning is the belief that the plan has the opportunity to succeed. If an organization enters the planning process with a negative perspective, then the potential for long-term success is severely strained. If an organization enters the planning process with the thought that economics will ultimately constrain the long-term success of the planning process, then this is likely to be a self-fulfilling prophecy. However, economics should not be abandoned as a factor in the planning process; rather the organization must alter the perspective on economics to one emphasizing global opportunity for industry growth. The following observations and facts reinforce this perspective.

First, it is important for organization leaders to recognize and remember that a significant investment in built facilities is made every year by both the public and private sector. As shown in Figure 2-1, the long-term spending trend for constructed facilities over the past 30+ years demonstrates a slow, but steady, growth pattern. The difficulty that exists for organizations to focus on this overall trend is that the total value of construction moves through a distinctive cyclical pattern. While the economic cycles support the previously identified sentiments at first glance, a closer look at the spending breakdown offers a greater outlook for economic opportunity. The first step in moving beyond these numbers is to note that the greatest influence on the cyclical spending curve is the residential market. More than any other component of the construction economy, residential housing influences the total value curve. Given that the majority of civil engineering firms work outside this market, it is important from a planning perspective to look beyond these numbers to those that have the greatest impact on the civil engineering industry. First, in the private, nonresi-

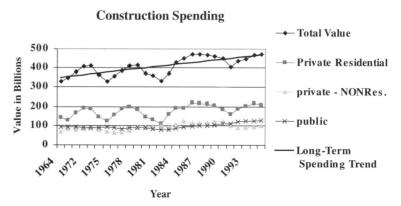

Figure 2-1 Economic analysis of long-term spending trends in both private and public building sectors (U.S. Census Bureau 1997).

dential market, the spending trend demonstrates a tendency to follow overall market trends with very small movements corresponding to short-term investment trends. However, the difficulty with this sector is the continued comparison to the artificial peak of the early 1980s. Rather than occurring from sound economic principles, this peak occurred due to tax considerations that encouraged developers to build commercial space even if it did not prove to be economically viable. From a planning perspective, such an artificial occurrence should not be counted upon to reoccur. Additionally, continued comparisons to this artificial level can be very damaging, since business success is always tempered by comparisons to a level that should never have existed. Reversing this comparison from a negative perspective to one that emphasizes economic opportunity requires an emphasis on the overall trend that shows continued growth and investment in the private market.

The second area of economic interest to the civil engineering industry is the public sector. The development of infrastructure facilities, including roads, water treatment facilities, and sewer lines, represents the backbone of business for many civil engineering organizations. This heavy construction sector represents a dominating economic component of the civil engineering industry, with the largest companies comprising the engineering-construction sector of the New York Stock Exchange. In addition to facing the economic pressures that influence private sector building investment, public sector companies face the unique challenge of predicting how local, state, and Federal elections will influence infrastructure spending. Based on road, water, and sewer conditions in many parts of the country, public and organization perception may incorrectly assume that spending in this sector is decreasing. In fact, spending in the public sector demonstrates few of the cyclical characteristics seen in the private sector categories. Rather, public sector spending has been consistent over the past 30+ years, with the latest time period demonstrating a pattern of sustained growth. The deterioration of existing infrastructure, combined with the expansion of metropolitan areas, provides a strong set of facts for predicting that opportunity will continue to grow in this sector as the long-established spending trend continues.

The underlying theme in each of these areas is an emphasis on perspective. Economics can be used by an organization as a barrier to planning or as an opportunity to develop long-term strategic positions. The annual investment in built facilities, whether it is in public or private sectors, continues to demonstrate a growth pattern. However, establishing successful business ventures in this growth period requires organizations to identify, analyze, and respond to the client sectors that are investing in facilities at any given time.

■ The Project Management Tradition

The emergence of catalysts related to the workforce, competition, information technologies, and economics represent a cross-section of the emerging business environment. From operations to administration, underlying assumptions held by civil engineering managers for decades are being threatened. In contrast to changes that have previously focused on narrow engineering operations such as the introduction of computer-aided design systems to replace manual drafting, these new catalysts are focusing on organization-wide changes. However, examining these issues from a strategic perspective raises an issue concerning the appropriateness of the project management tradition as a basis for long-term organization planning. The tradition of viewing business practices from the budget, schedule, and operations perspective must be challenged for its relevance to the civil engineering organization of the twenty-first century.

The project management culture is one that runs very deep through the academic and professional civil engineering communities. Questioning this tradition requires the development of a strong case. To build this case, the following analysis establishes the existence of the project management tradition and introduces the tools to break from this tradition and effectively compete in the new civil engineering environment.

The Existence of the Project Management Tradition

Research into current academic and professional practice provides strong indicators that the project management concept is the central focus for researchers, practitioners, and academic programs. Consider the following observations compiled from current literature, interviews with executives, professionals, and their clients, and the personal experiences of the authors:

1. In the civil engineering–based literature, an overwhelming number of paper topics focus on the technical and managerial issues that affect project success;
2. In graduate civil engineering programs, core courses are designed to teach a balanced combination of technical skills and project management techniques,

and very few offer more than one course in areas such as company management or strategic industry analysis;

3. The research reports funded by the Construction Industry Institute, a leader in industry-focused research, indicate support primarily focused on improving the cost effectiveness of projects; and

4. In civil engineering organizations, aspiring leaders are assigned a project-oriented career path, where extensive education and experience is provided in tasks that focus on the efficient planning, coordinating, implementing, and controlling of projects.

As seen by these examples, project-based topics form the majority of formal discussion items within the civil engineering industry. This focus should not come as a surprise to civil engineering professionals. Rather the project management focus is an extension of the education and professional paths followed by most organization personnel. On the academic side, university programs have followed and reinforced the project management tradition as they have prepared each succeeding generation of industry managers (Oglesby 1990; Betts and Wood-Harper 1994). In an attempt to respond to industry requirements for specific educational skills, university programs have slowly emerged as a mirror image of the industry itself. These programs instill a strong belief that the successful execution of a project, whether it is at the design, analysis, or construction phase, is the fundamental key to professional success. However, this line of analysis brings the civil engineering industry back to the longstanding argument concerning the appropriateness of topics for undergraduate and graduate programs (Long 1997; Singh 1992; Lowe 1991; Wadlin 1985; Hancher 1985; "Grinter" 1955). Professionals and academics have argued that the undergraduate level is inappropriate for the introduction of broad-scope management concepts. These arguments contend that students at this level should focus on the fundamentals of civil engineering design and project management. In terms of a management perspective, courses should focus on basic scheduling and estimating to provide students with the tools necessary to obtain entry-level positions. The arguments further emphasize that graduate school is the appropriate place to expand management knowledge. Based on these longstanding positions, undergraduate courses traditionally limit management focus to concepts of preconstruction planning and project cost controls.

However, the argument that broad management issues are being taught at the graduate level proves to be false. This conclusion is based on two facts: (1) Graduate civil engineering programs have failed to develop courses in organization management and strategic planning on a widespread scale, and (2) organization leaders are not obtaining graduate degrees in this area. Supporting the first fact is the curriculum of current construction management programs. Since construction management is traditionally where management practices are taught at the graduate level, the curriculum in these programs should reflect the demand for strategic-level education.

However, although an analysis by the authors revealed that approximately 40 graduate construction engineering and management programs exist in civil engineering with over 800 full-time students, few programs achieve the goal of providing knowledge beyond a project management level. Since most programs are designed for students to complete their studies in one year, many of these graduate programs do not have sufficient time to cover subjects beyond the project management level. Few programs offer more than a single course addressing strategic management topics such as forecasting emerging technologies, competitive industry analysis, business development, managing client relationships, and strategic planning. Thus the argument that graduate civil engineering programs are providing organization leaders with strategic knowledge proves false upon closer examination.

The second fact in the graduate education discussion focuses on whether organization leaders are returning to school to obtain advanced education in civil engineering management. In an effort to determine the answer to this question, the authors analyzed the education and professional statistics of 264 civil engineering executives (vice president level and higher), from a broad spectrum of midsize civil engineering companies (100–2500 employees) throughout the United States (Dun & Bradstreet 1996). If the education fact was holding up as stated, then a notable proportion of these executives should have obtained graduate degrees to assist them in successfully managing their organizations. In reality, the analysis found that only sixty of the executives (23%) held any type of graduate degree, and only fourteen of those executives (5%) held graduate degrees in an engineering discipline. These statistics demonstrate that the assumption that executives are returning to school to receive strategic management concepts is also false. In fact, the management exposure received by the vast majority of civil engineering professionals by the time they leave the academic environment is one focusing on project management concepts and practices.

Compounding the lack of strategic management knowledge provided by universities is the reinforcement of the project management tradition by the civil engineering industry itself. Whereas the trend in manufacturing-based industries and other service-based industries is to provide advanced management experience for rising executives, the trend in the civil engineering industry remains focused on slow advancement and long-term company loyalty. This focus reinforces project management as the cornerstone of management decision making by limiting managers to a single vantage point from which to approach corporate management issues. The evidence for this argument is found in the same analysis used to extract executive education statistics. The data provide an insight into the prototypical career path within the civil engineering industry. 63% of the executives spent twenty years or more with their organizations, while fewer than 10% spent less than 10 years (Fig. 2-2). While it cannot be concluded that all civil engineering professionals follow a similar path to the executive level, this finding does reiterate the notion that civil engineering organizations tend to create environments that facilitate slow, traditional career progressions.

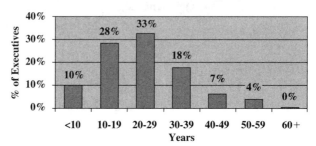

Figure 2-2 Statistical analysis of time that senior civil engineering managers stayed with a single organization (Goodman and Chinowsky 1997).

The Management Knowledge Gap Impact

The impact of the project management tradition on the civil engineering industry is difficult to put into a numeric value. It is unknown how potential business opportunities have been negatively affected, and it is unknown if this micromanagement focus has helped reduce costs at the corporate level. Rather than enter this monetary argument, the project management focus should be viewed from the perspective of its impact on the knowledge that is being used to make decisions in civil engineering organizations. Specifically, the project management tradition has resulted in the creation of a management knowledge gap between project-level perspectives and corporate-level responsibilities within the civil engineering industry. Understanding this gap and then bridging the gap must be a priority for organizations if an emphasis on strategic management is to be realized.

Defined, the management knowledge gap results when organization leaders address organization-level issues based on project management knowledge established through years of narrowly defined experience (Fig. 2-3). As reinforced by academic and industry tradition, the management knowledge gap is established through an extensive focus on project management responsibilities. This extensive exposure to a single topic establishes a reference point for future decision making. The extensive experience subsequently provides employees with a comfort zone from which to approach issues and day-to-day problems. The breadth of this comfort zone directly influences the breadth of perspective that the employee will draw from when approaching a problem. As illustrated in Figure 2-4, the balanced development of a personal, civil engineering knowledge base should reflect an inverse project–organization-level focus. At the beginning of the career, extensive focus on project management issues with a small complement of corporate knowledge introduces the employee to the terminology and issues that establish the context for individual projects. As the career progresses, the emphasis on project management knowledge begins to reduce as organization knowledge is increased to broaden the decision-making comfort zone. Ultimately, as the employee reaches a position of organiza-

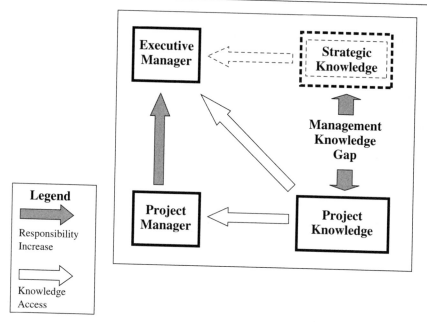

Figure 2-3 The management knowledge gap appears when senior managers continue to rely on project knowledge to make organization-level decisions.

tional leadership, the focus on personal knowledge base development should be a mirror of the early career stage, focusing almost entirely at the organization level with a small degree of project focus to ensure the employee remains current on project management practices. In this way, the decision-making comfort zone accurately reflects

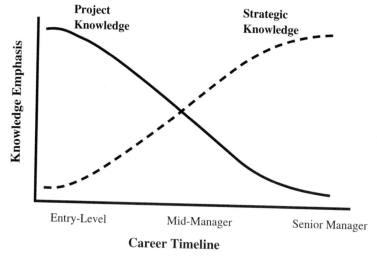

Figure 2-4 The idealized career timeline provides civil engineers with increasing exposure to organization-level knowledge in parallel to career advancement.

the responsibilities undertaken by the employee throughout the career progression. It is in this area of knowledge-base development that civil engineering organizations fail to bridge the management knowledge gap. The industry emphasis on establishing a firm foundation in project-related issues unintentionally contributes to a decision-making bias by limiting employee exposure to broad contextual issues.

To illustrate the formation of the management knowledge gap, an example can be taken that focuses on an employee that spends 15 years overseeing the structural design of high-rise structures. This employee progresses from a staff engineer to a senior project manager over the long term, successfully fulfilling design and project management duties at each career level. As the employee progresses through his career, he establishes an extensive personal knowledge base on how best to organize and control the project process. The goal in this process is to complete the design within both the design and budget constraints provided by the client or management. The employee learns to excel at this responsibility and develops personal methods for successfully achieving these goals. As the years progress, the employee is provided with emerging tools to manage individual projects efficiently. The success achieved from these methods is reinforced by senior managers, thus validating the appropriateness of this knowledge base for addressing and solving problems. In this manner, the knowledge base becomes a comfort zone that can be relied upon in day-to-day operations. However, throughout this career progression, the employee receives little, if any, exposure to issues outside immediate project responsibilities, the prevailing thought being that the employee will receive this knowledge when career advancement makes it necessary. As such, the employee continues to build a comfort zone around the project management knowledge base until it becomes an overriding perspective that the employee brings forward, even as career advancement requires a broadening of the management perspective. By the time the employee reaches a position of organizational leadership, the comfort zone has been skewed to emphasize project management perspectives to the point that leaving this comfort zone to rely upon newly obtained corporate knowledge becomes a daunting perspective. Thus it becomes easier to rely on the project management knowledge and reinforce the appropriateness of it by looking at other organization leaders who are also relying on this comfort zone.

The impact of the project comfort zone on organization-level responsibilities is emphasized in the perspective that organization leaders use to make operational decisions. The administration of any civil engineering organization requires leaders to focus on issues related to both daily operations and long-term organization viability. The extent to which organization leaders have the knowledge effectively to address everyday issues directly impacts operational success. Defining effective procedures for completing projects and controlling budgets is essential to achieving completed and profitable projects. However, these operational issues fail to address the need for the organization to understand its position within the overall context of the business environment. It is the responsibility of organization leaders to set policy in areas that will assist the organization in achieving long-term stability and growth that set leaders apart from midlevel project managers (Lih 1997; Betts and Wood-Harper 1994).

To illustrate the management gap concept, an example can be taken that focuses on the fictional B&B Construction firm that relies on Acme Development Company for 20% of its annual new construction business. Since the organization is profitable and continuing to receive new work from Acme, B&B's executives remain operationally satisfied with the condition of the organization. However, at the conclusion of a particular job, Acme expresses dissatisfaction with B&B's performance in managing the construction of a new regional headquarters. B&B's executives are thus faced with one of two primary options, attempt to placate Acme and retain the business, or attempt to diversify into new service opportunities. Ideally, B&B's executives can placate Acme and retain the business. However, this may not always be the case for many diverse reasons. Unfortunately, with limited exposure to client and market development, B&B's executives may not have the knowledge to envision a path to recover from the loss of Acme's business. Specifically, B&B's executives may not have the knowledge to remove the operational blinders that force them to revisit a path that may no longer be viable from a broader corporate perspective. This lack of corporate opportunity exemplifies the management knowledge gap. While an individual may be successful in operating an organization, if the individual is not aware of the broad responsibilities associated with the position, then the organization suffers from lost opportunity through the failure to establish a vision for long-term success.

Although the concept of the management knowledge gap may not be a popular argument within organizations grounded upon the project management tradition, the long-term impact of this gap on both academic and industry professionals cannot be ignored. The lack of strategic exposure by civil engineers at both the academic and industry levels is forcing both groups to examine long-held beliefs concerning the roles of universities and organizations in providing educational opportunities. At the academic level, growing sentiment is being voiced that students require a broader knowledge base to function successfully in the professional civil engineering domain (Lowe 1991; Russell et al. 1997). Industry is voicing concern that engineers are not being represented in positions of public leadership that impact the future of the industry (Hilton 1995; "Green" 1998). This concern is enhanced when the current academic trend toward reducing educational breadth is examined and it is realized that without a change in this trend, fewer engineers will leave school with the background necessary to enter these policy-making positions (Farr 1997). The concern is prompting some in the engineering community to urge engineering schools to rethink their mission, role, structure, intellectual foundation, and curriculum. Marshall Lih, the director of the National Science Foundation's Division of Engineering Education and Centers, exemplifies this thinking:

> . . . I believe that our schools are missing something crucial. We do not educate enough of our students with the broad perspectives and long-term aspirations to be decision makers, strategic thinkers, opinion shapers, and planners of our corporations—to be leaders of industry. The problem is that the preparation engineers receive is more of a training than an education. (Lih, 1997)

Education Impact

Although it is too early to predict the outcome of this management debate, the long-term impact cannot be underestimated. The university represents the starting point for civil engineering education. Establishing a lifelong learning expectation that encompasses a broad perspective of the industry is a critical step towards breaking the project management tradition. However, this perspective should not be obtained by transferring the responsibility to other academic departments such as industrial engineering or management. Students need to obtain a civil engineering perspective on the issues that affect the civil engineering industry. This perspective will not be obtained by exclusively studying cases from other industries. By delegating this education component to outside departments, the civil engineering industry risks losing students who at one time were committed to becoming civil engineering professionals. It is overly optimistic for the industry to assume that students who are exposed predominantly to manufacturing and financial examples will return to the civil engineering industry and transfer the concepts to the civil engineering domain. A principal example of this business bias is the lack of focus on civil engineering by the leaders in business publishing. For example, as of 1997, only 48 out of the more than 7,500 cases in the Harvard Business School Publishing library are directly related to civil engineering. Given this trivializing of civil engineering issues, academics must reverse the trend to increase specialization and focus on enhancing core engineering knowledge with contextual understanding of the business, political, and societal contexts in which the civil engineering industry operates.

The academic debate focusing on breadth versus depth of knowledge will inevitably bring into focus several positions on the issue. However, these discussions are not isolated to the academic community. Rather, renewed interest from the professional community is appearing in relation to topics of professional qualifications and continuing education and their relationship to long-term organization development. In terms of the former, universities are expending significant research effort to identify and develop student abilities that are desired in both new professionals and midcareer managers (Koehn 1995; Jennings and Ferguson 1995; Panitz 1998). These efforts and studies provide contradictory feedback for educators and continuing education planners. Consistently, studies by universities and professional organizations indicate that the primary concerns of engineering organizations are the lack of communication and leadership abilities demonstrated by engineering graduates and new professionals (Jansen 1998; Robar 1998; Hanna and Brusoe 1997). Organization leaders are becoming increasingly frustrated with the lack of nontechnical knowledge demonstrated by incoming professionals.

At a cursory level, the efforts reinforce the position that nontechnical knowledge deserves an equal standing with traditional engineering skills when evaluating future engineering leaders. However, these results present the civil engineering industry with a conflicting perspective of professional qualifications. Although research indicates an increased attention to nontechnical qualifications, hiring trends demonstrate exactly the opposite. Specifically, conflicting patterns emerge when discussing hiring preferences with civil engineering organizations. While organization leaders believe that nontechnical qualifications are an asset to civil engineering professionals, these same leaders are hesitant to diverge from traditional hiring patterns. The traditional focus on productive skills, or skills that can be effective immediately, such as software operators, estimators, or field assistants, remains the dominant focus in hiring decisions. This discontinuity leads to the question of what should be believed—the spoken thoughts or the organization actions. Under normal circumstances, this question would be a moot point; organization actions should always be the bottom line for educators and aspiring professionals. However, in this case the spoken thoughts may be just as important as the organization actions.

The discontinuity between the survey results and the hiring practices should not be viewed as a conflict that must be resolved prior to educators or organizations taking actions to support the move to greater breadth in education. Rather, the conflict should be viewed as a first step toward professional acknowledgement of nontechnical knowledge as a key component of future professional success. The conflicting data should be viewed as an opportunity to build upon current advances to introduce further breadth into the civil engineering domain.

Continuing Education Impact

The second education-related impact of the management knowledge gap lies in the area of continuing education. The potential impact of the management knowledge gap is forcing organizations to recognize the need for continuing education. Organization leaders are beginning to understand that the continuous enhancement of the corporate knowledge base is critical to the success and health of the organization (Fink 1998; Hecker 1997). Unfortunately, the prevalent choice available to civil engineering organizations is less than optimal. While civil engineering curricular decisions are moving inexorably toward the acceptance of breadth as a fundamental requirement for civil engineering education, organizations are faced with few options to enhance the existing corporate knowledge base. Although commercially sponsored seminars provide organizations with a sense of satisfaction that they are taking steps to provide their employees with up-to-date practices and management trends, interviews suggest ambiguous results. While employees may receive up-to-date information and bring this information back to the organization, most organizations do not have the infrastructure to ensure information is properly disseminated or the structure to determine the relevance of the information to the overall knowledge base.

The combination of these barriers places civil engineering organizations in a precarious position from which to develop the breadth of knowledge required for long-term planning. Although acknowledging the need to obtain knowledge is important, putting in place the structure successfully to integrate the knowledge is equally important. It is at this stage that civil engineering organizations stumble when attempting to bridge the management knowledge gap. Specifically, the rush to enhance the organization's knowledge base often obscures the need to plan for knowledge dissemination. As such, the issues of how to learn and what to learn become blurred in unstructured management directives. As stated by an executive of a leading civil engineering firm to his technology division in terms of expanding current organizational knowledge, "just identify *something* that will benefit the organization's work processes." This unstructured approach highlights one end of a continuum that can be used to characterize the current trends in continuing civil engineering education (Fig. 2-5). The "just learn anything" directive is an example of the far right side of the spectrum that the authors refer to as the Diffusion by Numbers approach. In this approach, organizations encourage as many staff members as possible to attend continuing education seminars with the hope that dissemination barriers will be overcome by extensive exposure, the intended result being that the greater the number of employees receiving exposure to continuing education, the greater the odds that the knowledge will be disseminated by informal communications. Although organization leaders recognize the shortcomings of this approach, the focus on education provides employees with the perception that the organization is concerned about knowledge development. However, with no formal dissemination structure, the ratio of dissemination becomes uneven, with selected employees receiving extensive exposure while others receive minimal exposure at best.

A contrast to the Diffusion by Numbers approach is the Stalemate Approach. Rather than spending excessive resources to ensure a minimal return, organizations adopting the Stalemate Approach fail to achieve organization learning due to inaction. While acknowledging that continuing education is important for the overall improvement of the corporate knowledge base, the stalemated organization fails to act since it understands that no effective dissemination method exists. Armed with this knowledge, organization leaders fail to send any employees to continuing education courses, since the benefit cannot be justified by long-term benefits. While these organizations save the expense incurred by the former approach, the failure to obtain any level of continuing education predisposes the organization to remain stagnant in developing the corporate knowledge base.

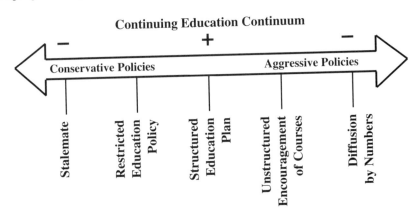

Figure 2-5 The continuing education spectrum encompasses approaches ranging from no emphasis on education to exaggerated attempts to infuse knowledge into the organization.

It is an oversimplification of the education issue to state that successful organizations will address the management knowledge gap by establishing a continuing education policy that falls somewhere in the middle of the two education spectrum extremes. Rather, organizations should approach the development of education structures, not by focusing on whether or not each employee should be attending continuing education courses, but rather on what each employee needs to know. The eventual success or failure of the organization will not hinge on the number of courses each employee attends. Rather, long-term viability depends on the ability of employees to understand the impact of emerging issues on organization goals.

■ Civil Engineering Knowledge Requirements

Establishing where the state of the industry is at a given point in time can cause an organization to either negatively react with disillusionment or positively react with enhanced motivation. In the negative reaction, self-analysis contains a notable downside by spotlighting the deficiencies and barriers that exist in the industry. In an organization experiencing a downward trend in business operations, an emphasis on barriers and deficiencies can provide the proverbial straw that breaks the camel's back. In this situation, disillusionment creates a self-fulfilling prophecy where organization leaders contribute to the continued decline by believing that no remedies are available to cure the downward cycle. Pointing to the barriers as proof of the unfavorable conditions that the organization is facing, leaders face the risk of falling into a holding pattern whereby no actions are taken to improve the current operating situation. In this mode, disillusionment grows as the organization continues to experience decline and sentiment grows that business opportunities are being lost to competitors. In the worst-case scenario, the organization eventually faces collapse from a combination of failed leadership and employee discontent. Although this picture paints a dire scenario, it is important to understand the ramifications of emphasizing negative self-analysis results.

In contrast to disillusionment, the desired outcome of industry self-analysis is organization motivation. Rather than interpreting deficiencies or industry barriers, organizations should receive these results as a wakeup call for immediate action. Deficiencies do not necessarily lead to collapse. As demonstrated by companies facing economic distress in diverse industries including Chrysler, Wang, and Unisys, the identification of deficiencies can lead to improvements that result in stunning economic successes (Hausman 1997; Williamson 1993). However, achieving these successes requires the proper knowledge from which to approach the improvement process. Using a project management example, a project manager would not be expected to undertake the building of a project without the appropriate budgeting, scheduling, and controls knowledge. Similarly, organization leaders should not implement improvements without assembling the knowledge to address organization deficiencies adequately.

The Strategic Knowledge Requirements

The advantage of knowledge lies in the collective breadth of its focus. Although an individual area of knowledge may be complete, the applicability of the knowledge is reduced unless an individual contains the breadth required to complete the given task. The individual areas of knowledge should not be looked upon as more or less important than any other area. Rather it is the collection of knowledge that enables individuals and organizations to evolve successfully from a project-based management approach to a strategic-based management approach.

Compiling this knowledge does not imply that an organization needs to hire outside experts to address specific knowledge areas. Rather, compiling knowledge requires commitment from the organization to either acquire the knowledge through an outside source or identify an individual in the organization who will be responsible for organizing an internal education effort. As illustrated in Figure 2-6, strategic knowledge is the top of a pyramid that builds upon the education and experience that characterizes a civil engineering career. Moving from science and math knowledge through engineering fundamentals, civil engineering specialty knowledge, and project management knowledge, the knowledge pyramid is completed with the acquisition of strategic knowledge. Specifically, knowledge in four areas combines to establish strategic knowledge: leadership, markets, technology, and economics. The four areas emerge from the need for civil engineering organizations to address four strategic concerns:

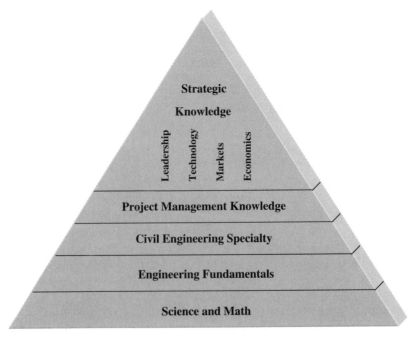

Figure 2-6 The knowledge pyramid contains the foundational and advanced knowledge areas required to compete in the emerging civil engineering environment.

1. How will the organization achieve the objectives set forth in the strategic management plan?

2. Where will the organization focus its limited resources?

3. What tools are available to assist the organization in working smarter, more efficiently, and with less overhead?

4. What economic resources are available, both internally and externally, to achieve the strategic objectives?

Leadership

As a foundational component, leadership represents one of the most widely documented elements of strategic knowledge (Tannenbaum and Schmidt 1973; Nadler and Tushman 1990; Farkas and Wetlaufer 1996; Kotter 1990). Educators, researchers, and professionals have promoted leadership as a cornerstone of the new management culture (Katzenbach 1997; Taylor 1995). Whether it is leadership in the form of a consensus builder or leadership in the form of a dynamic persona, leadership is the focus of a rising discussion that is crossing the barrier from the management domain to the civil engineering domain. As documented by civil engineers, including Hilton (1995) and Russell et al. (1997), the civil engineering profession requires a new generation of leaders to push the existing corporate envelope. Unfortunately, the development of a leader is not an overnight task that can be accomplished through a seminar or lecture series. Rather, as best exemplified by the military, building leaders takes years of incremental exposure to increasing responsibilities and positions of authority. Through this incremental process, individuals learn the roles, responsibilities, and boundaries associated with positions of leadership. Given this time commitment to preparing an individual for the responsibilities of leadership, the context of leadership in the pyramid does not focus on the ability to transform project managers into commanding generals. More important in the context of setting long-term objectives is the ability to create a core set of individuals that have the knowledge to act as custodians of the strategic management vision.

In this role of custodian, the organization is establishing a core group that commits to obtaining the knowledge and information necessary to establish and oversee the overall strategic management vision. Therefore, to bring leadership into the strategic pyramid, the goal should not be to identify a single individual that can act as a motivational figurehead. Rather, to establish the leadership core successfully, the organization should focus on giving the management team the responsibility to oversee the development of the strategic management direction and the information required successfully to translate the answers into operational objectives.

Markets

Every organization has a practical limit on its ability to apply resources to given projects. Coordinating these resources to leverage the greatest gain from their application is the goal of any organization. However, to obtain the greatest benefit, an organization must focus these resources within the context of a plan that emphasizes specific market goals. Developing responsibility for setting these goals is the focus of the markets component of strategic knowledge. Concepts such as market development, market segmenting, client identification, and service expansion are cornerstones of management writers. Companies such as 3M, CompUSA, and Citibank have built a large segment of their reputations based on their ability to identify market requirements and respond to them in an organized and efficient manner (Grant 1991; Heller 1997). In contrast, numerous examples exist where companies such as Woolworth, Eastman Kodak, and Apple Computers have experienced decline and even failure due to their inability to focus limited resources (O'Sullivan 1997; Pospisil 1997; Gelertner 1998). Although this topic has received secondary status within the civil engineering industry, the emergence of new markets and new competition requires organizations to reevaluate the importance of market knowledge. For a civil engineering organization to compete successfully in the diversified market of the twenty-first century, the organization will have to retrench and leverage its resources to respond to new market opportunities. The organizations that do this quickly and with focused effort will emerge as the leaders in the new market segments.

Does every civil engineering organization need to hire a marketing guru immediately to acquire market knowledge? No. However, organizations do need to elevate market development from the sidelines, where it is commonly relegated, to business development departments. Similar to the integration of leadership knowledge into an organization, the development of market knowledge requires a commitment to place market responsibility into the hands of a core group of individuals. However, the responsibility for market development cannot rest solely on the shoulders of these individuals. Rather, market responsibility will only be as successful as the leadership capability of the organization to respond to recognized market opportunities.

Technology

Commitment to leadership and markets provides the direction and focus for an organization. The third component of strategic knowledge augments this focus by providing alternatives to existing operational processes. The Information Revolution occurring within the civil engineering industry will inevitably impact every aspect of the architecture, engineering, and construction process. From design through construction and operations, information technologies are changing the manner in which both owners and professionals approach the development of capital projects. As evidenced by organizations such as Black & Veatch and Fluor Daniel, embracing the information revolution can lead to advances that provide strategic advantage in competitive markets (Calem 1995; Williamson 1993). Through streamlining of administrative procedures, sharing of knowledge, and the establishment of electronic communication capabilities, civil engineering organizations have the opportunity fundamentally to change their approach to project delivery. However, for this change to occur, organization leaders must understand that the introduction of technology needs to occur in the context of strategic placement, and not as part of a never-ending chase to keep up with the technology curve.

It is a fact that technology continues to change at a rate that makes it almost impossible for any organization to ensure that their resources are constantly at the cutting edge. As soon as technology is brought into the organization, it is guaranteed to be behind the leading edge shortly after acquisition. With this rate of advance, organizations that believe that constantly having the latest technology is the key to success are in for some very difficult lessons. Instead of this insatiable chase of the latest and greatest, organizations need to focus on acquiring and introducing technology to support operations over an extended period of time. For example, the introduction of videoconferencing technologies may enhance an organization's ability to communicate, exchange information, and reduce project errors by adding visual communication to the traditional voice and written media. With the focus on communication technology as a strategic decision, the organization can focus on the ability of the technology to fulfill its objective. In this way, the organization can acquire and upgrade the technology as it becomes necessary to improve its capability to meet the underlying objective, not because a newer or faster technology component has become available.

Establishing organization competence to ensure that technology is acquired as a component of a strategic vision is at the core of the technology knowledge requirement. For an organization to elevate the role of technology from automation tools to strategic assets requires a commitment to establish a knowledge base of technology capabilities. To benefit fully from this commitment, technology knowledge should not be isolated within the confines of a separate department. Rather, the responsibility for technology knowledge should reside in a cross-departmental group of individuals who can provide the operational, strategic, and project context for the acquisition and integration of emerging technology resources.

Economics

It would be unrealistic to develop any guidelines for civil engineering organizations without addressing the core element that many industry leaders believe is the primary driver in any organization decision. Economic knowledge is required to provide aggressive responses to opportunities and challenges presented by the operating environment. Civil engineering organizations must learn to rediscover and remake themselves to capitalize on untapped opportunity. The continued focus on reducing expenditures and lowering overhead has a finite limit in terms of its corporate benefit. Opportunity lies in the expansion of current income streams and the development of new income streams that were previously undiscovered.

Developing these undiscovered income streams requires analysis of several factors, including the identification of economically sound clients, rising economic trends, potential expansion into geographic regions experiencing strong growth patterns, and appropriate risk taking with organization resources. Although the identification of these opportunities and income streams is a significant endeavor, the eventual success or failure of the income growth objective will rest with the organization's willingness to take calculated risks. Assuming these risks without appropriate experience and knowledge is a foolish endeavor. The civil engineering industry has traditionally lagged behind other industrialized sectors in terms of consistent profits and returns on investment. The opening of new market opportunities provides the window to break from this underachieving trend. However, each organization must make an individual decision regarding its desire to pursue previously untapped income streams. Based on history, many civil engineering organizations will choose to watch these opportunities from the sideline. This leaves the door open for aggressive organizations to grab market share while others watch and wait. The focus of the final strategic knowledge component is the acquisition of knowledge that enables a civil engineering organization to pursue these new income streams aggressively based on the calculated analysis of risks, rewards, and unknown variables.

The commitment to acquire economic knowledge cannot be isolated to the business development department within an organization. Economic knowledge impacts every aspect of an organization from project execution to business development. As such, economic knowledge must reside in a cross-disciplinary group within the organization. Organization members responsible for issues including daily operations, project control, market development, and strategic planning each have a vested interest in the successful integration of economic knowledge into the organization. As such, each of these responsibility areas should be represented and given a mandate to elevate economics from a budget concern to a cornerstone of organization policy making.

■ Strategy—The Definitional Basis

In contrast to mathematics, physics, or material science, strategy does not contain universal truths that can be documented through scientific theorems and proofs. The civil engineer looking for the equivalent of Pascal's theorems or Newton's laws of physics will be disappointed to find that the elegance of science and mathematics is missing from strategic management. However, as illustrated through the extensive history of strategic management, scientific and management advancements have been integrally related to the field for centuries. From this development, strategic management encompasses principles from a combination of quantitative and qualitative fields. On the quantitative side, management and industrial sciences have formalized the domains of operations, logistics, and finance. Each of these specialties contains the rigor and scientific methodology equal to any engineering research effort. Complementing this quantitative rigor are the human dimensions of psychology, sociology, and human resource management. In these fields, the rigor of human study protocols contributes insights into the ill-defined areas of worker motivation, creativity, and organization purpose. When combined, these quantitative and qualitative elements address diverse organization needs, including professional, technical, and strategic demands.

However, similar to the difficulties that arise when architects, engineers, and constructors are unable to communicate due to incompatibilities in vocabulary, organizations cannot develop long-term plans when members are working from different definitional bases. Reducing uncertainty and miscommunication requires a common understanding and interpretation of foundational concepts. In the field of strategic management, these foundational concepts include strategy, strategic management, strategic planning, and strategic plans. While closely related, each concept represents a different component of the strategy domain. Understanding this difference and adopting a common vocabulary is the first step in the transition from project to strategic management.

Strategy Defined

Beginning at the highest level of abstraction, the first strategy concept is that of strategy itself. As illustrated in Figure 2-7, the basic concept of strategy is that of an idea, specifically, an idea that sets in place a path that responds to multiple internal and external influences (Porter 1979; Hamel and Prahalad 1989; Collis and Montgomery 1991). In contrast to the execution and control plans developed for individual projects, strategies are concepts that contain no intrinsic steps to achieve the final destination.

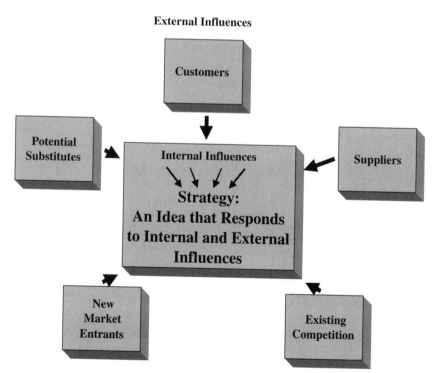

Figure 2-7 The basic strategy concept; an idea responding to internal and external influences (after Porter 1979).

Originally developed by rulers and military leaders attempting to broaden their empires, the concept of strategy can be traced to the beginnings of recorded history. Examples of strategies that remain a part of today's vocabulary or military foundation include the use of the Trojan Horse, the development of the warship, and the concept of the fortified castle. Each of these examples represented a new approach that responded to a current situation, predicted the needs of the future, or presented a new method for achieving a goal. The use of the Trojan Horse represented an advance in the concept of surprise and deception, the development of the warship represented an alternative to the traditional land-based campaign, and the fortified castle responded to both current and foreseeable future requirements for security, governance, and organization. Underlying each of these similarities is the fact that each advance was an idea or concept that focused on the achievement of a goal through innovation or long-term vision.

Similar to the expansion of political empires, the expansion of business entities requires organizations to take reactive and proactive steps for both existing and anticipated industry conditions. Some examples include the change from custom to standardized parts, the introduction of multiple product lines, and the formalization of the New York Stock Exchange. In each case, the idea encapsulated both a response and a vision for long-term organization advantage. Standardized parts reduced dependence on skilled craftsmen, multiple product lines provided the opportunity to build on existing name recognition, and the Stock Exchange formalized investment opportunities for growth and modernization. Once again, the underlying similarity is the introduction of a concept that focuses on the long-term viability of the organization rather than a short-term fix to an immediate problem.

Finally, the modern business interpretation of strategy is exemplified in industries as varied as telecommunications and grocery chains. The expansion of digital communications, global mergers, and the introduction of lean production have modified business practices to anticipate twenty-first-century scenarios. Digital communications have set in place the foundation for electronic services to any part of the world; global mergers are creating international alliances to address global economic conditions; and lean production is providing customization while retaining controls on production overhead. These ideas are not implementation plans, but rather, the visions from which specific action plans are being developed. In this context, strategy is defined as the underlying concept that responds to, or anticipates, industry conditions for the purpose of developing long-term plans.

Strategic Management—An Environment for Strategy

The development of strategic concepts, whether from a military perspective or a modern business perspective, does not occur spontaneously. The development of strategic concepts requires an environment that fosters strategic thinking and focus. The establishment, continuation, and enhancement of this environment is the focus of strategic management. In contrast to project management, strategic management is not limited to the concerns of individual projects. Rather, strategic management is the focus by organization leaders on the issues and topics that enable members to anticipate, identify, and respond to internal and external changes in the business environment.

The concept of strategic management is neither new nor unique to the civil engineering industry. As highlighted by the following section on the strategic management timeline, strategic management models have been evolving in the business domain on a continuous basis since the late nineteenth century. Combining input from these models with the results of interviews with organization leaders, the authors propose that strategic management in the context of the civil engineering industry comprises the following seven areas:

1. Vision, mission, and goals—The starting point for all organization endeavors; establishing a vision provides each member with a direction to follow in all business practices.

2. Core competencies—The business boundaries for an organization; core competencies establish what an organization does best and where its strength resides.

3. Knowledge resources—The combination of human and technology resources that provide the backbone for completing organization projects.

4. Education—A focus on the informal and formal requirements for lifelong learning and understanding of evolving business conditions.

5. Finance—A broad focus on monetary concerns beyond the project-to-project concerns of budget and schedule control.

6. Markets—The analysis of expanded business opportunities within domains that are related to core competencies.

7. Competition—A focused analysis and understanding of existing, emerging, and future competitors in both existing and potential market segments.

As illustrated in Figure 2-8, the seven areas of strategic management can be pictured as a series of segments within an overall structure. Rather than viewing the structure as a linear segment, the illustration is circular to indicate that strategic management activities are a constant process that return to the beginning at regular intervals to ensure that a constant focus is retained on the core purpose of existence. Underlying this entire structure is the understanding that the purpose of these focal points is to provide the environment that allows organization leaders to formulate strategic concepts.

Strategic Planning—The Implementation Side of Strategy

Strategic management provides the environment that encourages the development of strategic concepts. However, just as strategic concepts do not usually develop sponta-

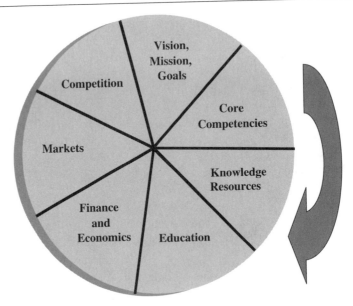

Figure 2-8 The seven areas of strategic management integrate within an overall feedback framework.

neously, the existence of a strategic management environment does not guarantee that organization members will focus on developing strategic concepts. To encourage this focus, numerous academic and business writers have proposed various strategic planning models (Thompson and Brooks 1997; Lemmon and Early 1996; Davis 1987; Mintzberg 1994; McCabe and Narayanan 1991). These strategic planning models provide specific instructions for approaching, executing, and evaluating the development of strategic concepts. For example, the model illustrated in Figure 2-9 emphasizes the need for an organization to: (1) build a strategic planning team, (2) set the strategic planning objectives, (3) gather member input, (4) synthesize the developed ideas, (5) develop an implementation plan, (6) execute the plan, and (7) evaluate the success of the ideas prior to the start of the next strategic planning time frame.

This structured approach is the primary exposure that the majority of organization members have to strategic concerns. The annual practice of setting and evaluating strategic goals is a ritual within many civil engineering organizations. Many business writers and engineering executives emphasize this annual process as the answer to the challenges facing organizations. The writers emphasize that the inclusion of employees at multiple levels in the planning process provides the foundation for employee ownership of the annual goals. However, as with any topic that focuses on procedural processes, the number of strategic planning methods is increasing at a rate that sometimes appears to be exponential. As such, the strategic planning process is slowly becoming synonymous with the entire field of strategy. This connection is incorrect. The strategic planning process is one element of the overall strategy topic. Strategic planning is the focused attention to the development of strategic concepts based on the inputs provided by the seven areas of strategic management.

Example Feedback Planning Model

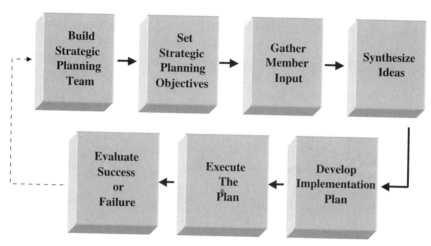

Figure 2-9 An example strategic planning model that outlines a structured process for formulating, executing, and evaluating strategies.

The Strategic Plan—Putting It All Together

The previous strategy elements combine to focus a civil engineering organization in a particular direction for a particular planning period. The strategic management areas set the foundation by focusing attention on the internal and external forces that impact everyday operations. From this foundational focus emerges a periodic focus on the development of strategies through a structured strategic planning process. Finally, the strategic planning process establishes a strategy, or set of strategies, that focus the organization in a specific direction for the duration of the current planning period (Schoemaker 1995). Although this strategic direction is a major milestone for the strategic planning process, it is not the final conclusion required for implementation purposes. Rather, a strategic plan is required to outline the goals, objectives, mileposts, and evaluation criteria that must be followed to achieve the developed strategy.

The development of this book revealed that strategic plans are not a foreign concept to civil engineering organizations. Almost all the organizations interviewed indicated that some form of strategic plan existed, was reviewed, and possibly was updated, each planning period (usually every year). However, translating this focus on long-term and strategic issues into a series of tasks that can be accomplished by individual departments is challenging. The time required to focus on broadening client bases, or examining new revenue streams, is often overridden by demands by projects for attention to budget, schedule, or personnel matters. In these scenarios, the immediate project demands gain management attention since the dissatisfaction of clients is a real penalty that exists as a tangible outcome if the immediate problem persists. Given this conflict for attention, a specific set of instructions is required to ensure that an organization remains focused on organization-level concerns. This set of instructions is the strategic plan. Encompassed within this plan are the measurable outcomes that both division- and organization-level managers can evaluate for progress and final achievement.

■ The Question–Answer Methodology

From an early emphasis on military planning to modern academic think tanks, setting a strategic vision has evolved from an operational necessity to a managerial philosophy. As seen from the strategic management timeline, this evolution is not limited to any one industry, country, or philosophical center. Strategic advances have emerged from great military leaders such as Napoleon, industrial giants such as General Electric, and internationally renowned academic institutions such as Harvard. The benefits from these strategic visionaries are evident in the advances that have occurred in diverse industries such as automobile manufacturing and hotel management. The questions that arise for the civil engineering community are why this body of work has not similarly impacted civil engineering organizations and why a new methodology should be introduced when such a diverse set of management practices is already in place.

The authors believe that both of these questions can be answered with a single response—perspective. When an organization adopts any tool, it is critical to evaluate whether that tool was designed for that type of organization. For example, when adopting a computer-aided design package, civil engineering organizations appropriately focus on the ability of the package to support civil engineering activities. It would be foolish to invest in a package that has been developed from an electrical engineering perspective with only generic drafting functions available for use in the civil engineering office. The same underlying concept exists in the management realm. While management theorists may disagree, it is important to note that the documentation and discussion of the majority of strategic concepts introduced over the past several decades fail to mention any component of the civil engineering industry as more than a passing footnote. In contrast, industries such as aerospace, automotive, computer technologies, and transportation receive extensive literature focus. The fact that the civil engineering industry is predominantly characterized by privately held, small and medium organizations, with executives whose primary educational background is not management, makes the civil engineering industry less attractive to management researchers. With few companies having to address shareholders, no product lines to manage, and very few companies that are familiar to the general public, the civil engineering industry does not fit the public profile desired by many management researchers and strategists. However, it is the existence of industry fragmentation, privately held organizations, engineers as organization leaders, and a strong project management focus that makes this industry unique in its management perspectives. As such, civil engineering managers deserve a strategic management tool that begins from their perspective, rather than one that has been developed for the manufacture of washing machines and as an afterthought introduced to civil engineers.

The question–answer approach presented in this book is the strategic management tool developed from this change in perspective. In this approach, organization analysis is undertaken through the use of questions that address the diverse factors that influence long-term corporate success. At the core of this approach are the two questions that were first presented at the beginning of the preface, "Where do you want your organization to be in five years?" and "How are you going to get there?" Although extensive evidence exists to support the position that these questions are the cornerstone of successful organization development and planning, many organizations have a difficult time successfully answering either of these questions. While organizations may have a clear picture of where they want to be in five years, achieving this vision is often left as an operational issue that will be determined as the time frame unfolds. History has shown that this approach rarely serves to achieve organization objectives fully. Rather, a comprehensive approach and commitment to set the organization on the path to implementing goals and objectives is required from the outset of the planning process. Setting this path and developing the foundation for the long-term commitment is the basis of the question–answer approach. Table 2-3 lists the questions that compose the question–answer methodology. Each of these questions has been developed to address a specific component that must be developed within a strategic corporate perspective. From the initial setting of an organization vision, to the development of human and technological resources, and the analysis of market and financial risks, the questions are designed to challenge organizations to analyze objectively the capabilities of civil engineering ventures. Collectively, the questions systematically provide a basis for implementing a long-term, strategic perspective to organization development. The focus in each category is the development of foundations for long-term growth. The emphasis on individual projects is reduced in favor of organization requirements and objectives. Individually, the questions serve as entry points into specific components of the strategic management perspective. For example, the initial question on the existence of an organization vision is intended to challenge the organization to determine if this vision exists. If it fails to have a positive response to the question, then an action item is required to develop this vision. The chapter associated with this question provides the background and proposed methodology to transform the identified need into a required action plan.

This question–answer approach was developed in direct response to the education and training that civil engineers receive in both the academic and professional environments. Civil engineers are trained to respond to facts and problems—the hard reality of a situation. Introducing a management perspective into this environment can take either an antagonistic or a complementary approach. It is the fundamental viewpoint of the authors that building upon the analytic foundation of civil engineers is the appropriate method to introduce strategic management concepts. Civil engineers possess a long-held comfort zone founded upon the derivation of answers to presented problems. When presented with a problem to solve, civil engineers will work tirelessly to develop

TABLE 2-3
The Questions Used to Transition a Civil Engineering Organization to Strategic Corporate Management Perspective.

Topic	Question	Challenge
Long-term goal	Where do you want your company to be in 5 years?	Is the organization prepared to start the strategic management process?
Vision, mission, and goals	What is the vision and the associated mission and goals that the organization has set as the foundation for long-term achievement?	Examine the fundamental purpose of the organization and set the forward-looking, almost out of reach goals that are the hallmark of innovative organizations.
Core competencies	What is the core business and strength of the organization?	Examine the resources and operations of the organization to determine if the greatest strengths are being identified as the emphasis for emerging opportunities.
Knowledge resources	What processes have been put in place to enhance knowledge and information resources within the organization?	Examine the integration of information technologies and human resources to determine if the work force is maximizing the knowledge-based opportunities presented by emerging technologies.
Education	What formal and informal procedures are in place for lifelong learning within the organization?	Examine the educational needs and opportunities within the organization and set the education objectives required to create a civil engineering learning organization.
Finance	What financial risk analysis is in place to forecast and protect the organization from economic swings?	Examine the strategic economic opportunities within the organization and set the economic objectives required to create long-term financial stability.
Markets	What market analysis procedures are in place proactively to identify new and expanded organization opportunities?	Examine evolving market opportunities and establish an expanded organization vision that exploration and innovation.
Competition	What positions are established to protect against new competitors and leverage the organization into new competitive positions?	Identify existing barriers to strategic positioning and establish a vision to aggressively pursue and defend market opportunities.

an answer that meets or exceeds the problem parameters. In the context of this analytical approach, the question–answer methodology starts at Question 1, developing a vision, and challenges an organization to analyze every facet of its operations until the final question is answered. The questions are designed to prompt an organization to resolve the problem of missing strategic management elements. The greater the ability of the organization to exceed the basic parameters of the question, the greater the ability of the organization to transition to a strategic management perspective. Through this comprehensive approach, an organization will progress along a structured path that leads to answering the question of how the organization will achieve its long-term goals.

Objectives, Development, and Context

The development of any product or service must be solidly grounded on a set of objectives that serve as the development guideline. Given this guideline, the development team can revisit the objectives throughout the project to ensure that the final product reflects the original project concept. While this concept of objective setting and project development receives extensive attention in the scientific literature, significantly less attention is paid to extending an understanding of these objectives to a target audience. Marketing to an audience often requires that project objectives be set aside as marketing hooks are developed to sell the developed product or service. Without entering into a lengthy discussion concerning the validity of this approach, the authors suggest that this method is inappropriate for the introduction of management concepts that are intended to have a fundamental organization impact. For a management concept to succeed, the adopting organization must have confidence that the concept matches their overall goals and objectives. As such, it is important for the reader to understand the underlying objectives that were established during the development of the question–answer methodology.

1. Cultural focus—Given the project management tradition within the civil engineering industry, the transition to a strategic management focus requires changes that extend beyond the office of the president. Organizations must be prepared to examine the underlying culture that promotes a project focus and be prepared to challenge long-held beliefs and unwritten cultural norms. Providing a structured approach to this cultural shift was the first objective set for the development of the question–answer methodology. Specifically, the methodology provides a path to examine current capabilities at every level of the organization, and provides the framework for collectively moving all phases of the organization forward along the path toward a strategic approach to corporate management.

2. Analytical methods—Civil engineers are educated from the day they enter a university to take an analytical approach to solving problems. The scientific method is emphasized as the golden rule for approaching, analyzing, and developing solutions to given problems. Moving beyond the absolutes of correct or incorrect answers takes years of field experience. However, even when shades of gray are allowed into the problem-solving process, the ability to evaluate answers based on quantitative or logical approaches remains as a fundamental component of the engineer's behavior. Rather than ignore or compete with this behavioral tendency, the second objective of the question–answer methodology is the establishment of a framework that allows civil engineers to establish a strategic approach to management based on objective, rather than subjective, criteria. The framework encourages engineers to build upon academic and professional education by extending problem-solving concepts into the strategic management domain.

3. Comprehensive focus—Concepts such as TQM and strategic resources are important components for organizations to consider when establishing a strategic vision. However, when taken as individual components, they address only a limited segment of the organizational requirements for strategic management. For example, a segment of existing management concepts focus heavily on the identification and development of strategic market segments while ignoring the importance of emerging technologies to support these marketing initiatives. Similarly, other concepts focus heavily on the need for organizations to implement comprehensive education plans while failing to address the need to establish a strategic vision that this education is supposed to support. Although this limited focus may be acceptable in organizations where an overall strategic management approach is already in place, it is insufficient to assist organizations making the shift from project- to strategic-based management. Thus the third objective of the question–answer methodology is a perspective that comprehensively addresses the management requirements of an organization and provides a framework to develop all phases of the organization's operations.

4. No assumptions—Although significant progress has been achieved by management theorists and consultants in terms of developing strategies for management responsibility, the majority of these developments have emerged from management schools or management consulting organizations. Although this may seem to be an obvious and preferred development environment, it actually serves as a disservice to the civil engineering community. To illustrate this disservice, it helps to use an example from a civil engineering perspective. When civil engineers develop an industry advance such as new fluid analysis computer programs or spatial analysis techniques, the developers of these systems make a fair assumption that the users of the technology have a background in civil engineering from which to use the technology in the proper perspective. When the user of the technology does not have this background, the consequence is often that the tool is used improperly or the tool fails to achieve its intended benefit because the developer has not provided the user with the complete perspective required to leverage the capabilities of the technology. The same issue exists in the development of management tools. When a management academic or a management consultant with a pure management background develops a tool, assumptions are made concerning the background of the user. Whether explicitly or implicitly placed in the tool, the developer makes the assumption that the user of the tool has a background that is related to the perspective of the developer. As seen by the statistics on the background of civil engineering executives, this assumption is highly incorrect. Civil engineers require management tools that do not assume that they have a management degree. Rather, the tools must be developed from a perspective that acknowledges the traditions and historic tendencies of the civil engineering organization. As such, the final, and overriding, objective of the new question–answer methodology is to approach the framework development process from a civil engineering perspective rather than a management theorist perspective.

Development

The combination of the preceding objectives provides the guidelines for the question–answer approach to strategic management. However, objectives alone should not provide a basis for civil engineers to adopt this management tool. In addition to the establishment of appropriate objectives, a development process that brings both industry and academic perspectives together into a coherent management framework is required. To achieve this coherent framework, the authors undertook a multiperspective approach to establish the requirements, tools, and questions embodied within the framework. Specifically, the question–answer methodology builds upon three critical inputs: industry case studies, established management theories, and external forces of change.

The core of the development effort focuses on industry case studies. Interviews conducted with organization leaders and staff provided an industry perspective on current business practices, organization concerns, and visions of the future. Both public and private civil engineering organizations involved in all areas of the industry were included in the studies to ensure a broad representation of viewpoints and industry practices. To ensure diversity, organizations of various sizes were included in the study with the acknowledgment that fundamental differences exist in the resources available to the organizations. The case study process provided the foundation to ensure that the question–answer methodology was solidly grounded in the requirements of the civil engineering industry. In contrast to many case study exercises, the process was not restricted to well-known organizations. Rather, a cross-section of the industry was examined to ensure that the strengths and weaknesses identified within the organizations were relevant to the broad civil engineering community. As such, several of these case studies have been summarized in the following chapters to illustrate organizations that have built upon individual strengths to implement successful plans in areas such as education or marketing without the benefit of a multi-billion-dollar budget.

Complementing this industry input are management theories that provide relevant concepts and strategic approaches. Emerging and classic management concepts were combined in an attempt to bridge the management gaps found during the case study analysis. Relevant concepts from managerial scholars including Peter Drucker, Henry Mintzberg, and others have been included where they offer direct benefit to the civil engineering industry. Theories such as organizational learning, core competencies, and strategic intent offer strong promise for civil engineering organizations. These management theories provide organization leaders with unique approaches to solving unfamiliar challenges. However, the lack of focus on the civil engineering industry has limited the benefit of these theories for civil engineering managers. Therefore, to break this disciplinary boundary, identified management concepts were layered upon case study results to create a bridge between the identified industry needs and the opportunities afforded by management theory.

The final area in the development effort addresses the need to look to the future. The emergence of issues and forces that are changing the face of civil engineering practice requires answers that adequately address the issue of potential impact. In response, the question–answer methodology integrates emerging issues as a check against the theoretical and industry inputs into the process. To ensure the question–answer methodology focuses on guiding civil engineering organizations into the future, emerging issues served as a reality check for each component of the framework. Where a question was proposed that did not reflect future responses to emerging needs, the question was reevaluated to either alter its focus or eliminate it from consideration as part of the framework. In this manner, the final question–answer framework combines the insights of industry managers, the theory of management analysts, and the context of emerging industry requirements.

Precautions

The concept, objectives, and development background establish the context in which the question–answer methodology was created. With a focus on comprehensiveness and objective analysis, the questions presented at the beginning of each chapter provide the opportunity to examine every facet of an organization's operations. However, it is important to add some cautionary qualifications to the question–answer context.

1. The question–answer methodology is not an immediate cure-all—When any organization is in a difficult situation, the drive to find a quick fix is a strong one. Putting out fires with temporary solutions is often a necessary component in the day-to-day world of project management. In contrast, putting in place a strategic management perspective requires analysis of the systemic processes within an organization. Effectively analyzing these deeply ingrained processes take time. The question–answer methodology does not replace this requirement; it structures the process in which the analysis takes place.

2. The question–answer methodology is not a substitute for additional organization learning—The strength of any organized group of individuals is only as strong as the weakest link. In the context of strategic management, the weakest link appears when individuals responsible for implementing the strategic plan are not provided with the education necessary to understand fully the issues for which they have been given responsibility. If any section of the plan is weak in terms of its development, then it will serve as the breaking point within the vision. As such, it is necessary for an organization to commit to education not only for technical competency, but also for strategic advantage.

3. The question–answer methodology is a first step along the strategic management process—Many professionals have experienced the supervisor who practices the "management by the latest chapter" philosophy within an organization. Each month this individual enters the meeting room with a new management philosophy or direction, appearing to have just read a chapter in the latest how-to management guide. Lack of commitment to an idea ensures its failure before it has the opportunity to impact existing operations. Strategic management cannot succeed in this context. The questions and associated issues presented in this book are intended to serve as starting points for individuals and organizations to pursue solutions in areas requiring enhancement. Failing to pursue this path of knowledge and organizational enhancement will result in the strategic focus vanishing as quickly as a management trend of the week.

The underlying philosophy behind each of these contextual warnings is the fundamental concept of commitment. Strategic management focuses on long-term success and growth. Achieving this will not be accomplished with short-term solutions. With this as a precursor, it is now appropriate to focus on taking the first step in the strategic management process—establishing an organization vision.

◼ Taking the Next Step

The most difficult step in any endeavor is taking the first step on a previously untried path. The transition from a project-focused management perspective to a strategic-focused management perspective is no different. The reliance on project management principles has provided civil engineering organizations with a comfort level that is difficult to ignore. Breaking from this comfort zone requires a commitment to change that is very daunting to some organization leaders. However, the recognition for change is the first step in moving an organization forward. The following chapter builds upon the foundation for change introduced in this chapter by leading civil engineering organizations through a step-by-step approach to building visions, missions, and goals.

DISCUSSION QUESTIONS

1. How may the career path of knowledge workers differ from traditional members of a civil engineering organization?

2. Organizations grow and sustain through the identification of new markets. Given the increased competition in the civil engineering marketplace, where are the potential new markets for civil engineering organizations, and what are the challenges associated with these market opportunities?

3. Given the economic trends of the past 30 years, for what type of economic conditions should civil engineering organizations plan as part of their strategic management process (e.g., length of positive or negative trends, severity of cycles, susceptible industry sectors, etc.)?

4. Historical records indicate that most civil engineering executives stayed with a single organization for at least 20 years. Does this trend continue today, and what are the factors reinforcing or threatening this tradition?

5. The management knowledge gap concept states that project knowledge differs from strategic knowledge. However, some individuals disagree with this assertion, stating that strategic knowledge is a direct extension of project knowledge. What are the differences between an executive's responsibilities and a project manager's responsibilities?

6. The strategic management knowledge pyramid identifies four areas of knowledge that are critical for civil engineering executives. How does the need for this knowledge impact existing graduate and continuing education programs?

7. Seven areas of strategic management are presented for civil engineering organizations. Which of these areas currently receive greater or less attention by civil engineering organizations?

8. What is the greatest barrier that prevents civil engineering organizations from implementing a full strategic management focus in all seven areas? Is it possible to develop strategies and strategic plans without a strategic management environment?

PART II

INTERNAL STRATEGIC MANAGEMENT ISSUES

THE FIRST STEPS 3

Visions, Missions, and Goals

What are the vision and the associated mission and goals that the organization has set as the foundation for long-term achievement?

The start of any venture begins with a single step. The transition from a project-based management focus to a strategic management focus is highly dependent on a series of successful first steps. Of particular concern is the establishment of the roadmap that employees, managers, and executives can each look to for guidance in daily and long-term decisions. In contrast to the prevailing thoughts of many managers, the hierarchy of position does not provide adequate guidance for employees. Although senior and middle managers may provide direction during individual projects, these directions fail to provide a strong foundation for long-term organization success. Specifically, management directives do not provide the vision that motivates organization members to strive beyond project goals to achieve organization goals. To move beyond this barrier requires long-term goals that each member can point to as the overall purpose for expending the energy necessary to succeed in today's global economy.

Enabling civil engineering organizations to break from the project management paradigm requires a specific roadmap to support the pragmatic civil engineering tradition. However, vision alone does not ensure organization success. Although vision provides an organization with a basis for long-term goals, vision does not necessarily create motivation. Establishing this fundamental organization element requires a layering of vision, mission, and goals—the heart of a successful long-term plan. As evidenced by organizations as diverse as Wal-Mart, Disney, and Boeing, the successful establishment of these elements provides individuals with the sense of ownership that lies at the core of organization cohesiveness (Collins and Porras 1996). Providing a reason and a common focus for achieving long-term goals is essential for establishing the spirit of cooperation that is the hallmark of great organizations (Riggs 1995). The lack of this common purpose leads to divisional fractures since the divisions lack the purpose for cooperation other than they belong to the same organization.

In this chapter, the development of strategic visions, missions, and goals is examined in the context of the question posed at the beginning of the chapter. The chapter emphasizes the development of these elements as building blocks for establishing organization roadmaps. A methodology to create these building blocks is introduced that emphasizes breaking from pragmatic traditions and embracing forward-looking strategies. Combining existing strengths with visions for obtaining new resources is emphasized as the first step in achieving a transition to a strategic management focus. Often discussed, but quickly dismissed in the heat of day-to-day operations, the setting of these strategic elements diverges from traditional civil engineering education. The lack of budgets, calculations, and tangible conclusions creates a viewpoint that these elements have secondary importance in the practical administration of a civil engineering organization. However, breaking through these management barriers sets the challenge for this chapter—*examine the fundamental purpose of the organization and set the forward-looking, almost out of reach goals that are the hallmark of innovative organizations.*

■ The Organization Vision

The giants of international business are recognized beyond the barriers of industrial sectors. The names of Gates, Eisner, and Buffett are synonymous with success, power, and wealth building. Of greater significance is their image as visionaries (Collins and Porras 1991). These individuals are seen as having the capacity to see beyond the horizon to detect, influence, and direct the future of entire industries. Their individual companies rose from the industry segment pack to achieve the status of blue-chip companies. In this role, the companies are bellwethers of economic, political, and social stability within a rapidly changing world. From this foundation, reputations were built, expanded, and firmly entrenched into the annals of industrial myth and lore. The days of solely emphasizing completion of projects were erased as issues of investor returns, international trade, and rapid customer response became entrenched into the daily vernacular of the leading organizations. With this image of success, the industrial world has slowly become divided into the very few large-scale players that make up a fraction of the overall number of organizations, but which dominate the overall collected revenues, and the rest of the players that comprise the vast majority of the industry sectors, but which account for significantly less accumulated income. In this split, the question arises as to why some organizations are successful in crossing the bridge to growth opportunities and some organizations remain tethered to the constraints of size and traditional customers.

There is no simple answer to this question. A calculator, formula, or theorem does not exist that ensures individual organization success. However, when looking at the common traits of successful organizations, one common element emerges as a strong guiding force, the organization's vision (Pearce 1982). Taken at its bare form, the vision statement embodies the direction, values, and long-term goals of the organization. The vision provides the outside world with a discrete statement that captures the essence of the organization and its outlook on the industry. Similarly, the vision provides the internal organization with a message that consistently portrays the reason for the organization's existence. As Figure 3-1 illustrates, the vision statement can encompass a broad range of statements, ranging from an environmental focus to a customer focus, to a pure business emphasis. However, in each case, the statement provides an insight into the organization's outlook on its individual industry and the role it plays in that industry and society as a whole. From this single statement, the organization displays its attitude and intention to competitors, investors, clients, and future employees.

In addition to these constituents, the vision statement addresses the most critical constituent in terms of long-term success, current employees. Long overlooked in the civil engineering industry for their potential to transform an organization from an industry player to an industry leader, the individuals within an organization represent a collective resource that can be energized to create a unique organization identity. From this identity emerges a unique organization loyalty that is often preserved from generation to generation. In one well-documented example, Proctor & Gamble has long prided itself on its ability to create an environment where P&G people strive to push the organization to levels of excellence beyond those of its competitors (Collins and Porras 1991). In this environment, P&G people understand and exhibit the underlying vision of respect and quality that has been the core of the organization for

---

Example Vision Statements

CDM will be a global consulting, engineering, construction, and operations firm committed to exceptional client service to improve the environment and infrastructure.
Camp Dresser & McKee Inc.

An unrelenting commitment to quality and a service philosophy that seeks to always treat customers in an honest and straight forward manner.
Rudolph and Sletten

Our goal is to be the best builder in America.
McCarthy Construction

To become a national home building company with the financial and organizational strength to provide quality homes in distinctive neighborhoods throughout the nation.
U.S. Home

Develop high quality products and reduce time-to-market for our clients.
Battelle

A commitment to helping our clients reinvent their organizations, enhance their capacity for learning and change, and create lasting value for their customers, employees and owners.
Arthur D. Little

Committed to being the best provider of computer products in the business markets we choose to serve.
Dataflex

Singularly positioned and strategically committed to global leadership as the preeminent financial management and advisory company.
Merrill Lynch

To be the recognized global leader and preferred supplier of high performance coatings on flexible substrates.
Southwall Technologies

Figure 3-1 Vision statements set the foundation for every decision and practice addressed by an organization.

over a hundred years. The result of this common focus is a business environment that is both respected and emulated by competitors.

Taking the impact of the vision statement one step further, the ability to create a vision that is both inspiring and respected by peer organizations provides employees with a purpose for being a member of the organization that transcends everyday responsibilities. The successful vision provides the employee with the opportunity to view their position as a component of a team that has a unified direction and purpose, rather than viewing their role as one who is responsible for an individual task. Given this purpose, employees have been found to exhibit greater loyalty, productivity, and enthusiasm for individual tasks (Leavitt and Lipman-Blumen 1995; Katzenbach and Smith 1993). In this environment, organization leaders can spend significantly greater amounts of time focusing on external issues such as identifying new client requirements, new market opportunities, and defending moves by competitors, rather than facing internal personnel issues (Kotter 1990). This focus on employee motivation and loyalty may seem unusual by many in the civil engineering industry. The industry tradition that focuses on long tenures and consistent promotions to managerial status often diverges from personnel issues such as motivation. Specifically, many senior managers spend all or most of their careers working their way up from field or entry-level design positions to achieve leadership positions within the organization (Goodman and Chinowsky 1997).

With this tradition, the obvious question is "Why is there a need to focus on vision statements?" The answer to this question is the difference between loyalty and passion. Many organizations of different sizes and emphasis have great employee loyalty. Their employees are happy to remain with the organization for their entire careers. These employees believe in the methods and practices, and will defend the organization against claims from competitors and outsiders. However, this loyalty should not be confused with visionary passion. The loyalty that these employees feel is inspired less by the intrinsic belief that they are contributing to the benefit of others, but rather by the comfort that they enjoy being part of a team. Transforming this comfort into an environment that encourages achievement, success, and purpose is the difference between loyalty and passion (Jeffords, Scheidt, and Thibadoux 1997; Ritner and Taylor 1997). In an environment where employees are passionate about their vision and purpose, such as the Proctor & Gamble example, the individual strives to make the organization succeed since there is a belief that the success of the organization serves a greater purpose for the community and industry as a whole. In this mode, the organization strives for success and growth beyond the level of a sector player to a sector leader.

The effort required to achieve this success is not a trivial task. As a first step, a cohesive vision must be developed that provides a focal point for every organization member. Successfully establishing and implementing this vision requires several elements:

· An understanding of the organization beliefs
· An understanding of the organization purpose
· The establishment of a long-term goal
· The establishment of a comprehensive vision statement

Organization Beliefs

As described by Collins and Porras (1996), organization beliefs are intrinsic values that the organization holds as unalterable tenets that guide the practices and activities of all members. Examples of these beliefs include respect for the individual, preservation of the environment, honesty and integrity in all business arrangements. Although vague from a traditional project management perspective, the beliefs represent principles that bind the organization together. These beliefs are not transitory sentiments that are intended to change in response to every shift in the economic winds. Rather, the emphasis is on the identification of principles that were intrinsically and unarguably appropriate when the organization was founded, and will continue to be the correct things to believe throughout the existence of the organization. To use a civil engineering illustration, a large southeastern civil engineering organization was founded on the belief that high-quality service with a focus on advancing a customer's position in their business sector is the key to business success. Today, that belief still stands as a cornerstone of all activities in which the organization engages. In fact, during the development of a new division, the senior managers specifically focused on the vision to ensure that the activities of the new individuals emphasized the common beliefs of the original founders.

Although it is difficult to argue with organization beliefs, many individuals, especially those who have been exposed to the less than noble side of business through highly publicized cases, are understandably cynical when the topic of organization beliefs are raised. When the topic is first introduced, students often raise the thought that these beliefs are solely for customer consumption. Similarly, employees who are exposed to organization beliefs through a poster or memo are often less than convinced of the organization's commitment to these ideals. With this skepticism as a background, is there a tangible benefit to establishing a set of organization beliefs? The best answer is seen in the organizations that have strong organization beliefs. The roll call includes business leaders from a diverse set of industries including Wal-Mart, Southwest Airlines, Proctor & Gamble, 3M, and Disney. Individuals from these organizations overwhelmingly speak of feelings that include pride, enthusiasm, and family. The connection between these feelings and the financial results that place these organizations among the elite of United States industries cannot be overlooked.

Identifying Organization Beliefs

To assist in developing a set of organization beliefs, a structured process can be used to aid in the documentation of the underlying ideals that form these beliefs. As illustrated in Figure 3-2, a series of questions capture the underlying focal areas for organization beliefs. The intent of the questions is to prompt organization leaders succinctly to record beliefs that are related to each focal category. The beliefs should be identified independently of the perceived expectation of clients or employees. Specifically, the beliefs must originate from the fundamental beliefs of the individuals. With-

---

Summary of Organization Beliefs

Instructions:
For each category, enter a one-to-two sentence description of a belief that is fundamental to your approach with issues emanating from the category.

Category 1: Employees—What do you believe should guide all interactions with organization employees?

Category 2: Customers—What do you believe is the fundamental element that should guide interactions with clients of any size or potential opportunity?

Category 3: Practices—What is the basic tenet that guides all organization activities whether the organization is earning a profit or absorbing a loss?

Category 4: Environment—As the developers of the built environment, what is your belief about managing the natural environment?

Figure 3-2 A set of four questions should be asked in identifying the core beliefs of the civil engineering organization.

out this level of commitment, it will be difficult to establish the unique environment that separates the organization from its competitors.

Employees: The first belief question focuses on organization employees. This question identifies the employee relation principles that reside at the core of organization practices. For example, principles in this category may include respect for individuals at all times, equal opportunity in all circumstances, unbounded freedom to create and motivate, or maximum job satisfaction. In each of these examples, the organization leaders record a succinct, but clear, statement that evokes a deep understanding within the reader. Since employees may be skeptical of the belief statement, sincerity is of the greatest importance. If the organization is not ready to create an environment based on the belief statement, then it should not be developed since the failure to commit to the belief can be more damaging than not having the statement at all. The bottom line is that employees will respond to an organization that is sincere about their concerns, but will quickly abandon this loyalty if they feel that the organization is insincere in its stated beliefs.

Customers: The second element of the belief structure switches from an internal employee focus to an external customer focus. For this purpose, customers are defined in the broadest context to include private or public clients, partners, or agencies. In each case, the customer receives an output that is developed by the organization. As part of developing this output, the organization should identify an underlying belief that guides every aspect of the transaction. This belief may be very simple, such as "honesty in every project," or more directed, such as "quality above all else." In either case, the belief is a fundamental guide for every member of the organization to follow during his interaction with clients. As such, the belief must have a powerful impact on each employee to ensure that the belief becomes an ingrained part of every decision-making process. If achieved, the statement establishes a culture that is conveyed between veteran and new employees. However, if the statement does not create the required culture, then the danger arises that the statement becomes a buzzword phrase that is repeated, but never fully believed. For example, adopting a statement that ties to current trends such as "sustainable engineering" will never fully be adopted because it is seen for what it is, an attempt to capitalize on current trends.

Practices: The third element of the belief structure emphasizes the professional and business practices that are conducted by the organization. In contrast to the people-oriented beliefs that are addressed by employee and customer beliefs, practice beliefs are the code of conduct for the organization. Examples of these beliefs include "fairness in dealing" or "corners will never be cut." A statement is developed that every member of the organization can follow during the completion of his or her assigned tasks. The statement should be strong enough to leave no question as to its meaning, but simple enough to be easily remembered and practiced. Once again, the central key is the visible practice of the beliefs within the organization at all times. Creating a statement that organization leaders fail to follow will ultimately have a negative backlash from employees, customers, and partners.

Environment: The final element in the belief structure focuses on the interaction between the organization and the physical environment. As professionals that directly or indirectly derive primary income streams through interactions with the physical environment, this element should be considered as a potential component for every organization. However, it is critically important that an environmental belief be a statement that has intrinsic value to the organization. This is in contrast to the vastly overused environmental statements that civil engineering organizations have adopted to jump on the environmental bandwagon. For example, an intrinsically valuable environment-based statement is "examination of environmental options in all design solutions," while a bandwagon-type statement might be "green and lean." Although the latter statement is catchy, it fails to convey the underlying belief that the former statement conveys. This failure results in providing the organization avenues to deviate from the statement since it is considered to be a marketing slogan rather than a core belief. As the creators of the built environment, civil engineers continue to face increasing pressure from customers to protect this environment. The organization that demonstrates core beliefs in this area will have a fundamental advantage over those that fail to demonstrate this belief.

The completion of the belief identification process results in a minimum of four statements that reflect the internal and external practices that guide each organization practice. Upon completing these statements, the organization must decide whether all these statements will be retained, or if a subset will be incorporated into the final vision statement. Although that decision does not need to be made at this point in the process, each belief statement can be questioned at this time as to whether or not it is of such an intrinsic value that the organization members can never see it being changed. As an initial filter, this determination separates the marketing jargon from the core beliefs.

Organization Purpose

If the organization beliefs provide the guidelines for conducting operations and business, then the organization purpose provides the foundation on which all organization decisions are based. As documented in many industries over many years, the existence of a strong organization purpose can be directly correlated to the existence of a strong organization culture (Kotter 1990). Loosely defined, the organization purpose is the reason why the organization exists. It begins as the reason for starting the organization, progresses to the focus of the initial growth decisions, and emerges as the guideline for expansion and external acquisitions. In each stage, the organization purpose provides the unvarying statement to which every member can return when doubt emerges as to the appropriate direction in which the organization should move. Additionally, the purpose provides the point of reference around which these same members can congregate to achieve long-term objectives that the organization has set out to achieve. Finally, the purpose provides the single statement that leaders, members, and customers can focus upon to determine the fit between anticipated projects and organization purpose.

With this emphasis on organization purpose, the statement that is developed in this process can either serve as a solid foundation for long-term success, or as an insincere statement that serves no purpose other than to grace the pages of an annual report or business card. The difference between the two outcomes rests on the depth of belief that the organization demonstrates in the purpose. If an organization sets in place a statement of purpose that sincerely reflects the underlying reason why the organization was created, then the organization has the potential to build upon this statement and successfully create an organization culture. This culture emerges as continuity is reinforced from a purpose statement that not only reflects a fundamental purpose that was true when the organization was initiated, but continues to remain consistent in today's operating environment. As such, the statement instills a sense of pride and unity in the organization that transforms the members from a group of individuals into a single, organized team.

Is there a fundamental reason why this unity and pride is not seen as often within civil engineering organizations? There is nothing inherent to the industry that prohibits employee unity. Rather, the element that the authors have found to be missing within the industry is a sense of need by the organization to create this culture. Overwhelmingly, organization leaders interviewed by the authors were found to believe that civil engineering professionals are satisfied simply with the opportunity to practice their skill. Consequently, the need to create an organization culture is considered secondary to daily operations. Compounding this belief is the feeling by organization members that profits are the sole reason that the organization was founded. The organization members rarely give statements advocating the opportunity to achieve anything other than profits. The combination of these perceptions by both ends of the organization spectrum establishes a barrier that needs to be broken by a strong statement that speaks to the competitive nature of every member within the organization.

Specifying the Organization Purpose

Capturing the organization purpose into a single statement that reflects the intent of the organization requires the leadership to examine the true reasons for the organization existence. While this may seem to be a trivial statement, the existence of any contradictory perception within a civil engineering organization can make it difficult to retain a central focus. Therefore, to minimize the impact of this barrier, Figure 3-3 illustrates a method that can be used to create an analytical approach to specifying the organization purpose. The method addresses the basic elements that drive the establishment of an organization in a layered approach that examines the root purpose for existence and not simply the obvious surface reasons. The layers progress from the broad external concerns of the global market down to the personal satisfaction of the individual founders.

At the broadest level, the global market, the emphasis of the statement is on the intent of the organization to have an impact on the world at large. This may be a focus on sustainable engineering, assisting underdeveloped countries, or capitalizing on a worldwide demand for civil engineering services. In any such case, the statement reflects an organization that started specifically to address a broad, global issue. While this type of statement may be commonly associated with new business ventures developed by large organizations, there is no fundamental reason why any size organization, new or old, cannot be established on the principle of a global focus. If this is in fact the basis of the organization, then it should be documented and defended with the same passion that was required initially to prompt the idea.

Moving one step inside the global ring is the emphasis on customers. At this level, an organization is established based on the intent to assist customers to the greatest capacity of the individuals within the organization. An example of this type of orga-

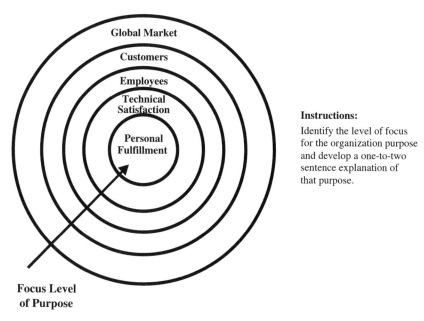

Instructions:

Identify the level of focus for the organization purpose and develop a one-to-two sentence explanation of that purpose.

Figure 3-3 The identification of the organization purpose can focus on a number of layers depending on the emphasis of the founders and current leaders.

nization would be a new Army Corps of Engineers office that is opened with the specific purpose to assist the general public through the project permitting process. The organization is conceived and implemented with an external customer focus that must be retained by all members at all times. The caution here is that the customer is an overused statement of purpose that creates skepticism in many individuals. Due to this fact, it is imperative that an organization that emphasizes the customer as the organization purpose instills in all members that customer satisfaction permeates all business practices. Any deviation from this focus will result in the erosion of confidence in the statement of purpose and reduce its effectiveness to serve as a unifying element within the organization.

Moving one step inside the customer ring transfers the focus of the purpose to an internal concern for employees. In this focus, the intent of the organization is to provide an environment where employees have the greatest opportunity to reach their individual potential. When placed in an environment where employees do not have the opportunity to be creative or work outside standard procedures, some employees will eventually build enough desire to change the environment that they will leave and establish a separate organization. The employee focus recognizes this potential by establishing a purpose that provides employees with the greatest opportunity to reach their individual creative and technical potential. Given this stated purpose, the organization is guided by the concern for the employee as the overriding determiner in any future organization decisions.

If the purpose of the organization has not been identified at this point, then the focus of the examination turns introspective to the individuals who originally founded the organization. Specifically, the development of the purpose statement focuses on the internal drive that led to the original idea for the organization. This focus on internal drive is divided into two levels, technical and satisfaction. The outer ring is the technical focus. At this level of purpose, the organization is founded based on a strong desire by the individuals to have the opportunity to utilize their unique knowledge to the greatest extent possible in the context of projects that they choose. The emphasis at this level is on the opportunity to select challenges and work under self-imposed, rather than externally imposed, guidelines. The individuals who initiate this type of organization are extremely self-motivated and have very personal ideas about the operation of the organization and the broader definition of success beyond profits. The potential difficulty in this situation is translating this to a statement of organization purpose that transcends from the individual founders to the organization as a whole over an extended period of time. In this situation, the greatest opportunity for success is to provide a statement that reflects the initial intent, with an extended focus on providing this opportunity for individual technical achievement to every member within the organization. Given this opportunity, the organization will attract individuals with a similar desire to demonstrate their unique civil engineering knowledge. However, of equal importance is remembering that freedom for one individual can turn into strict guidelines for another if every member of a team is not provided an equal opportunity to utilize their unique problem-solving skills.

If personal drive to achieve technical goals does not accurately reflect the personal desire to initiate an organization, then a second type of personal drive exists as the innermost basis for the organization purpose. In this second component of personal drive, the emphasis of the organization purpose is on personal fulfillment. Considered by psychologists such as Maslow to be the highest form of personal needs, fulfillment can best be understood by contrasting it to personal satisfaction. Personal satisfaction is achieved when individuals believe that they have completed a task to the best of their ability. The individual is proud and can proceed to the next task, comfortable with the knowledge that the task was completed to the highest personal standards. While this state of mind is desirable, it can actually be achieved in any setting. An individual can achieve this satisfaction from completing almost any task. In contrast, personal fulfillment moves beyond this level to address the need by individuals to achieve a goal that is beyond normal task-level activities. In the context of an organization, this translates to creating an organization based on the drive to fulfill an inner desire to achieve a level beyond that possible as an employee in another organization. The individuals who join together to create this organization share a desire to prove to themselves that they can achieve this goal, and will work beyond their normal capabilities to achieve personal fulfillment. When translated to employees, this type of organization has the purpose to create an environment where every employee has the opportunity to achieve the highest goals that each individual can set. Of greatest importance here is the opportunity for the employee to achieve at an organization level. Specifically, to attract employees with this motivation, there must be an opportunity for the employee to ascend to the ranks of ownership or directorship. If the employee does not have this opportunity, then it is likely that the employee will leave the organization after a period of time to create his own organization. Therefore, to avoid this loss of a unique individual, the purpose must reflect the opportunity to achieve individual and team goals without barriers or preconceived limits.

The identification of the organization purpose cannot be taken lightly. Of the three components that comprise the vision, the organization purpose serves as the philosophical anchor for all activities that the organization will undertake. The purpose must be an unwavering statement that guides every member to continue the tradition on which the organization was founded. Far too often, this is forgotten within an organization, resulting in members who have no sense of commitment or loyalty to the ideals that formed the basis and purpose for establishing the original entity.

Long-Term Goal

The third component that drives the development of a comprehensive vision statement is the establishment of a long-term goal for the organization. In a great divergence from the project-to-project mindset of the civil engineering industry, developing a long-term goal forces an organization to confront the distant future. In this future, the organization is setting a milepost that appears almost out of reach. Coined as a BHAG (Big, Hairy, Audacious Goal), the long-term goal represents everything that could be achieved if the organization could fulfill its ultimate plans (Collins and Porras 1996). Examples of this goal include statements such as, "To become the largest civil engineering organization in the Western United States," "To have annual earnings exceeding one billion dollars," "To have regional offices in every area of the world," and "To be recognized as the best structural design organization." As these examples illustrate, the BHAG can either be quantitative or qualitative. In either case, the goal is large enough that it appears to be out of reach to everyone outside the organization.

The obvious question for this component of the vision statement is why such a goal is required when so many uncertainties exist in the business world. Answering this question requires two components, a psychological component and a business component. The first of these factors emphasizes employee motivation. As outlined in the previous components of the vision statement, the need to provide employees with a direction and purpose lies at the core of successful, team-oriented organizations. Taking this concept one step further, the core element on which organizations can build cooperative, long-term relationships with employees is the ambitious goal. The key term in this statement is "cooperative," as in cooperation between organization leaders and employees. Cooperation between employees and organization leaders is a concept that is consistently highlighted within traditional civil engineering organizations, but acted upon as a core commitment to a far lesser degree. Specifically, the common sentiment voiced in organization interviews is the "need to know" perspective. This sentiment emphasizes the belief that employees outside the senior staff have very little need to know the long-term perspective of the organization. These employees should spend their energy and focus on specific tasks, while relinquishing strategic goal setting to the organization members who have the responsibility of keeping the organization a viable entity. Although this sentiment has been a common, although largely unvoiced, opinion within the civil engineering industry, management research has demonstrated that an alternative approach provides greater opportunities for organization success.

The psychological component of the long-term goal contributes to organization success by focusing directly upon the motivation of employees (Dreyer 1996; Brown, Cron, and Slocum 1997; Carr, Pearson, and Provost 1996). The limitation imposed by project-based goals is the lack of opportunity they provide organization members to extend themselves beyond short-term expectations. Given the narrow perspective of a short-term project, individual members do not have the opportunity to build an internal perspective of the broad opportunities that exist for the organization. Taken to the extreme, individual members rarely have the opportunity to explore the long-range (20–30 years) future. As such, the opportunity is limited to build a cohesive team that jointly strives toward a goal that is broader than an individual project. Reversing this trend requires a motivational tool that breaks this barrier and reestablishes the fundamental concept of teamwork. From this basis, the psychological component of the long-term goal emerges as a driving concept.

The contrast between short-term and long-term goals in terms of employee motivation is striking (Billingsby 1992; Kotter 1990). Short-term goals provide employees with a motivation factor that subsides at the conclusion of each project. When combined with the fact that a project incorporates a segment of the organization rather than the entire organization, a situation such as that illustrated in Figure 3-4 is created. In this scenario, the organization experiences multiple, out-of-phase motivation cycles where cohesiveness is substituted with individual pursuits. In contrast, the establishment of a long-term goal presents a scenario such as that illustrated in Figure 3-5. In this scenario, a primary motivation cycle is set in place that incorporates the entire organization and exists over an extended period of time. The establishment of this cycle provides a common goal that each member can reference as an ultimate achievement. Given this overarching objective, individual project goals serve to reinforce rather than substitute the primary motivation factors. An example of this scenario would be the establishment of a goal by a heavy equipment provider to have their equipment located on every jobsite in the United States. If a startup or small equipment provider establishes this goal, then individuals outside the organization may consider the goal unobtainable. However, to those inside the organization, the goal represents a significant challenge that can serve as a cohesive element designed to motivate each individual to act as a team in pursuing the final objective. Although each project will provide individual incentives for the organization, the long-term goal provides an underlying fabric that establishes an overriding objective that is stronger than any individual goal. In this manner, the psychological impact of the long-term goal is stronger than any single economic benefit.

Complementing the psychological benefit of the long-term goal is the business benefit. Specifically, the long-term goal provides the foundation for an organization to establish objectives and milestones that exhibit a consistent pattern over an ex-

Motivation Levels-Competing Divisions

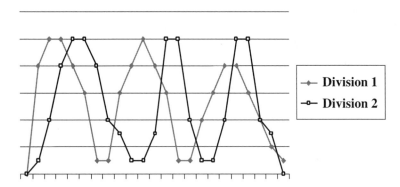

Figure 3-4 Individual objectives can lead to out-of-synch motivation cycles and employees experiencing highs and lows as projects start and complete.

Motivation Levels-Cooperating Divisions

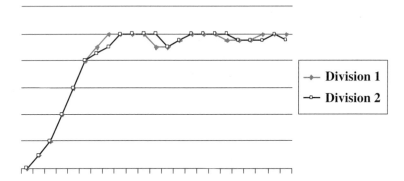

Figure 3-5 A coordinated approach to goals and objectives will lead to stability and long-term motivation focus as employees focus on organization accomplishments over individual group accomplishments.

tended period of time. In the same way that short-term goals limit cohesiveness within an organization, short-term business goals have the adverse effect of emphasizing short-term gains rather than long-term viability. Elements of this viability include:

· The ability to set consistent marketing plans
· The ability to hire contributing staff members
· The ability to establish consistent operational structures
· The ability to establish coherent and motivated teams

The common thread within each of these elements is the need to focus on a goal that reaches beyond the project barriers. Given this thread, individual project goals merge and define a broader business objective. In terms of the example elements in the list above, the establishment of the long-term goal provides the opportunity to guide human resources, operations, and marketing among all business elements. Each of the elements in the list represents a business component that requires either monetary or manpower commitments. The establishment of a long-term goal that remains beyond the project time frame provides organization leaders with a guideline for investing resources to obtain the goal. For example, the decision to enter new markets can be influenced by numerous market factors. However, when placed in the context of a long-term goal, the decision to enter a market can be evaluated against a consistent objective. To illustrate, the decision to enter an infrastructure rehabilitation market may be enhanced by a long-term goal that guides an organization to be the dominant player in the field of earthquake engineering. However, the organization that has set the goal to be the dominant player in recreational facility design should reexamine the decision to enter the infrastructure rehabilitation domain since it does not support the long-term goal.

This same goal also creates a unifying force that establishes an insider versus outsider competition within the organization. In this competition, organization members strive to achieve the long-term goal against apparently insurmountable odds. When confronted with doubtful statements of individuals outside the organization, organization leaders can answer these criticisms with strategically placed monetary and personnel investments consistently to challenge these doubts. For example, in returning to the equipment provider, the establishment of the audacious goal to have equipment on every site within the United States requires concerted organization decisions. From the location of rental facilities and repair yards, to the establishment of marketing and sales staffs, the organization must retain the ultimate focus as a guiding element in decision making. Questions regarding investments in technology, human resources, and infrastructure can each be evaluated against a common foundation. Additionally, while retaining this goal, the business decisions reinforce that the organization is dedicated and determined to prove that it can accomplish what many consider impossible. The business of heavy equipment thus transitions from an organization endeavor to a challenge that every individual can become committed.

The final component of the business perspective on long-term goals is the opportunity to achieve financial results that may not be possible when the organization is restricted to a project-based perspective. The establishment of the long-term goal provides organization leaders with a broader context in which to place financial results and goals. Rather than focusing entirely on the profits and losses that result from a particular project, the existence of a long-term goal provides the organization with the opportunity to address the financial results in terms of the relationship to the final objective. For example, a structural design firm entering the infrastructure rehabilitation field may lose 2% on an initial job. However, this loss can be viewed from different perspectives. From a project-based perspective, the loss appears as a significant strike against the management team engaged in the project. However, from the perspective of a long-term goal that explicitly states that the firm aims to become the dominant player in the infrastructure market, the existence of a loss can be justified from a strategic positioning perspective. Although a loss is never desired, if the completion of the job provides the firm with a solid foothold in the marketplace, then the loss may be recouped many times over in the long term. This business lesson has been proven and demonstrated internationally in diverse industries including personal electronics, automotive parts, and retailing. From the perspective of long-term gains, the existence of a short-term loss can be justified as the price of achieving the ultimate organization goals. However, without the existence of the long-term BHAG, the organization is relegated to justifying the loss from the traditional project perspective with little opportunity for long-term business analysis.

Comprehensive Statement

The final component within the vision statement summarizes the preceding elements into a single, comprehensive statement that reflects the ideas, sentiments, and goals of the organization. Although the elucidation of the values, purpose, and long-term goals is an essential component of strategic management, the practical side of the strategic management process requires a succinct representation of these elements. Specifically, a succinct statement is required to convey the organization vision to both internal and external constituents. Internally, the constituents are the employees that the organization desires to develop into a cohesive team. Externally, the constituents are the clients, suppliers, and agencies that interact with the organization on both a regular and one-time basis. Although each of these entities may study the broader vision statement components, the initial focus will center on the comprehensive statement. If the vision conveyed by the statement piques the interest of the constituents, then further analysis of the vision will occur. However, if the statement fails to convey the underlying vision, then the opportunity to sell the constituent during a first impression is irreversibly lost.

With the importance of the comprehensive statement originating from both internal and external constituents, the successful development of the statement cannot be overstated. This importance sets the stage for the synthesis task required to generate the comprehensive statement. In this task, a component of each vision element is combined into a single, representative sentence. To assist in completing this task, the following method can be used to structure the approach. In this illustration, a general civil engineering organization has decided that its current management practices are inadequate due to insufficient income.

1. Start with the purpose—Since the organization purpose represents the underlying reason for the organization existence, the purpose provides a backbone on which to build the comprehensive statement. For the civil engineering organization, the organization leaders identified an organization purpose that emphasized employees. Each of the founding members believed that the fundamental purpose for their firm was to provide every employee with the opportunity to use their individual knowledge in creative ways. Therefore, their organization purpose was stated as, "To provide an environment where every employee can demonstrate knowledge and creativity while solving unique civil engineering problems." Given this statement, the first step in the process is to identify the element that needs to anchor the comprehensive statement. In this case, two elements are potential candidates for the final statement, "demonstrate knowledge and creativity" and "solving unique civil engineering problems." Each of these elements represents a fundamental ideal for the organization's purpose. The former emphasizes the internal employee goal, while the latter represents the external professional goal. Selecting either or both of these elements for the final statement depends on the remaining selections.

2. Add the values—The organization values represent the beliefs of the organization that are intrinsically good. The only significant variation between the values of one organization and the values of the next is going to be in the focus of these values. Given the universal acceptance of intrinsically good values, these values play a supporting role in the comprehensive statement. Returning to our example, the organization leaders identified two overriding values from which the organization never diverges; "Organization actions must recognize the need to preserve the environment for future generations," and "Every employee should be treated with respect in every task." The key issue here is to abstract these values as a supporting element in the final statement. The options in this case focus on two possibilities, "preserve the environment," and "employees treated with respect." The values emphasized in these statements are subtle but clear to anyone who reads them concerning the intent of the organization. As with the organization purpose, the selection of the final element will be chosen in the context of the entire statement.

3. Add the goal—With the purpose and values providing the foundation for the vision, the long-term goal completes the statement by providing the motivational emphasis. When extracting this component, the long-term goal must be conveyed without giving away information that the organization would prefer to keep as an internal motivation tool. For example, although the organization may internally focus on overtaking an industry leader as the number one service provider in a specific niche category, it may not be strategically beneficial to state explicitly that goal to the external world. This may be due to marketing, industry, or professional concerns that could result in negative publicity or perceptions of the organization's intent. Therefore, the incorporation of the long-term goal must be made in a manner that accomplishes both the motivational and strategic goals. Returning to the illustration, the firm has set the long-term goal to become the national experts in the design of environmentally sensitive watershed areas and to become the dominant player in that specific industry segment. While this is a goal that the organization holds as the ultimate direction for operations, it may not serve the best interest of the firm to emphasize the dominant player aspect of the goal to their external constituents. Therefore, the candidate for the comprehensive statement is the role of the organization as "national experts in the design of environmentally sensitive watershed areas." This segment of the statement represents both vision and professional dedication while de-emphasizing the focus on becoming a dominant player.

4. Synthesizing the statement—With the identification of the three underlying components complete, the final step in the construction of the comprehensive statement is the synthesis of the components into a comprehensive vision. The goals in this final step are to create a statement that conveys the essence of the underlying components, serves as a marketing tool, motivates internal constituencies, and sets the foundation for long-term stability. Given this diverse set of objectives, it is understandable that the statement must combine specifics with generalities. The specifics are required for near-term issues that the organization chooses to direct through interpretation of the vision, while the ambiguity is required for flexibility of interpretation over an extended period of time. Putting this together in the context of the current example, the first step is to collect the potential elements. The result of this collection is as follows:

"demonstrate knowledge and creativity"

"solve unique civil engineering problems"

"preserve the environment"

"treat with dignity and respect"

"be national experts in the design of environmentally sensitive watershed areas"

Combining these statements into a cohesive statement requires the organization to focus on the underlying theme and then augment it with descriptive phrases. In this example, the underlying theme is the environment. Augmenting this theme are the concepts of respect, creativity, and problem solving that are expressed in the remaining statements. Combining this primary theme with the secondary concepts leads to a comprehensive statement such as the following:

> The vision of this organization is to demonstrate national leadership in the preservation of environmentally sensitive watershed areas through the application of unique environmental knowledge and creativity.

The decision in this case is to focus exclusively on the environment and knowledge components of the vision. Although the value of employee respect is not explicitly included in this version of the comprehensive statement, it does not mean that it has been eliminated from the organization's vision. Rather, the value is included as part of the entire vision package with the purpose and long-term goal. In this manner, the comprehensive statement serves as an invitation for both internal and external constituents to continue their exploration of the organization through examination of the full vision statement.

■ The Mission Statement

If the vision statement provides an organization with a general roadmap of an area to pursue, then the mission statement provides the directions on how to get to a specific point (Pearce 1982; Sunoo 1996). Whereas the vision statement addresses both internal and external constituents, the mission statement focuses inward to provide guidance to organization members concerning how the organization anticipates achieving these goals (Fairhurst, Jordon, and Neuwirth 1997; Rigby 1998). Figure 3-6 illustrates a diverse range of mission statements that exemplify the common focus on action and strategy implementation. As seen in these examples, the emphasis of a strong mission statement focuses on the actions that will be undertaken by the organization to transform the vision statement from a goal to a reality. Similar to the motivational role of the vision statement, the mission statement provides the organization with the focal point around which employees can rally for a common objective (Moore 1995). However, in contrast to the vision statement, the mission statement appeals to the desire of the individual to take immediate actions.

From an internal constituency perspective, the mission statement provides an opportunity for the organization to set in place the tasks required to put the mission statement to work. With a common goal set forth in the mission statement, the responsibility of organization leaders is to translate the mission statement into a series of tasks that are definable in terms of responsibility, outcomes, and mileposts. Through this definition, the mission statement transitions from an organization directive to a series of mileposts that serve as a framework for the development of individual action plans. Equally significant, the mileposts establish the competitive arena for the

---

Example Mission Statements

Parsons will be the premier integrated source of consulting, engineering, design, construction, operations and program/project management services. We will provide these services globally to a diversity of markets consistent with responsible social and business practices. Our work will be performed to the highest standard of quality, and we will exercise absolute integrity in our dealings with employees, clients, business associates, subcontractors, suppliers and the general public.
Parsons Engineers and Constructors

To help our clients make substantial and lasting improvements in their performance and to build a firm that is able to attract, develop, excite, motivate and retain exceptional people.
McKinsey & Company

To assist client s in attaining a competitive advantage by delivering quality services of unmatched value.
Fluor Daniel

Be the preferred supplier of open computer graphics solutions providing the highest quality hardware, software, and services to the complete satisfaction of our customers and partners.
Intergraph

To provide construction services in a professional manner to exceed the needs of our clients.
J.E. Dunn Construction Company

Provide professional construction solutions through the application of estimating, scheduling, inspection, project management, and dispute resolution services to owners, development teams, design teams, lending institutions, and the legal profession. Our goal is to deliver high quality services, on time, which are profitable to our client and ourselves.
Boyken International

To help clients change to be more successful. To work with clients from a wide range of industries to align their people, processes and technology with their strategy to achieve best business performance.
Andersen Consulting

Figure 3-6 Mission statements set the direction for achieving broad visions and establish a roadmap for setting specific objectives and goals.

organization. By translating the mission statement into specific mileposts, the organization sets the parameters to evaluate both groups and individuals. In this way, the abstract concepts associated with strategic management are brought back to the tangible reality often desired by engineering managers.

Similar to the internal constituency perspective is the value of the mission statement to the external constituency. Rather than limiting the roadmap concept to the organization members that will be traveling along the route, organizations can choose to include the external constituents in the proposed plans. This inclusion focuses on both clients and competition. For current and prospective clients, publishing the mission statement provides a security net for their investment of resources with the organization (Lammers 1992). No client wants to risk expending resources on an organization that it believes will not be a viable entity for an extended period of time. Given this understandable reluctance, the organization must provide some tangible evidence that strategic thought has been given to the long-term survival of the organization. The mission statement provides one component of this evidence. By providing the vision and mission to the client, the organization is stating that a long-term goal has been developed and the roadmap is in place to achieve this goal. With this foundation, the opportunity exists to build a relationship with the client that is based on a shared vision of goals and objectives.

Related to this shared client vision is the use of the mission statement to convey the organization's roadmap to the competition. In a competitive industry such as civil engineering, the organizations that emerge on top are the ones that take the opportunity to use both explicit and subtle advantages. In terms of the mission statement, an opportunity exists to use the statement as a subtle advantage in the continuing push to establish competitive positions. This opportunity emerges from the fact that the greatest threat to an organization that invests the time to develop visions and missions will be from other organizations that invest similar efforts. As part of this effort, organizations invest significant time evaluating the capabilities of the competition, including the development of these vision and mission statements. Where this becomes an advantage to an organization is in the psychological advantage that can be achieved by the mission statement. Specifically, if the mission statement provides a bold roadmap that can be translated into a set of aggressive actions, then the organization immediately puts the competition into a defensive position since they must respond to the potential actions. In this sense, the mission statement serves as a strategic component similar to a chess game. Since the competition is limited to viewing what is on the table, it must anticipate the next move that is implied by the current situation. In this mode, the competition can either act defensively to guard territory or respond by organizing a new offensive strategy. Either way, the competition is reacting to the organization's movements rather than focusing energy on establishing proactive positions. The value in this strategy is twofold. First, the organization is placing all competitors immediately in the position of playing catchup to an offensive position. Second, the competition is placed in a position where it must always consider the direction of the proactive organization in its future strategy sessions, creating a concern that requires additional resources. Although neither of these elements ensures success for the organization, they contribute to the overall positioning that serves as the foundation for the battleground within which successful organizations must maneuver to achieve competitive advantage.

Creating the Mission Statement

The development of the mission statement can be approached in a structured framework similar to that of the vision statement. The basis for this framework is the successful response to the following mission statement components:

- Who is the target group for the mission?
- What is the emphasis of the mission?
- How will the mission materialize?
- When will the mission be put in place?

Target Group

The first component within the mission statement is defining the target group that the mission statement will address. There is no limit imposed on who this target group can be. The only constraint is that the target group be a realistic consumer of the intended service of the organization. For example, if the organization decides that it will provide construction management services to Fortune 500 organizations, then the target group for the organization is not small, family-owned businesses that have small-to-medium capital expenditure budgets. Rather, the organization must explicitly focus on Fortune 500 organizations as the core recipients that will determine the success of the initial vision. Similarly, a public agency that works exclusively with individual developers on issues related to land use and environmental management should not focus on the institutional developer as the primary target group. Although this may appear to be an obvious sentiment, the misappropriation of resources to target groups that are not the core business recipient of the organization is a costly mistake to incur.

Returning to the example of the general civil engineering firm, the firm's emphasis on the preservation of "environmentally sensitive watershed areas" sets the boundaries for the target group. The vision explicitly emphasizes the focus of resources on projects that encompass issues related to watershed areas. The target group in this arena is extremely large, since it could encompass public works projects such as highways and bridges, recreational projects such as resort hotels, commercial projects such as shopping malls, or residential projects such as apartment complexes or subdivisions. Given this diversity, the organization must decide whether it will focus on a subset of these groups or retain a generalist point of view that emphasizes service throughout the target group. The two possibilities result in potential mission statement components such as "The organization will provide commercial developers with . . ." or "The organization will provide watershed developers with . . ." The former statement establishes a focused component of the potential target population, while the latter statement retains a flexible target population that can be narrowed or expanded, as the circumstances require.

The Mission Emphasis

The second component of the mission statement is the emphasis of the service to be delivered. Whereas the vision statement provided the broad outline for the organization, the mission statement provides a guideline for the specific services that will be offered. For example, if the vision statement refers to being the leader in the field of highway design, then the mission statement must elaborate on that sentiment to define what design services will be provided. In this translation process, the mission statement serves as the interpreter of the vision statement. Since it is difficult to build an action plan around the generalities espoused in the vision statement, a solid footing is required to build an action plan. The purpose of the mission emphasis is to provide this footing. However, to create this emphasis, the organization must thoroughly examine the potential services that it intends to offer. Although every service the organization may offer during the course of its existence does not need to be explicitly stated, the statement is a public forum in which a statement of services may be publicized.

The important emphasis in the preceding statement is the public focus. Anything that gets placed in the mission statement will be exposed to public scrutiny and review. Since the public is the ultimate consumer of the civil engineer's service, the emphasis placed in the statement must reflect the actual intent of the organization rather than an assumption of what the client desires to read. Similar to the vision statement, the longevity that accompanies the mission statement implies that the elements within the statement be avenues that the organization intends to pursue as part of a long-term strategy rather than a short-term market reaction. The alternative to this approach is a mission statement that must be updated every time a market alteration occurs. In this reactive mode, the stability, motivation, and confidence that the mission statement is intended to project is undermined as consistency fails to materialize.

Returning to our example, the emphasis of the mission statement must reflect the services that the organization intends to offer as part of its commitment to preserving environmentally sensitive watershed areas. In this example, the firm is composed of civil engineers with diverse backgrounds, including permitting and policy development, water resources and hydrology, project management, and general site design. This diversity converges in projects that require regulatory approval for developing watershed areas. Within these projects, the firm has the capability to assist in all phases, from permit applications to design and regulatory compliance. Given this broad set of capabilities, the choice that the organization must make at this point is determining the level of specificity that will be included within the statement. At one end of the spectrum is the generalist perspective that advocates a broad description of capabilities such as ". . . provide a full spectrum of civil engineering services for the entire project life-cycle. . . ." This perspective permits the mission statement to be interpreted as broad or as narrow as the potential customer requires for a given project. In contrast, the specialist perspective advocates a narrow description of capabilities such as ". . . provide full-service civil engineering services, including permitting, design, compliance, and project oversight . . ." This perspective emphasizes specific organization capabilities with the belief that the customer desires a clear description of the organization prior to considering it as a potential service provider. Either approach is an acceptable avenue for an organization to pursue. However, the defining factor rests on the profile of the potential customer. Since the mission statement is used both internally and externally, the constituents are the guiding factors for determining the appropriate specificity to include in the statement. In the example, the recommendation is to include a greater level of specificity. Given the technical character of the intended customer, the greater appeal to this customer will be in the specific rather than the general.

Mission Implementation

The third component of the mission statement builds upon the previous two by providing the framework in which the service will be conducted. In this framework, the mission implementation answers the "how" component of the mission after the previous sections have answered the "who" and "what" questions. In contrast to the previous elements, the mission implementation can be a challenging concept to convey in a few words within the overall mission statement. However, the mission implementation has a critical role to play in the overall mission statement. The brief statement that emerges from this element provides the first concrete indication of the path that the firm intends to follow to put the abstract vision statement into practice. Therefore, this brief indication needs to have an immediate impact on the reader in terms of both short-term action and long-term organization development. For example, a mission implementation element that states ". . . within in an environment that emphasizes individual achievement, thought, and drive . . ." immediately creates an image within the reader's mind of the professional environment in which the service will be provided. The reader can interpret from the element that the individual in charge of the project has the opportunity to develop action plans independently from a central control group. Similarly, long-term goals have been put in place that emphasize individual advancement through the demonstration of individual achievement. The combination of these goals conveys an environment that is progressive, action oriented, and responsive to the customer rather than traditional and process oriented.

Given the potential impact of the mission statement, the selection of the words to include in the mission implementation must be considered carefully. Both internal and external constituents will analyze each word in the element for the underlying meaning. From the internal perspective, the constituents will be looking for consistency between the stated objective in the mission statement and the actions taken by the organization to provide a commensurate work environment. Similarly, from an external perspective, the constituents will be looking for consistency between the stated objective and the level of service provided to the customer. In either situation, if a constituent believes that an inconsistency is detected, then the result can be a loss of trust between the individuals responsible for establishing the organization direction and the target constituent. Taken to its extreme, if the target constituent believes the mission implementation element is an empty statement representing no true belief, then the trust between the organization and the target constituency could be permanently damaged.

In the example, the mission implementation element needs to respond to the vision statement claims that the organization will demonstrate national leadership through the application of unique environmental knowledge and creativity. Given a limited number of words to express how this vision will be implemented, the mission statement must focus on the core component behind this statement. In this specific case, the national leadership is intended to emerge from the practices demonstrated by the firm's projects. Therefore, the emphasis of the mission implementation should focus on the latter part of the statement. But what does it mean to apply unique environmental knowledge and creativity? The vision sounds appealing, but lacks the specifics to create a tangible direction. However, this ambiguity also provides multiple opportunities to create a mission implementation element. The words "unique," "knowledge," and "creativity" each provide opportunities to expand upon the underlying concept as follows:

Word	Equivalent Phrase
Unique	Progressive, technology-based environment that combines emerging technologies with traditional civil engineering practices
Knowledge	Organization structure that combines project, theoretical, and experimental experience and knowledge into an integrated, organization-level knowledge repository
Creativity	Environment that emphasizes individual achievement in producing one-of-a-kind solutions to unique problems based on organization experience and tradition

As each of these examples illustrates, the mission implementation element walks a thin line between abstractness and specificity. Each phrase includes enough detail to provide a foundation for implementing organization action plans, while retaining the abstract quality of the vision statement to allow for flexibility and long-term adjustment.

Mission Timing

The final component of the mission statement places the statement in a temporal context. Providing grounding for the mission statement, the mission timing establishes a timetable in which the mission statement will be pursued. In contrast to the previous mission statement components that provide a framework irrespective of evolving time, the mission timing provides both internal and external constituents with mileposts that can be evaluated in terms of mission completion. In contrast to the abstract focus of the vision, and the previous mission statement components that provide no definitive plan for achievement, the mission-timing element contributes a definitive time-based goal to the long-term plan. Given this specific requirement, every employee and customer can evaluate the organization in terms of its progress towards achieving the stated deadline.

Although a definitive time frame for achieving a mission statement is preferable by some organizations, others shy away from this element due to its success or failure emphasis. In these cases, an alternative approach can be adopted where the mission statement focuses on an indefinite time period. Specifically, a descriptive time frame can be included in the mission statement that clearly defines a time goal, but without the limitation of a numeric value. For example, returning to the example firm, the phrases "by the beginning of the next decade," "prior to the competition," "while the natural resources remain" each establish an implicit time frame for their mission statement without the constraint of a specific date. With these implicit time-frame variations available, the firm can adopt one of two general approaches to including a mission-timing component. The first approach is to focus on a relative time frame relating to the existence of potential danger to environmentally sensitive watershed areas, such as "with the explicit purpose of establishing a national awareness of existing environmental concerns." The second approach is to focus on the ascension of the firm to national prominence in a given time period, such as "prior to the emergence of competitors attempting to duplicate our efforts." Each of these approaches provides the firm with a specific goal, but eliminates a definitive date that leads to the redevelopment of the mission statement on an annual basis.

Putting It Together

With the identification of the target group, mission emphasis, mission implementation, and mission timing, the pieces are in place to create a synthesized mission statement. Similar to the synthesis of the vision statement, the objective of the synthesis process is the development of a statement that serves the organization over an extended period of time. Specific to the mission statement, the synthesis of the parts must establish a statement that serves as an intermediary between the abstractness of the vision statement and the specificity of the organization goals. The preferred result creates a statement that provides both a degree of abstractness for internal and external constituents to interpret and a degree of specificity from which specific goals can be established.

The first step in obtaining this result is to gather the potential mission statement elements. For the example, the synthesis process results in the collection of the following elements:

"The organization will provide commercial developers with . . ."

"The organization will provide watershed developers with . . ."

". . . provide a full spectrum of civil engineering services for the entire project life cycle . . ."

". . . provide full-service civil engineering services, including permitting, design, compliance, and project oversight . . ."

". . . progressive, technology-based environment that combines emerging technologies with traditional civil engineering practices . . ."

". . . organization structure that combines project, theoretical, and experimental experience and knowledge into an integrated, organization-level knowledge repository . . ."

". . . environment that emphasizes individual achievement in producing one-of-a-kind solutions to unique problems based on organization experience and tradition . . ."

". . . with the explicit purpose of establishing a national awareness of environmental concerns . . ."

". . . prior to the emergence of competitors attempting to duplicate our efforts . . ."

Combining these diverse elements into a cohesive statement requires the organization to focus on a singular theme that serves as a binding thread throughout the statement. In this example, the binding thread follows the environmental theme established in the vision statement. Each of the statements explicitly or implicitly emphasizes environmental awareness or the use of creative problem solving techniques to achieve environmental protection. Given this emphasis, the following synthesis represents one potential mission statement.

The organization will establish a national awareness of environmental concerns based on experience and tradition while providing developers working with environmentally sensitive watershed areas full-service civil engineering services including permitting, design, compliance, and project oversight in an environment that emphasizes individual achievement and one-of-a-kind solutions to unique problems.

The final statement is a reflection of the thoughts, purpose, and ambition embodied within the organization. The statement provides direction without specifics, long-term goals without explicit deadlines, and action without micromanagement. Through this combination of abstractness and explicitness, the mission statement provides the entry point to the final component of the vision–mission process, the setting of specific goals.

The Organization Goals

Managers require goals and milestones to measure progress effectively within individual groups and projects. Management research in diverse fields including electronics, appliances, and athletics has demonstrated the value of specific goals in developing action plans, evaluating employees, and transforming vision and mission statements into tangible directives (Clark and Koonce 1995; Rogers 1993; Kaplan and Norton 1996; Mintzberg 1994). The underlying theme behind each of these research efforts is the necessity to have tangible directives that each member of the organization can review during the execution of individual projects. In this transition, the organization goals provide the first steps away from the common focus established by the vision and mission statements. In this divergence, the establishment of organization goals translates the mission statement into specific objectives that emphasize the role each group will play over the next planning period. Depending on the size of the organization, this could translate into a set of goals for a single group, or sets of goals for dozens of organization teams. In either situation, the development of these goals follows a similar pattern that emphasizes placing the vision into the domain of numbers and measurements.

The establishment of goals does not completely focus on establishing numeric objectives for individual managers. Rather, the establishment of goals provides organizations with both specific objectives and intermediate mileposts from which to evaluate the progress toward obtaining the overall vision and mission. In these objectives and milestones lie the individual instructions for reaching the destination laid out in the vision and mission roadmap. From specific checkpoints to time and cost estimates, the goals set the tangible measurements that both executives and entry-level employees can objectively review during progress evaluation.

Goals as Objectives and Milestones

The development of organization goals that serve as specific objectives and milestones for managers and staff-level personnel can be accomplished in a structured process that facilitates a focused approach to goal generation. In this approach, goals are developed that focus on specific directions that put the mission statement into an achievable action plan. The goals focus on items such as new market exploration, new revenue opportunities, employee education, and technology enhancement. In each case, the emphasis is on the incremental enhancement of existing conditions in an effort to achieve the broad objectives outlined in the mission statement. For example, an organization that sets a mission objective to demonstrate technological superiority in all operations may set an individual objective to investigate and acquire the latest visualization technologies available for the civil engineering industry. This is a specific action-oriented goal that can be evaluated at a specified time increment in terms of its completion of lack of completion.

Developing these objective goals follows a similar process as that followed for the vision and mission statements. At the core of the process is the need to emphasize the underlying purpose that the organization has established. Similar to the mission generation process that elaborated on the organization purpose established in the vision statement, the process for setting goals focuses on the elaboration and refinement of the elements outlined in the mission statement. Specifically, the process requires the organization to:

- Develop ownership of the goals
- Integrate the goals into a cohesive objective
- Demonstrate commitment to the goals

Ownership of Goals

The first step that an organization must take to enhance the odds of success of achieving the stated goals is to ensure that every member of the organization believes in the goals. Research has demonstrated that the greatest impact of organization goals is realized when each member of the organization is committed to the attainment of those goals (Riggs 1995; Rao, Schmidt, and Murray 1995; Reichers 1986). Additionally, the greatest opportunity to achieve this commitment is through the collective generation of goals. In contrast to many traditional organization processes that focus on a management team establishing a set of goals for each individual group or division, the preferred method to establish goals is through inclusion and exchange. Specifically, the generation of organization goals should focus on a bottom-up approach that encourages individual group and division members to participate in the goal generation process.

The reasoning behind this bottom-up approach is twofold. First, individuals who directly participate in the generation of goals will have a significantly higher commitment to the goals since a sense of ownership in the goals is established. Second, the goals that the individual groups establish will be more ambitious than those set by upper management since a sense of competition emerges between groups and divisions. The first issue addresses the basic desire of individuals to achieve personal goals. The challenge of meeting these goals is significantly higher when the goals are self-imposed rather than externally imposed. This phenomenon of rising to the occasion can be seen time and again in organizations that permit individual groups to set an initial path to success. This does not exclude the organization leadership from modifying the goals, but emphasizes the role of the individual in serving as the impetus for the goals. Similarly, the second issue builds upon the fact that individuals are more committed to goals that are self-imposed than goals that are externally imposed (Goodson and McGee 1991). Given these self-imposed goals, a strong desire exists to not only achieve these goals, but demonstrate to other groups the ability to achieve at a higher level than is expected. Although this competition must be controlled to eliminate the possibility of the organization digressing into a series of independent operations, the desire to excel beyond expectations is a human desire that can be focused into constructive energy. The combination of these two factors creates a compelling argument for broader employee inclusion in the goal development process. However, to achieve this inclusion, the organization must present the members with a sincere request for involvement. The organization cannot approach the topic by presenting the members with a set of specific goals and then ask the members to approve the list. Rather, the organization should request the groups to take the initial steps in developing the goals and objectives that will directly affect their members.

The development of these goals should begin with an open request that retains only two primary restrictions, the goals must reflect the emphasis of the mission statement, and the goals must be stated in a manner that can be evaluated. Given this broad interpretation capability, the individual groups can proceed to establish goals that focus on any topic that is considered a core component of their operations. Returning to the example firm, one group within the firm is responsible for project oversight tasks. In the past, this group has been a service group for the organization, providing oversight services as part of a larger package provided to customers. However, in this strategic management phase, the group believes that it can package project oversight services as an individual service directly to outside customers. As such, the group establishes the following goals:

1. Obtain outside customers that account for 30% of group revenue—This goal emphasizes a measurable revenue amount that establishes the potential market for independent project oversight services.
2. Identify public and private organizations that potentially will require project oversight services over the next three years—This goal emphasizes a market analysis activity to establish a baseline study for long-term viability of the project oversight concept.
3. Build technological capabilities to demonstrate support for independent project oversight—This goal emphasizes the need to enhance the infrastructure of the group to pursue the project oversight goal.

The group submits these three goals to the senior management team as an initial foundation for the ensuing year's activities. The goals are limited in number, but powerful in scope. When analyzed in their entirety, the goals establish an ambitious roadmap for the group. The combination of goals creates financial, marketing, and infrastructure goals that demonstrate the opportunities available for an independent project oversight group. Of equal importance, the goals establish a measurable set of actions that the organization can evaluate at the end of the current planning period.

Integration of Goals

As illustrated in Figure 3-7, the development of individual goals may lead individual groups in directions that move in directions away from the common mission. Although this divergence from a single set of goals is necessary to establish the ownership of the goals, the organization must integrate these goals into a cohesive plan to reduce the possibility of the organization splitting into a number of groups that remain under a single umbrella in name only. As illustrated in Figure 3-8, the integration of goals refocuses the organization on the overall goal of achieving the long-term elements outlined in the mission statement. For example, if three different groups within the organization are each proposing that they establish independent revenue streams, the organization leaders must examine the goals to ensure that the groups are not proposing that they compete for the same customers. In this situation, the greater benefit to the organization is to approach the customers with a comprehensive package of services rather than a series of independent proposals. In this manner, the organization appears to the customer as an integrated organization that can utilize diverse resources to provide comprehensive services during the entire project life cycle. However, the responsibility to integrate goals such as these must be re-

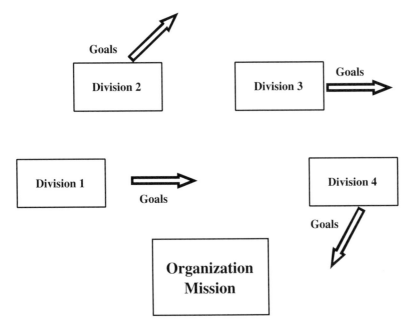

Figure 3-7 Individual goals can diverge from the central organization mission if they are not coordinated in a central effort.

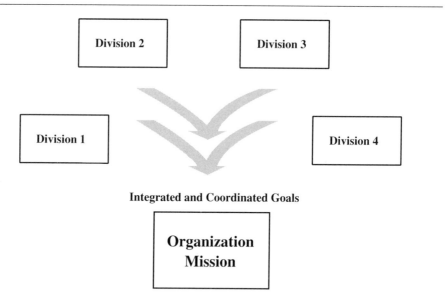

Figure 3-8 The integration and coordination of goals reinforces the organization commitment to a central mission.

tained at the organization level rather than the individual group level since an overall perspective is required to identify the potential conflicts.

The integration step requires senior managers to retain a balance between constructive input and control. Although this may appear to be two independent concepts, the relationship between the two is significant in this stage of establishing goals. In terms of the former, the emphasis of the senior managers should remain on the refinement of the submitted goals to ensure alignment with the mission statement. The ownership established by the individual groups can easily be lost if the senior personnel eliminate or fundamentally alter the submitted goals. The task assigned to these personnel is to review the goals for a realistic chance to succeed and then modify the goals to bring them back to a level where success is possible. Additionally, the review team should emphasize relationships between the submitted goals. Wherever possible, the goals should be modified to build upon the strengths of related goals from different groups. For example, if two groups are submitting goals that dovetail together in terms of the project life cycle, then the review team can modify each goal to include language that emphasizes cooperation between the groups in achieving each goal. Finally, if a submitted goal is outside the domain of the mission statement, and the organization does not wish to pursue this service area, then it should be removed with the explicit statement that extending into this area is inappropriate at this time. Each of these review tasks emphasizes refinement and structuring to achieve common purpose and advantage.

The opposite approach to the integration task occurs through either lack of control or overcontrol. The lack of control approach can be referred to as "integration by the stapler." In this mode, senior managers accept the goals from each group, combine them into a notebook without revisions, and reexamine them at the conclusion of the planning period to determine if the groups have achieved their stated goals. The lack of integration leads to the strengthening of divisions to the detriment of the overall organization, since each group understands that their measure of success is the ability to meet individual goals rather than goals set by the organization. In contrast, the overcontrol approach allows the review team to rewrite the submitted goals to a point where they reflect the ideas of the review team rather than the submitting groups. This approach is possibly the most destructive approach that an organization can adopt. By rewriting the goals to achieve a different result, the organization is saying to the groups that their input is an exercise in wasting time. The inevitable result will be the lack of commitment that is seen in organizations during the strategic planning process. Since the groups believe that their efforts are looked upon as a formality by the organization, the amount of effort in the process is reduced to a minimum, and the identified goals are forgotten as quickly as they are submitted. However, of greater importance is the lack of ownership developed by the review team for the goals. These are not the goals of the individual groups, and as such, it is unreasonable for the organization to assume that the groups will fully commit to achieving an externally enforced set of benchmarks.

Returning to the context of the example, the original project oversight group had established the following goals:

- Obtain outside customers that account for 30% of group revenue.
- Identify public and private organizations that potentially will require project oversight services over the next three years.
- Build technological capabilities to demonstrate support for independent project oversight.

Independently, these goals provide the group with a set of defined, measurable objectives for the upcoming planning period. Although these goals may be acceptable to the group, the organization must ensure that the goals do not conflict with those set by sister groups. For example, a sister group that is responsible for policy analysis consulting has set a goal that emphasizes technological improvement to assist in expanding the core business of the group. At the organization level the congruence of these goals can be, and should be, detected. Once detected, the project oversight group can modify its technology goal to emphasize integration at both an organization and group level as follows:

> Build technological capabilities in conjunction with the organization to demonstrate support for complete project development and oversight.

Demonstrate Commitment to the Goals

Although an organization may take initial steps to require the development of goals, the simple establishment of goals does not guarantee success. Similarly, the integration of the goals is not sufficient to ensure the successful pursuit of the objectives. Although these two steps are necessary components of the goal process, the final component of the process that is required for successful achievement is the demonstration of commitment to the goals. Trust between management and staff can either be developed or shattered due to the perception by the staff about the level of commitment that management has for supporting, committing, or enhancing group goals. Although psychological and industrial management research efforts have documented the subtle effects of mistrust between management and staff, the element of trust can be reduced to a demonstration of explicit commitment (Klein and Kim 1998; Cocco 1995; Tubbs 1993; Locke, Latham, and Erez 1988). An organization demonstrates commitment to goals by providing the resources required to achieve the individual objectives set by the groups. Illustrating this idea is best achieved by returning to the example firm. As illustrated below, each of the goals requires organization resources to establish the greatest opportunity for success.

Goal	Organization Resource
Obtain outside customers that account for 30% of group revenue	Commit marketing and business development resources for customer identification
Identify public and private organizations that potentially will require project oversight services over the next three years	Commit cross-group resources to identify customer leads from previous projects and known leads
Build technological capabilities in conjunction with the organization to demonstrate support for complete project development and oversight	Commit organization financial and human resources to design, implement, and train personnel for an integrated project system

Although each of the goals has a realistic possibility for success without organization assistance, the opportunity to succeed is significantly enhanced when the organization commits to assisting the groups in their pursuit of annual objectives. If organization leaders elect to remove themselves from the pursuit of group-level goals, then the groups will understand that the responsibility for success or failure rests completely within their domain. The subsequent result of this independent responsibility contains two elements, resentment and competition. The first of these elements refers to the break that occurs between the group and the organization. When the group believes that the organization is not committed to the stated goals, then the trust between the organization and the group quickly changes to resentment. This resentment in turn can evolve into division between the group and the organization that results in a decline in productivity. Since the organization demonstrates no commitment to the group, then the group understandably experiences no commitment to produce results for the organization. The resulting downward spiral in productivity assumes a momentum that is difficult to reverse.

The second element that can develop from a lack of organization commitment is the rise of internal competition. Given a scenario that leaves groups responsible for achieving their own goals, the tendency of the groups will be to erect virtual boundaries around their individual domains. The virtual barriers may consist of varying levels from simple lack of cooperation with other groups to hostile inter-actions and information obstruction during group communications. Left to build over an extended period of time, this competition results in deep divisions devel-oping between organization personnel. With the increasingly deep divisions be-tween groups, organization leaders are placed in the positions of arbitrators. At this point, the focus of organization discussions diverges from the core issues of objectives and goals to a focus on equality of resource commitment and commu-nication barriers.

It is important at this point to emphasize that the transfer of visions and missions to successful goals and objectives is not destined to result in destructive competition and resentment. Rather, the generation, integration, and commitment to goals pro-vides a strong foundation for an organization to transform abstract concepts of vi-sions into tangible financial results. However, the final component required to ensure the success of this transition is the human element that is often generalized under the topic of leadership.

Change in Goals over Time

The introduction of goals and objectives provides an organization with measurable milestones that every member can personalize to their own task. However, an orga-nization cannot rest on the success of this goal identification process. Although these goals may serve as the immediate requirements for the organization, the changing market forces require the organization to reevaluate goals on an annual basis (or more frequently depending on the business environment). This reevaluation process is an integral component of the strategic management process.

Whereas vision and mission statements serve as long-term roadmaps, goals must serve as reminders of the need to reevaluate the position of the organization in its larger market sector. Specifically, goals serve as checkpoints for achieving the vision statement. In this role, the organization must accept that goals will change and wa-ver over time. This change can occur in one of two ways. First, goals can be expanded or contracted but retain the original intent. In this mode, the organization takes an original goal such as revenue and reevaluates the revenue target in response to stronger or weaker market conditions. This change represents an alteration, but falls short of a change in direction. While this form of modification can be beneficial, organiza-tions need to resist making a regular practice of setting a single goal and then rely-ing on alteration as the main source of annual reevaluation.

In contrast to the alteration approach to goal reevaluation, organizations can elect to modify the goals at a fundamental level. In this approach, the organization ac-knowledges that the strategic goals are not providing the path required to achieve longer-term missions and visions. In this case, organization leaders must demonstrate leadership by moving the organization in a new direction. While significantly more difficult from a leadership perspective, the modification of goals is often necessary fully to accept the conditions of the changing marketplace.

■ Leadership in Strategic Management

The establishment of the organization vision, mission, and goals provides the foundation on which to build a directed management plan. Aggressive or conservative, the organization can choose a path that reflects the philosophy of the organization founders. In either direction, as long as the organization is consistent with the statements developed during the previous processes, then the organization is ensured of a roadmap that follows the agreed-upon goals. However, progressing along this path presents an organization with a strategic issue that transcends industry segment, public or private status, size, and length of existence. Specifically, the issue of leadership underlies the ability of each organization to transform the statements placed on paper into strategic actions.

The mantra of leadership has rung clearly through management research, literature, and popular books since the inception of business as a formal discipline. From Barnard in the 1930s, to Herzberg in the 1960s, and on through Porter in the 1980s, a succession of influential scholars, industry leaders, and writers have placed their personal perspective on the leadership topic. Concurrently, military tacticians throughout the centuries have focused upon the art of leadership as a necessary requirement for successfully carrying out complex military missions (Hart 1930). The common link between these diverse scholars and practitioners is a focus on the importance of the human element in the successful translation of formal plans and visions into tangible results (Kotter 1990; Tannenbaum and Schmidt 1973; Katzenbach 1997; Farkas and Wetlaufer 1996; Rosener 1990). In contrast to the analytic and numeric rationales conveyed by the accounting and finance professionals, the leadership writings rely primarily on the psychological and sociological aspects of human nature to convey arguments. Although potentially difficult for civil engineers to accept due to the analytic basis of the industry, the focus on leadership is transcending the business domain into the realm of civil engineering.

Leadership as Caretaker

The identification of leadership as a necessary element within a successful organization provides a basis for developing leadership skills, but this identification begs the question of what skills need to be developed to establish leadership qualities. This question becomes even more difficult to answer when the spectrum of leadership writings is studied. In these contexts, leadership is discussed at every possible level of detail, leading one to conclude that leadership implies that a single individual can be responsible for every aspect of organization behavior. Within the civil engineering industry, this diversity of definitions and thoughts is reflected in the growing number of professional education courses emerging on the industry stage. Proposing to educate civil engineering professionals on topics ranging from team dynamics to motivation and dynamic speaking, the introduction of leadership as a central education theme is rapidly bringing leadership to the forefront of industry concerns. Additionally, statements in journals such as the *ASCE Journal of Management in Engineering* are bringing the leadership topic to the attention of a broad audience by highlighting the observed lack of leadership in the civil engineering industry (Hilton 1995). When these statements focus on the lack of leadership training that civil engineering students receive during their formal education, the topic transcends the professional domain into the educational domain. This crossover powerfully brings the leadership topic home to educators as well as practitioners. However, leadership should not necessarily be equated with experience. A civil engineer with 25 years of experience does not have any greater intrinsic advantage in the leadership arena than an entry-level engineer. Rather, leadership is a quality that can be demonstrated at any stage in an individual's career. As listed below, the emergence of leadership can encompass a broad range of roles that are not necessarily related to the length of time an individual spends within an organization structure.

- Leadership promotes the development of visions and strategies for future organization direction.
- Leadership promotes the development of collective consensus for achieving long-term goals and objectives.
- Leadership promotes the motivation of individuals to exceed perceived limitations and boundaries.
- Leadership promotes the development of an organization culture that encourages creativity, individual achievement, and the promotion of future leaders.

Fully discussing leadership and its role in the civil engineering firm justifies an entire book by itself. Of greater importance to the current discussion on leadership in strategic management is a significantly narrower emphasis, leadership as the caretaker of vision, mission, and goals. As stated earlier, the development of the vision and mission statements, together with the accompanying organization goals, provides the organization with a roadmap for pursuing short-term and long-term objectives. However, the value of the roadmap is reduced to virtually nothing if an individual chooses not to refer to the map. In the context of the organization roadmap, the vision, mission, and goals may be visionary, inspiring, and motivating on paper, but the organization requires an individual to lead the transformation of these statements from words to focused work efforts. In practical terms, the information and direction may be available, but the lack of reference to the direction will result in the organization veering away from the plotted course. Preventing the organization from veering away from the intended path is the role of the civil engineering leader in the context of strategic management.

From this perspective, leadership is placed in the context of a caretaker because of the relationship to a caretaker who has been granted the critical responsibility of protecting a vulnerable individual. In the latter role, the caretaker is responsible for protecting the individual from harm or threats that may originate from internal or external forces. Adopting methods that are appropriate to the circumstance, the caretaker is required to maneuver through these threats to either arrive at a predetermined destination, or reinforce the defensive position in which the individual resides. In either circumstance, the underlying objective remains the successful fulfillment of a planned set of goals that are meaningful to the protected individual and the organization in which the individual operates. When compared to the caretaker role of the civil engineering leader, this description can be applied with minimal alterations. The leader as caretaker has the responsibility to protect the vision, mission, and goals from threats by internal and external constituents who have the intent to alter the direction of the planned objectives. The leader must adopt the roles and procedures necessary to ensure that the effort placed in developing the strategic statements is rewarded with the successful transformation of the statements into tangible actions.

Given the diversity of individuals who enter the civil engineering industry, the role of caretaker can be assumed by a broad spectrum of personalities and management styles. Rather than attempt to mute these individual differences within the context of an ideal caretaker, these differences should be recognized within the multiple methods that may be adopted for successfully achieving the organization objectives. Provided that the organization does not allow the caretaker to veer from the intended course, the caretaker may adopt one of several approaches to achieving the organization goals:

> *Caretaker as motivator*—The first, and recommended, approach that a caretaker may invoke to achieve organization goals is the role of motivator. The motivator approach emphasizes the caretaker providing the individuals with opportunity to achieve beyond their believed limitations. In this role, a caretaker demonstrates leadership by speaking to the ambitions, dreams, and loyalty of the members. For example, in the case of the example firm, the leader can motivate the organization by focusing on the stated objective to produce "one-of-a-kind solutions." From this one statement, a motivating leader appeals to the organization to approach client projects with a combination of proven solutions and progressive alternatives to deliver the unique, one-of-a-kind solution. However, the motivator does not rely solely on organization-level appeals. Rather, the caretaker who adopts a motivator approach spends equal or greater amounts of time speaking to team members to inspire within them a desire to achieve results that are greater than the individual believes can be achieved. Through this individual motivation, the leader recognizes that the achievement of large, organization-level visions is dependent on each individual achieving a small component of the vision. As such, the motivator may need to adopt the role of friend, confidant, or executive depending on the requirements of each organization member. The motivator recognizes that each member of the organization responds to different internal drivers, and subsequently adopts a motivating style to address these drivers. Finally, the motivator recognizes that the key to achieving organization objectives is

the delicate balance between encouraging group achievements for the organization, and the encouragement of individuals for personal advancement.

Caretaker as exemplar—The second approach to strategic management leadership recognizes that not all civil engineers feel comfortable in the role of a motivator. In an industry where the majority of members are introverts, it is unreasonable to advocate that every leader adopt the motivator role as the method for achieving organizational success. In the second approach to caretaking, the leader approaches the achievement of organization goals by adopting the role of exemplar. In this role, the leader encourages the members to achieve beyond their assumed limitations by setting an example for personal effort and achievement. Rather than relying on words to convey the achievement message, the leader conveys the point by personally striving to excel at required tasks. For example, in the general civil firm, the exemplar leader would focus on achieving "one-of-a-kind" solutions to client issues by personally working to develop solutions that were unique from anything presented by the competition. By combining a small number of organization speeches with a great deal of personal effort, the exemplar leader adopts a "do as I do" approach to achieving organization goals. The exemplar leader recognizes that individuals relate to examples of excellence and effort, and attempt to achieve at these levels in an effort to live up to the example set by the organization leader. However, it is important for the exemplar leader to recognize that a delicate balance is also faced in this leadership method. If the leader uses the exemplar role to demonstrate that organization achievement can only be reached through superhuman efforts, then the approach can rapidly backfire. Specifically, if the leader pushes the staff increasingly to raise the number of hours at their jobs by consistently adopting the first one in, last-one-out attitude, then significant numbers of organization members may turn on the leader. The exemplar leader cannot cross the line that separates professional and private lives. If organization members believe that professional responsibilities are replacing personal responsibilities as the primary role in daily existence, then the example set by the leader will be seen as manipulative rather than motivating.

Caretaker as celebrity—The third approach to the caretaker role swings 180 degrees from the exemplar approach to an emphasis on celebrity and charisma as the keys to organization success. The most difficult of the four caretaker roles, the celebrity approach to achieving organization goals relies on the small percentage of individuals who can push an organization to succeed solely on the power of their individual message. As demonstrated by the long-term references to celebrity leaders such as Chrysler's Lee Iacocca and Microsoft's Bill Gates, the number of individuals who can equate the organization with their personality is scarce. In this role, the caretaker relies on inspiring speeches to the organization as well as the perception that the leader has the individual capability to lead the organization to success through whatever challenges may be encountered. This personal strength is subsequently transmitted to the achievement of tasks through the belief by organization members that the failure to achieve stated goals is a failure to satisfy the personal request of the leader. This personalization changes the abstract goals into challenges reflecting the personal commitment of the charismatic leader. Placing this leader in the context of the example firm, the celebrity caretaker would approach the "one-of-a-kind" solution statement as a personal belief that the organization has the ability and responsibility to provide clients with solutions that meet this criterion. Through the use of passionate appeals at both group meetings and individual sessions, the celebrity caretaker infuses the objective with personal belief. However, similar to the previous caretaker methods, the celebrity approach has fundamental flaws. In particular, the celebrity approach fails if the organization finds a reason to doubt the personal integrity of the leader. Whether the indiscretion is real or perceived, celebrity leaders rapidly lose the capability to achieve grand visions if the staff believes that the invincibility of the leader is in doubt. Since the leadership is based on actions that reinforce the charismatic message, the existence of an action that counters the perceived strength can inflict a fatal blow to the leadership mantle. Therefore, the existence of a celebrity-based caretaker should be approached with great care by an organization.

Caretaker as superior—The final role for a caretaker to adopt is the role of superior. In this role, the caretaker builds upon the formal authority of the leadership position by conveying personal authority over the organization members. Through the use of speeches that emphasize the expectations of individuals to achieve the stated goals of their positions, and personal interactions that reinforce these expectations, the caretaker reinforces the leadership position and sets the requirements for every member. Depending on the individual in the leadership position, these requirements may include an implied consequence if the organization does not fulfill these expectations, or the leader may explicitly state the repercussions that may befall the team if the goals are not reached. In either case, the members are made aware of the expectations of the superior and the commitment that the organization has made toward achieving the vision, mission, and goals. Placed in the context of the example firm, the caretaker that assumes the position of superior approaches the "one-of-a-kind" solution by personally calling each member of the team to convey the importance of breaking from previous problem-solving approaches to achieve an innovative solution. The one-on-one meeting is friendly, but clearly conveys the message of expectation. Similar to each of the leadership positions, this approach to achieving organization goals has its potential pitfalls. Specifically, the caretaker as superior approach can transition to caretaker as intimidator or enforcer with minor changes. Given the authority of a leadership position, the individual who has internal drive to ascend to a leadership position can turn that drive outward and attempt to project the drive onto other members. This projection can result in unreasonable expectations of team members, who may not have the same goals as the current caretaker. Similarly, the belief by the caretaker that he alone has the best focus on the final goal may result in the caretaker not accepting external advice. Either of these movements may result in the organization losing its team-based focus and adopting a traditional hierarchical approach that reacts to leader requests rather than proactively generating drive from the individual members.

Caretaker Responsibilities

The diversity of personalities that exist within the civil engineering industry guarantees that each of the caretaker approaches will appear in any cross-section of organizations. Although each of these personality types has its intrinsic advantages and disadvantages, the fundamental need for the caretaker should not be forgotten. Specifically, the caretaker is responsible for providing the strategic-level perspective that binds the organization together to meet the stated vision, mission, and goals. Achieving this level of continuity requires the caretaker to retain a focus on the overall objectives and avoid the micromanagement trap of delving deeply into the concerns of individual projects. Given that civil engineers are trained as project managers, project-level concerns should be the strength of the individuals managing the projects. The caretaker needs to focus on blending these strengths into a cohesive focus on overall organization achievements.

The enumeration of leadership tasks is a popular topic within leadership research. A broad spectrum of leadership educators, consultants, and researchers has developed numerous leadership frameworks that outline the essential tasks that must be performed by a successful leadership position. However, retaining the current focus on the leader as a caretaker of the organization vision, mission, and goals, the leadership role can be narrowed to a subset of leadership tasks that emphasize the achievement of strategic organization goals. These tasks include the introduction of the strategic objectives, the reinforcement of the objectives, and the evaluation of the objectives. Together, the tasks are a linear process that moves the organization from the initial stages of abstract strategy formulation to the annual process of strategy reevaluation.

Communication of strategic objectives—The first requirement of the caretaker is to communicate the strategic statements to the organization members. If the organization is in its initial stages of formation, then this responsibility includes the communication of all three levels of statements. However, if the organization is in a mature stage of development, then the responsibility focuses on the reintroduction of the foundation-level vision and mission statements and the introduction of the annual goal statements. Both of these scenarios emphasize the role of the caretaker to focus the organization on the task that lies ahead for the current planning period. Since it becomes easy for the organization to focus inward on project-level concerns, the caretaker must remind the organization of the overall objectives that remain as the cornerstone of the organization existence. In this role as organization conscience, the caretaker must forcefully outline the requirement for the organization to stay close to the original vision. By reminding each member of the responsibility to retain a focus on the overall objectives, the caretaker serves as the focal point for all decisions relating to the overall vision, mission, and goals.

Reinforcement of objectives—If the communication of strategic objectives serves as the traditional "tell them what you are going to tell them" of speechmaking strategy, then the reinforcement of objectives is the "tell them what you want to tell them" component of the strategy. After introducing the strategic objectives to the organization members, the caretaker must transition to the role of objectives monitor. Periodically, the caretaker must reinforce the objectives to the organization to ensure that every member remembers the goal that each project is intended to serve. This reinforcement may take different forms depending on the role that the caretaker has assumed. Individual conferences, group discussions, or organization-level meetings may each serve the purpose of reminding the members of their responsibility to the objectives laid out for both long and short-term duration. In each case, the caretaker must focus on two specific objectives, reinforce the direction toward achieving the overall vision and mission, and reinforce the need to meet the annual planning goals that serve as stepping stones toward the fulfillment of the long-term vision and mission. Although a magic formula for achieving this goal does not exist, as long as the caretaker remains focused on reminding each member that their role is equally important in enabling the organization to achieve its strategic goals, then the opportunity remains to progress forward along the strategic roadmap.

Evaluation of objectives—The last step in the push toward achieving the stated objectives is a final evaluation to determine if the objectives set forth at the onset of the planning effort are the same ones that were achieved at the conclusion of the planning period. In terms of the speech framework, the evaluation of objectives is equivalent to "telling them what you told them," or returning to the initial message to remind the organization what was expected at the beginning of the planning period. This evaluation process is required to determine if the short-term mileposts are being achieved on the road to the long-term organization vision. This evaluation process diverges significantly from the familiar project evaluation process. Rather than focusing on the budget and schedule of a specific project, the evaluation of organization objectives requires the caretaker and the senior staff to determine if the projects and procedures adopted over the previous planning period are consistent with the long-term goals. Since these goals may emphasize intangible elements such as the example firm's "one-of-a-kind" solutions, the evaluation process emphasizes analyzing the strategic achievements of the organization. Specifically, the caretaker must focus on elements such as market penetration, publicity, profit gain, quality, and business growth to determine the advancement made toward the long-term goals. At times, these advancements may conflict with the objective numbers obtained from the traditional project analysis. However, the strategic management emphasis for an organization must retain the long-term perspective as the overriding viewpoint of success.

Each individual who assumes the responsibility to complete the tasks may undertake these caretaker tasks in a different manner. However, the underlying purpose remains the same for each caretaker. For each individual, the goal remains to evaluate whether or not the organization is making progress toward the long-term objectives stated in the vision statement. If the evaluation process determines that satisfactory progress is being made, then a continuation of the process can occur for another planning period. However, if satisfactory progress is not being achieved, then the organization must reevaluate its generated goals. In returning to these goals, the organization must determine if either the proper goal was actually set, or if an attempt was made to pursue goals that did not coincide with the strategic path. In either case, the organization can either choose to return directly to the goal determination phase set forth previously in this chapter, or elect to examine the overall strategic management objectives that complement these goals as set forth in the following chapters.

DISCUSSION QUESTIONS

1. Many organizations choose to develop mission statements without developing vision statements. Why would an organization choose to follow such a plan of action?

2. One argument that is prevalent in the civil engineering industry is that setting a long-term goal is almost impossible due to the cyclical nature of the industry. What are the ramifications of this statement on organization members?

3. Mission statements are often seen in civil engineering offices on a wall or on cards on members' desks. However, individual members are often unaware of the specifics of the mission statement in these same offices. Is it important for organization members to know the organization mission statement, and if so, what can the organization do to enhance the knowledge of the mission statement?

4. Goals are an acknowledged part of any successful organization. However, in a client-driven industry such as civil engineering, how can an organization set goals that they intend to pursue independently of client influences?

5. Leadership is a popular topic in the management and civil engineering literature. Books focusing on leaders such as Bill Gates and Jack Welch are commonplace. However, books and articles on civil engineering leaders are few. Why? Is this a reflection of the leadership qualifications of civil engineering leaders?

6. What are the required leadership characteristics of today's civil engineering organization leader?

Vision, Mission, and Goals in Business Creation

The value of vision, mission, and goal statements in successful business development extends beyond the establishment of statements that internal and external constituents can analyze. These statements also serve as unifying concepts for a civil engineering organization. An example of this unifying purpose is illustrated by Metric Constructors. Metric Constructors is the Southeastern division of Smith Construction, a national construction organization with offices in every region of the United States. In turn, Smith Construction is owned by an international holding company that has interests in engineering, materials, and international construction. In 1935, Metric Construction was founded in Atlanta as a division of Smith Construction. Today, Metric has grown to the number two ranking construction company in Atlanta, with an annual revenue of $217 million. Geographically, Metric has extended its operations from Texas to Florida and north to the Carolinas. Within this geographic region, Metric has established four main divisions: Commercial Buildings, Multifamily Residences, Hospitality, and Corporate Services.

The Players

Metric Constructors is managed by Jerry Taylor, Vice President of Smith Construction. Serving as Metric's director for over twenty years, Jerry Taylor has established Metric's reputation for diversification in project development and long-term experience of senior project managers. Jerry Taylor has developed Metric into a close reflection of his personal beliefs on the construction industry. Specifically, Metric reflects Taylor's strong beliefs that a successful construction organization should emphasize diversification as the key to long-term success. Anticipating that economic swings will alternately favor different components of the industry, a diversified organization will have the opportunity to absorb these swings by placing itself in a position to capitalize on the current building expansion. Additionally, by recruiting project managers with strong leadership skills, the organization establishes a personal relationship with corporate clients through these managers. The successful completion of projects results in return business with personal requests for specific project managers. Combining these two beliefs, Metric has built its foundation and reputation on the development of diverse projects, led by long-term employees with strong connections to the local corporate environment.

Supporting Jerry Taylor in his development of long-term strategy for Metric is Mark Foster, senior project manager, and director of operations. Mark Foster focuses exclusively on setting strategic policy and direction for Metric through the development of annual goals and objectives for each of the four operating divisions. Representing the business influence within Metric, Mark Foster faces the challenge of extending the project-to-project focus of the operating divisions into a cohesive organization strategy. Mark emphasizes the challenge of this responsibility in areas such as translating business concepts including core competencies, strategic planning, and annual objectives into the vernacular of the project manager. Given the role of integrator and business development oversight, Mark stresses the need to acquire personnel who have the foresight to envision innovative income opportunities, as well as the need to retain a cohesive team through common goal development.

A primary example of the innovative personnel recruited by Mark Foster is Charles Mayfield, director of the Corporate Services division. Charles represents a divergence from the Metric standard of long-term organization employment. Employed for only seven years within Metric, Charles was brought into the organization specifically to develop the Corporate Services division. Focusing on the development of close relationships with corporate clients that require facilities improvement projects on a regular basis, the Corporate Services division has established itself as a principal revenue generator within Metric. Pursuing projects with rapid turnover and strong potential for success, Charles has developed a strong reputation within Metric and built a close relationship with Jerry. Never satisfied with the success of the Corporate Services division, Charles epitomizes the Metric emphasis on diversification by approaching both Jerry Taylor and Mark Foster with a regular stream of ideas for new revenue development. Ideas for new Metric services have included in-house subcontracting, material recycling, and strategic alliances with national partners.

The Issue

The success of Metric Constructors has included both positive and negative ramifications. From the positive perspective, the success has enabled Metric to build on its reputation for excellence and superior personnel through continued recruitment of experienced project managers from competing organizations. According to Metric employees at all levels, the emphasis on Metric people has never been stronger. However, this success has resulted in an accompanying downside. The drive to demonstrate increased revenues has placed the four primary divisions into a state of competition that has grown from healthy internal challenges to potentially destructive internal conflict. Specifically, the healthy competition qualities of obtaining clients and demonstrating quarterly revenue increases has slowly deteriorated into competitive barriers between the divisions. Instances of personal resentment and questions of internal favoritism have emerged quietly among senior personnel within the divisions.

In interviews with personnel in each division, feelings of resentment have surfaced as each division believes that disproportionate resources have been allocated to ensure the success of one division over another. Additionally, questions of leadership have emerged as division leaders question the desire of Jerry Taylor to reduce the internal conflict. Some personnel have anonymously indicated that Jerry Taylor is promoting this conflict as a method to determine which division leaders have the leadership potential to build their division into a leading position within the organization. However, Metric personnel indicate that the strong personality of Jerry Taylor has dissuaded Metric personnel from bringing this matter to the attention of either Jerry Taylor or Mark Foster due to fears of being labeled as a non-team player.

The Solution

With this move toward internal conflict as a backdrop, both Jerry Taylor and Mark Foster recognized the need to bring the divisions closer together under a unified Metric objective. With this as a goal, Jerry Taylor directed Mark Foster to lead a vision- and mission-building initiative. In this initiative, the Metric organization was directed to develop a new vision and mission statement that reflected the direction of the organization as it prepared to enter the twenty-first century. At the core of this effort was a directive to unify the organization through a single vision and mission that each division could translate into a set of objectives that collectively benefit Metric. This broad directive was given to Mark Foster to develop into an implementation plan that would address the entire Metric organization.

The implementation plan developed by Mark Foster in response to this directive was simple and straightforward. A team was developed that reflected each division as well as the various levels of seniority within the divisions. Each member of the team was directed to meet with their divisions to develop a core set of ideas that reflected Metric's commitment to its customers, employees, community, and investors. Additionally, the team was directed to develop the ideas from the perspective of Metric as an overall organization rather than the individual viewpoints of the divisions. The team was given two weeks to complete this task prior to a meeting where the individual contributions would be combined into an overall vision and mission statement. To ensure that his personality did not influence the outcome of this effort, Jerry Taylor removed himself from the development process, leaving Mark Foster to coordinate the effort.

Although the initial phases of the process ran into difficulties due to the internal division conflicts, the personal attention by Mark Foster kept the team moving forward toward its stated goal. The final step of the process occurred with the team convening in a meeting to consolidate its input into a unified vision and mission statement. This meeting emphasized the hesitance of the team to commit to the effort as each emphasized their doubts concerning Metric's commitment to the new vision and mission statements. In turn, each team member voiced his division's concerns and their belief that internal competition was Jerry Taylor's intended avenue to improved organization performance. Faced with this reluctance, Mark Foster was forced to act as the organization mediator. In this role, Mark Foster concentrated upon bringing the team to a state of cooperation where the members agreed to focus on the immediate task. With this as a less than optimum environment, the team proceeded to combine their efforts into an overall vision and mission statement. The resulting cooperation generated the following:

Vision Statement

Metric Constructors will achieve growth in financial, professional, and community pursuits by providing our customer community (including economic and owner developers, designers, consultants, and subcontractors) with individual attention in building services from project inception through project warranty and beyond.

Mission Statement

Metric Constructors will implement our mission by leveraging our reputation, tradition, knowledge, people, and depth of resources (financial, technical, political, etc.) to provide clients with a competitive advantage in their industry through award-winning service that exceeds expectations.

To Provide:

Clients with quality services and craftsmanship in order to achieve growth in financial, professional, and community pursuits.

Employees with a safe, empowering, enriching environment that fosters and rewards education, growth, and performance with financial and professional security.

The Community with enhancements that leverage industry knowledge and resources to create greater visibility, enhanced company image, and promotion of opportunities.

Stakeholders with a top performing return on invested capital while achieving a sound fiscal structure, stability, long-term growth, and security within the framework of the building industry.

QUESTIONS

1. Given the reluctance of the vision and mission team, how should Mark Foster proceed in transforming the statements into a unified set of organization objectives?

2. Given the strong indicators by the team regarding the influence of Jerry Taylor in the direction of Metric, how should Mark Foster relate the team's concerns to Jerry Taylor?

3. Placed in Jerry Taylor's position as Metric executive, what is his role in translating the vision and mission statement into twenty-first century objectives?

4. Placed in Charles Mayfield's position as a division manager, how do you reconcile the stated goal of developing common objectives with the knowledge that Jerry Taylor prefers individual achievement?

4

CORE COMPETENCIES OF THE CIVIL ENGINEERING FIRM

What are the core business and strength of the organization?

The second question in the question–answer methodology transitions from the visionary focus on organization missions and visions to a hard focus on organization strengths. Although the development of the vision, mission, and objectives provides a roadmap for future direction, the lack of an appropriate vehicle can severely hamper progress in this direction. In practical terms, the vision may be admirable, but if the organization personnel do not fit the profile required to implement the vision, then either the vision or the personnel need to be altered. Although these elements are often viewed independently, the proper combination of personnel and mission statement can provide the motivation required to achieve the organization vision. In an era where considerable concern is raised about declines in productivity, the establishment of motivating elements can provide opportunities to reverse this trend. However, in establishing these far-reaching visions, an organization can overlook motivational opportunities by downplaying the fact that visions have to be implemented by individuals possessing definite spheres of knowledge and abilities.

The need to focus on these specific areas of knowledge has been repeatedly demonstrated during the twentieth century. This time period ushered in the great expansion into organization diversification. The middle of the twentieth century saw organizations attempt to enter 10 or more markets that had little, if any, practical connection. From automobiles to food to paper goods, the drive to diversify dominated private organizations. For example, Mobil Oil bought Montgomery Ward department stores and Great Bear restaurants, while GE owned Kidder Peabody (Dutton 1997). However, the latter part of the century witnessed a retrenchment by many of these organizations back to original business sectors. Similarly, public organizations underwent a streamlining of operations to reduce overlap and strengthen focus. In both cases, the realization emerged that the long-term organization success is highly dependent on the ability of the organization to focus on what it does best. These underlying strengths, or core competencies, provide the greatest opportunities for long-term success by providing a foundation for all activities the organization attempts to pursue (Hamel and Prahalad 1989). In other words, embarking on an activity with a weak foundation is destined to fail, as the exterior appearance can only mask the underlying failure for a finite period of time.

In this chapter, the development and refocusing of core competencies is examined in the context of organization strengths and abilities. The chapter emphasizes the need to build and develop these strengths as avenues to achieving long-term missions and visions. Reexamining these strengths requires organizations to move beyond the thoughts of current, short-term profit centers, to a focus on the objectives the organization can potentially achieve over a sustained period of time and effort. Leveraging human resources into core strengths provides the organization with the confidence, knowledge, and ability to pursue objectives with the greatest possibility to succeed. Similarly, investing organization resources into a core area provides the opportunity to leverage existing investments rather than investing in an area that lacks the fundamental strength to provide the anticipated return on investment. However, identifying these core competencies sets the challenge for this chapter—*to examine the resources and operations of the organization to determine if the greatest strengths are being identified as the emphasis for emerging opportunities.*

■ Examining Core Competencies

The establishment of an organization vision provides members with a specific goal to pursue. The larger and more ambitious the vision, the greater the challenge becomes for the members to build upon their existing position to achieve the vision. However, after an organization successfully builds on this vision statement to develop a mission statement and specific objectives, the time arrives when management must ask, "How are the objectives going to be put into actual practice?" Although this is a proper procedure in evaluating the feasibility of the objectives, the question must be altered. Specifically, the management team should initially focus on whether the objectives and the organization strengths are a proper match. Subsequently, the organization must ask a related second question, "Are these objectives focusing on what the organization does best?" Although both of these questions serve as checkpoints for the development of visions and missions, they diverge in their emphasis.

The former question focuses immediately on implementation. It makes a logical leap to assume that the objectives are appropriate for the organization as it currently exists, and the organization should pursue these objectives as central components of long-term strategy. Immediately, the implementation focus forces the organization to examine the personnel and resources available to achieve the objectives. Strategic plans are developed that outline needs, objectives, and evaluation criteria that should be followed during the implementation phase. Slowly, momentum will begin to move the organization in the specified direction. Given a direction and a starting point, the members slowly build belief in the direction until momentum is developed in a particular direction. At this point the organization has committed itself to achieving the specified interpretation of the vision and mission as it was laid out in the short-term goals and objectives.

Focusing on the positive side, this focus on implementation is laudable from the point of view of establishing a management objective. With a specific objective in place, the organization can rally around a set of goals to achieve. The acquisition of resources, the creation of business development materials, and the development of evaluation criteria can each focus on a unified endpoint. As discussed in the previous chapter, the development of common objectives is a central component of the strategic management process. However, this immediate focus on implementation contains a negative side that must be weighed against these positive elements. Specifically, the immediate focus on implementation ignores the second question, "Are these objectives focusing on what the organization does best?" While the development of objectives is a necessary component of strategic management, the development of objectives that are inconsistent with the organization focus can have a greater detrimental effect than failing to develop objectives. The development of objectives that build momentum in a direction that is incompatible with organization strengths can send an organization down an implementation path that has greater odds of resulting in lost revenues than resulting in bright new horizons. Management history has demonstrated that the investment in personnel, resources, and related items that do not focus on organization strengths will ultimately result in monetary losses (Dutton 1997). Although the initial idea may be positive and seem limitless in its potential, all ideas eventually rest on a solid foundation. The organization strengths represent this foundation. Building in an area where the organization has never established this foundation is a risky effort that has a greater chance to find insufficient support than a solid footing.

It is this concept of focusing on what an organization does best that resides at the center of the core competency philosophy. Originally referred to as distinctive skills or distinctive competencies (Selznick 1957; Learned et al. 1969), the emphasis on building on existing strengths as foundations for future expansion is a concept that is neither new nor unique. However, it was not until 1990 when Hamel and Prahalad published their seminal work on core competencies that a groundswell of management focus developed to emphasize the need to build on these core competencies (Hamel and Prahalad 1990). The essence of this focus is on the relationship between the long-term vision and the core capabilities of the organization to achieve the vision (Schoemaker 1992). While it is admirable for an organization to attempt to capture a new business sector, if the sector fundamentally strays from core competencies, then history tells us that the endeavor has a greater chance of failing than succeeding.

Relationship to Civil Engineering

The core competency concept appears obvious when taken to the extreme. The appropriateness of department stores buying financial institutions or oil companies buying restaurants appears questionable to the majority of observers. Of course, viewing this diversification from the perspective of several decades of experience provides a greater opportunity to evaluate these acquisitions. Placing the acquisitions in the context of 1950s diversification may not validate the mergers, but it places the incompatibilities in the context of the times where diversification was the apparent key to organization success. Moving forward to current times, the lessons of failed diversification are strong indicators of boundaries that organizations must recognize during the quest for growth and expansion. However, the question emerging at this point is, "What relevance does this have to civil engineering?" Based on the limited publications focusing on civil engineering organizations, the core competency perspective would appear to be misplaced (Goodman 1998). The civil engineering industry has very few players that match the economic power of a General Electric or IBM. Furthermore, the strength of individual civil engineering organizations appears to be quite obvious. Although the industry is diverse in its focus, the specialization and fragmentation that dominates the market appears to have forced companies to narrow their focus to their strengths whether they intended to or not.

The reality of the industry is actually quite different from this perception. A random sampling of civil engineering organizations finds organizations of varying sizes branching out into areas including mining, software, telecommunications services, construction equipment development, and the ownership of storage facilities. In some instances, organizations are starting new operating divisions to oversee the ventures, while in others, existing departments are being expanded to undertake the diverse business ventures. In some instances, the diversification emerges from the acquisition of, or merger with, an existing organization. However, in each situation, the common element is an attempt to either build on an acquisition or enter an emerging market that is misaligned with core competencies. This fact returns us to the base question of what is an appropriate extension of a core competency and where should the line be drawn in terms of establishing goals that do not coincide with these competencies.

While specific lines and recommended boundaries can be developed as organization guidelines, the value of these recommendations ultimately resides in the restraint developed by an organization to resist entering a business segment that diverges from its core strength. With this as a precursor, the development of core competency boundaries starts with a basic understanding of core competencies in the civil engineering domain. To assist with this understanding, core competencies will be placed in the context of an example construction firm, Mountain Regional Builders (MRB). MRB is a small firm consisting of 20 office staff and 15 field personnel. The firm specializes in light commercial construction, although they have undertaken a few projects in specialty areas such as health care and detention facilities. Over the 25-year history of the firm, the partners have slowly changed the focus of operations from self-performing as many operations as possible to subcontracting almost 95% of operations. The reason for this change in focus was predominantly based on economics. The cost of retaining skilled field personnel during lean times was greater than the overhead that was charged by subcontractors for the same tasks. Therefore, MRB retained field supervisory personnel to oversee operations, but slowly eliminated positions related to activities such as site development, concrete placement, or masonry work.

With this as a background, Figure 4-1 illustrates MRB's core competencies and the relationship of these competencies to additional business opportunities. At the center of the radius is the core competency of the organization, or what MRB does best. Having divested itself of most construction operations, MRB has built its business on the ability to oversee projects. This project oversight strength establishes the core competency of MRB, the ability to direct and oversee complex assembly operations. Although the first instinct may be to focus on MRB's strength as a constructor, the competency is actually on assembly oversight since the vast majority of project work is implemented by subcontractors. Given this strength in assembly oversight, MRB could build upon this core competency at several levels. First, MRB can look at applying the same strength in a related area. For example, the oversight of projects can be expanded to the oversight of facility management services. In this role, the same management expertise that is used initially to construct the facility can

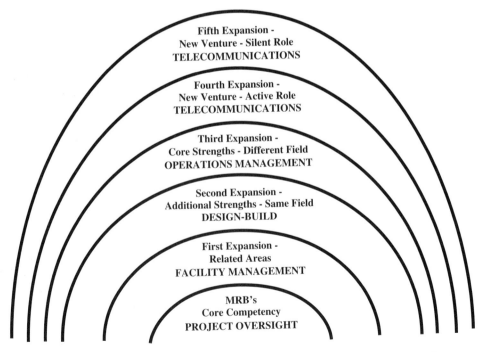

Figure 4-1 Organization expansion should initially focus on internal strengths and gradually expand to new market sectors.

be transferred to the management of facility expansion, reconstruction, and renovations. Through these corporate service activities, MRB can expand its client services without expanding its core employee strengths. Additional examples of related area expansion for MRB could include construction management consulting, preconstruction consulting, and contract administration. Each of these areas builds upon the core strength by building a bridge to a related area that uses the same strength currently existing within the organization.

Expanding beyond the boundary of applying the same strength in a related area, the second area of expansion focuses on appending additional strengths in the same field. For MRB, this expansion emphasizes the analysis of the AEC project life cycle to identify expansion opportunities. Given MRB's strength in managing complex construction processes, MRB can examine the possibility of managing a greater proportion of the project development process. Following the emerging emphasis on design–build projects, MRB could develop alliances with architecture and engineering firms to provide clients with a turnkey project solution. However, MRB must be careful in this expansion not to abdicate its control to another member of the project team. Far too often, civil engineering organizations fail to realize their potential in project development, since they believe that their role is that of an implementer rather than a leader. A greater number of civil engineering companies need to examine the traditional role of civil engineering organizations in the industrial sector, where the engineering organization plays the lead role in managing the project development and implementation processes. The management strength of MRB places it in the position to follow these examples and develop the alliances necessary to create a design–build or similar approach to alternative project delivery.

Moving beyond the boundary of its related strengths, the organization begins to accept a greater element of risk as it analyzes additional revenue opportunities. Specifically, the organization is making the decision to expand their core competencies. Undertaking this expansion encompasses several risks including the acquisition of knowledge that is unfamiliar to existing employees, the possible entry into a field where competitors are unknown, and the development of business strategies that address operating conditions that could be vastly different from those experienced in the core business. With this as a caveat, the next area of expansion is the application of core strengths to different industry environments. For MRB, this transfer of knowledge focuses on the identification of industries that undertake assembly processes that are similar to the contractor–subcontractor relationships found in the construction industry. The limitation to this identification process is bounded only by the risk aversion adopted by the MRB leadership. Similar assembly relationships exist throughout the manufacturing domain. From small components such as microprocessors to large artifacts such as airplanes, it is conceivable that the underlying assembly knowledge developed over 25 years of construction experience could be transferred to a manufacturing operation. Through the acquisition of personnel with knowledge in the product domain, the management processes developed in MRB's core operations could be transferred to the manufacturing operation. Although MRB personnel would not act as direct floor supervisors, the project and strategic management knowledge would allow MRB personnel to act as operation managers and senior managers.

While many individuals may hesitate to adopt the core strength transfer approach to business expansion, from a core competency perspective it contains less risk than the expansion opportunities that follow. In the next option, the organization elects to enter a completely new business venture that has little or no relation to the core business strength. This expansion can occur through a number of avenues, but the most common entry point into new ventures will either be through the acquisition of an existing organization or the initiation of a new division with in-house resources. Examples of these ventures are boundless and include those mentioned earlier, including telecommunications, software, and equipment design. The distinguishing feature of this category is the intent of the current organization executives to take an active role in the management of the new venture or acquisition. In this mode, the civil engineering executives elect to guide the new venture operations through direct policy, management, or strategic objectives. Although this may seem the prudent approach given the investment that the organization is making in the venture, the challenges and opportunities that exist in these new ventures may be entirely different from those experienced by the civil engineering executive during normal operations. For example, the emphasis in civil engineering organizations on job charges as the basis for daily operations is incompatible with the research and development requirements of a software venture. Software requires a heavy up-front investment that cannot be directly billed against an existing client. The organization must accept the fact that economic benefits will occur through postdevelopment sales rather than man-hour reimbursements. Attempting to oversee the software development process through traditional engineering processes will ultimately encounter difficulties due to incompatibilities in the required operations. From this perspective, the organization entering a fundamentally different market domain while retaining an active management presence that includes traditional engineering perspectives is taking a significant economic risk.

The risk associated with the active management approach is significant. However, the risk is easily matched by the risk associated with a final expansion option, the entry into a new business domain with the decision to act as a silent owner or partner in the operation. The lessons of the 1950s and 1960s diversification emphasis should ring loud in the ears of civil engineering organizations that attempt to take this avenue to new market opportunities. Acquiring or starting a business venture in a domain where the civil engineering management team has little or no direct knowledge opens the organization to numerous unforeseen challenges. However, compounding this challenge with the decision to take a silent role in the endeavor places a greater potential barrier in the path to success. Specifically, the organization that elects to take a silent role may not be aware of the strategic opportunities being developed by the new venture. Potential clients, business opportunities, or new technologies may emerge out of the new venture. However, if the organization is not aware of these opportunities, then the overall strategic value of the endeavor may be compromised. Therefore, to heed the lessons of history, the organization that elects to diversify as a silent partner accepts the greatest risk of all expansion avenues since the organization cannot build a unified framework of business units within an overall strategic vision.

It is important to note at this point that all organizations that choose to pursue one of the expansion paths highlighted in the outer rings have a high probability of experiencing failed business ventures. However, the understanding that must emerge from the core competency perspective is that the further the organization moves away from its core competency, the greater the possibility that the endeavor will fail to meet expectations. From this perspective, an organization that enters into one of these outside ventures should enter it with the understanding that failing to build on core competencies places investment capital at a higher risk than core business efforts.

Learning from Outside Examples

Placing the core competency concept into a tangible perspective can be accomplished through many negative examples. For every success story such as General Electric, which has operations ranging from jet engines to media ownership, many more exist that do not result in such success. The instances of organizations moving beyond their core competencies to enter fields where they had no experience and then suffering large financial losses are numerous. In fields ranging from civil engineering to electronics and finance, organizations have suffered losses from entering business sectors with little domain knowledge. However, rather than focusing on the negative examples, the opportunity exists to highlight well-known success stories where organizations have used a single core competency to build international reputations in diverse services and products. The following two examples provide strong lessons for today's strategic management team that is looking for avenues to expand revenue and market opportunities.

Honda

The first example is one of the most discussed examples of core competencies being leveraged into multiple products and international success. Two partners who were interested in developing cost-effective motorcycles established the Honda Corporation in postwar Japan (Pascale 1996). Mr. Honda was the engineering innovator who served as the focal point for product development. At the core of Honda's research and development effort was the development of a lightweight and affordable engine that could be used as a basis for producing affordable motorcycles. Initially focused on racing as a means to develop name recognition, sponsorship, and research results, the corporation built on the results obtained from the track to introduce their first lightweight, commercial motorcycle in 1958. Powered by a lightweight, high-horsepower 50cc engine, the new motorcycles were an immediate success in Japan. Within one year, Honda was the leading motorcycle manufacturer in Japan. By 1964, Honda was the leading distributor of small motorcycles in the United States. Much has been written about this rapid entry and success in the United States market (Mintzberg

1996a; Goold 1992; Boston Consulting Group 1975). However, the story of interest is the expansion of a single core competency into an array of internationally successful markets.

The success of Honda's small engines in motorcycles led the company to examine the possibility of pursuing the same strategy in automobiles. In 1963, Honda built upon their competence in small engines to release their first compact automobiles in Japan. Continuing to improve upon this technology, Honda introduced the N600 sedan to the United States seven years later. Although the United States auto industry had remained focused on large-car production, Honda introduced the American consumer to an alternative. In 1972 this alternative coincided with the OPEC oil embargo. Building upon the need for fuel-efficient cars, Honda introduced the Civic to the United States market. Using its small engine technology, the Civic was affordable, fuel efficient, and a statement about changing environmental perspectives. The popularity of the Honda automobiles was enormous. By 1980, production of the Honda Accord reached 1 million units. By 1982, Honda became the first Japanese auto manufacturer to open a plant in the United States. Within a few short years, the Honda Accord gained the title of Most Popular Car in the United States.

Although the expansion of the Honda Corporation from motorcycles to automobiles is impressive, the success of the corporation is not limited to transportation. The same expertise in small engines allowed Honda to expand into additional markets. In 1984, Honda expanded into the United States consumer product arena with the opening of the U.S.-based Honda Power Equipment Mfg., Inc., plant. Focusing on an array of small-engine-based products such as lawnmowers and generators, Honda once again built upon its core expertise rapidly to build market share in an established market domain.

Hewlett Packard

In another of the more popular business success stories, Dave Packard and Bill Hewlett started a small electronics business in the garage of the Packard's house in 1938. Educated as electrical engineers, Bill and Dave started their business with $538 in working capital and an innovative idea on the development of electronic test equipment for sound equipment. Producing a breakthrough in negative feedback technology, HP's first product was released in 1938 and received eight orders from Walt Disney to assist in making the movie *Fantasia*. By 1940, HP expanded its electronic test expertise into eight products and attained a net revenue of $34,000. World War II provided the opportunity to expand into related fields of microwave signals and generators. Becoming the acknowledged leader in this field during the war, HP built upon wartime demand to increase product development until revenue reached $30 million in 1958 with 373 products and 1,778 employees.

Although HP could have remained within this market domain and flourished as an international leader in electronics, HP decided to branch into related fields in the 1960s, including medical electronics. However, it was the introduction of the first desktop scientific calculator in 1968 that leveraged the core electronics expertise into a new market leadership position. Entering the field of computing, HP introduced minicomputers, scientific workstations, and timeshare operating systems. Building upon this expertise, HP introduced the first handheld calculator in 1972. By 1980, net revenue had increased to $3 billion, with an international workforce of 57,000. However, the greatest success for the company was still to be achieved. Based on expertise in electronics, computing, and desktop electronics, HP introduced the Laser-Jet printer in 1984.

The LaserJet printer and its successors became the company's single most successful product and evolved into the international standard for laser printing. By combining document technology with electronics, HP developed a new market position in computer peripherals, including plotters, printers, scanners, and fax machines. Focusing on its core expertise, HP continues to grow while expanding research and development efforts into related product areas. Preparing itself for the next generation of technology requirements, HP enters the twenty-first century by expanding into electronic commerce and the handheld organizer markets with a workforce exceeding 120,000.

Lessons Learned

The success of both Honda and HP are unusual in the scope and longevity of the organizations. The scope of the success is beyond what most organizations are able to achieve. However, the approach that was adopted to achieve this success is not limited to these organizations. In each case, the organizations focused on a single core competency, small motors for Honda and electronics for HP, and then expanded market domains based on these competencies. In neither case did the organization attempt to break from this competency to follow a new path. If a new market segment emerged such as luxury cars for Honda or computers for HP, then the organization analyzed the role its competency could play in developing new products that responded to the market. The lesson that civil engineering organizations should take from these examples is that adherence to a core set of competencies provides the greatest opportunity for leveraging core business activities into greater market opportunities. Developing prominence in a given area, together with expertise and vision, provides the foundation for expanding into areas that directly benefit from this expertise.

■ The Need for Core Competencies

Demonstrating the value of core competencies is a manageable task when the opportunity exists to reflect back in time on the operating philosophy. However, in the middle of daily operations, the luxury of this reflection is generally not available. The focus on deadlines and milestones often overrides a focus on a central operating philosophy. Given this short-term perspective, the argument for core competencies must be developed to a point where the concern for emphasizing these competencies is equal to or greater than daily operational concerns. If the case for competencies is not developed to this point, then the demands of daily operations will quickly outweigh a central focus on core competencies. To establish this importance, an organization must examine core competencies as more than an operating philosophy. Rather, core competencies must be established as a foundation for the entire business enterprise. Including the opportunity to solidify the vision statement, the work force, and the market position of the organization. However, to arrive at this point, the organization must clearly understand how core competencies diverge from current business practices and how to reevaluate current operations to fit this change in perspective.

How Core Competencies Diverge from Current Business Practice

The emphasis on core competencies demonstrated by Honda and HP represents only two of a much larger number of organizations that have achieved success by following the core competency strategy. Organizations as diverse as Sharp Electronics, Kodak, and U.S. Surgical have each benefited from a focus on core competencies (Gallon, Stillman, and Coates 1995). Achieving this level of success is not a trivial matter. There is no magic core competency wand that will transform an organization from a losing proposition to an instant cash cow. However, as one component of a larger strategic perspective, an emphasis on core competencies is a valuable step toward achieving broader vision and mission statements. In this case, the critical first step is an understanding of how core competencies diverge from current business practices. Specifically, an organization must understand how to reevaluate current business practices and acknowledge the unique difficulties that face civil engineering organizations who attempt to implement a core competency strategy.

Reevaluation of Business Practices

The first of these issues, the reevaluation of business practices, requires organizations to question the validity of current operational procedures at several levels. As illustrated in Figure 4-2, one method for conducting this operational analysis is to divide operations into internal and external concerns. Within the internal sector, several levels focus on the breadth of operational concerns, including activities, projects, and systems. Similarly, the external sector is divided into levels that emphasize the breadth of revenue focus including customers, specialties, and markets. Within the internal sector, the reevaluation of business practices begins at the activities level. An emphasis on core competencies requires individual activities to combine into products or services that reflect an overall emphasis area. Viewed in isolation, individual activities can appear to be disjointed with very little in common. However, it is at the activity level that the stage is set for overall organization success. Specifically, if the personnel undertaking an individual activity do not understand the context of that activity within the overall emphasis on core competencies, then the personnel

Business Practice Analysis

External Concerns
Market - Does a market anchor exist from which to expand?
Sector - Is the organization focusing on a specific sector?
Customers - Are all customers aligned with strategic objectives?

Internal Concerns
Systems - Is the organization in balance with each division?
Projects - Are all projects reinforcing strategic strengths?
Activities - Are all activities focused in a common direction?

Figure 4-2 Dividing business practices into internal and external concerns provides a structured method for reevaluating business practices.

will approach the activity as if it were an isolated task. In contrast, if the personnel understand that all activities are related in an overall context, then opportunities emerge to identify relationships between diverse activities. The 3M corporation is an example of this concept. Within 3M, organization members understand that innovation is a central tenet that underlies every endeavor. To facilitate this innovation, small areas exist throughout the buildings that allow individuals to sit down at tables and exchange ideas on new products (Bartlett and Ghoshal 1997). At any time, employees may be found exchanging ideas or describing current activities with the idea that an activity that is part of one project may spawn an idea that assists, expands, or changes the direction of an activity on another project. This exchange does not occur if the personnel do not possess a greater understanding of their role within the "big picture" of strategic direction. Therefore, the first task in reevaluating current practices is to question whether or not the organization has successfully conveyed the strategic direction to every employee at every level. If this has been done, then organization members will be actively comparing approaches and ideas to activities in which they are involved. If this has not been accomplished, then organization members will be working as isolated components with the sole purpose of finishing their assigned task.

Moving to the next level, the focus changes from individual activities to overall projects. Taken individually, projects reflect the current market demand for products or services. A boom in commercial property development will be reflected in the civil engineering organization by a comparable increase in commercial property projects. This reflects basic economic concepts of cyclical economies and demographic fluctuations. However, this emphasis on the individual project prevents an organization from developing strategic connections between projects. From the core competency perspective, every project undertaken by the organization should be a direct reflection of organization strengths. Taking the commercial development example one step further, if an organization has core competencies in designing heavy civil projects, then it should hesitate before undertaking a commercial design task. Although the underlying design task is similar, the emphasis is fundamentally different. Unlike a transition from multifamily housing to high-rise commercial, everything in the heavy-to-commercial transition, from the eventual user to building codes and oversight agencies, is focused on different project needs. As such, the organization must obtain a new set of resources to address these specific requirements. Rather than reinforcing organization strengths, these resources dilute the strength. Of greater importance, these resources require the organization to continue an emphasis away from its core strength because additional projects are required to keep the resources profitable. In this manner, a single project can lead to a long-term emphasis away from core competencies. Avoiding this dilution of strength requires an organization to examine its project decisions. If a project does not reinforce the strategic strengths, then it should be questioned. If this questioning does not occur, then the organization could be committing itself to a long-term divergence, and not simply a one-time change in direction.

The final internal level is the systems level. The concept of organizations as systems is one that has been promoted by organization theorists for decades (Drucker 1974; Miles and Snow 1986; Nohria and Eccles 1994). In the systems perspective, activities, projects, and overall operations are viewed as an interrelated whole that is influenced by every component. The activities undertaken by an organization member in one part of the organization may not have an obvious relationship to a member in a different sector, but an effect exists at a strategic level. The organization is a set of interconnected relationships, some stronger and some weaker, that are affected by every event that occurs. As such, the organization must examine its operations from the perspective of ensuring that the entire system is working in balance. Specifically, organization leaders must ensure that a set of strengths is identified, a vision has been conveyed, and that each part of the system is aware of the activities of the other parts. Placed in simpler terms, organization leaders must examine whether the organization is working cooperatively on a common set of objectives, or is it operating as a set of independent components. Operating in this latter mode, the system is prone to imbalances, since each part is pursuing its own objectives. The system must retain in balance to ensure that each part contributes to the overall health and growth of the organization.

Developing an organization with the capacity to remain balanced through the conflicting demands of daily operations is a significant achievement. However, the strategic use of core competencies does not end at the boundaries of the organization. Rather, an organization that uses core competencies as an avenue to focus growth opportunities must also focus on the external sector. In this sector the strategic focus transfers from keeping the organization members directed toward a balanced system to retaining a cohesive focus in relationships with customers. At the first level of this external focus is the individual customer. Paralleling the focus on internal activities, the external focus on customers is the need for organizations to pursue individual customers or projects that fit within the organization's strengths. Retaining this customer focus is a central component of the core competency concept. Although diverging from the core competency by pursuing a single customer may appear to be trivial in terms of retaining an underlying philosophy, the implications of chasing noncore customers is significant. First, from an employee's perspective the pursuit of a noncore customer represents a break from specified objectives. Given this break, the organization begins to lose credibility in terms of its commitment to strategic objectives. Once this credibility begins to erode, organization members will have doubts regarding other organization statements. Similarly, from a potential customer's perspective, the pursuit of noncore projects results in the organization appearing unfocused. The existence of these diverse customers will be in direct conflict with stated strategic objectives. Contrary to the desired image of strength in a given market niche, this contradiction makes an organization appear vulnerable to competitors. Therefore, while the individual customer may appear to be a small component of the overall strategic vision, the impact of a single customer can never be underestimated.

Combining these individual customers into a specific market specialty represents the next level of the external sector. Strategic focus requires organizations to strengthen core competencies through experience and personnel. However, attempting to enter multiple market specialties diminishes the experience factor. An organization cannot effectively be all things to all people. Organizations must focus their strengths on a specialty and then build on that specialty as a foundation for growth and expansion. From this basis, organization leaders must focus their efforts on building a reputation within a specialty that is marketable in future project efforts. Establishing this specialty reputation is essential to broadening client opportunities. For example, ABB Engineering is known worldwide for its work in the power industry. Similarly, Tishman Construction built its reputation in high-rise construction. These reputations cannot be purchased from a vendor or supplier. Rather, specialty reputations are built through repeated examples of excellence in a given area. For example, returning to the MRB organization, the strength of MRB is in the management of complex processes. Each of the customers that MRB pursues should conform to that strength. However, to build a reputation of strength and knowledge, MRB should limit these customers to a specific specialty area. The focus of this area is flexible. It can either be a vertically segmented area such as applying construction management expertise to healthcare facilities, or horizontally divided such as applying construction management expertise as a generic construction manager. In either case, it is important that organizations such as MRB identify this specialty and stay within its boundaries until a reputation has been established.

Establishing a reputation within a given specialty provides an organization with the strategic focus to expand its operations within an overall market. Depending on the specialty developed, this expansion can either occur along the project lifecycle to include greater interactions with project participants, or deeper into a specific area to emphasize knowledge of a specific market niche. In either situation, the organization is returning to the emphasis on strategic balance. Just as the organization can lose balance in terms of internal operations, an organization can lose balance in a given market niche. Each market niche is characterized by the organizations that compete within its boundaries. To retain a balance within the market, an increase in the strength of one member must be offset by a weakening of another. Similarly, the entrance of a new participant requires other participants to either reduce their strength proportionately or expand the market boundaries. The focus on strengthening the core competencies in a given market domain provides an organization with the resilience to absorb these fluctuations. Of greater importance, strength in one market segment allows the organization to emerge as the new factor in a new market niche. By leveraging strength in one market niche, the organization can attempt to alter the balance in a related area. However, if strength in a given area is not established as an initial starting point, then the broadening of market focus sends the organization out of balance since no strength exists as an anchor point. Therefore, the last component of operations analysis is to emphasize the development of a market strength that will serve as the anchor for endeavors into compatible market domains.

The Difficulty in Civil Engineering

Reevaluating business practices provides an organization with an opportunity to identify areas that can be improved, modified, or enhanced in support of long-term goals. As outlined in the previous section, this analysis covers the complete spectrum of business activities, from the individual level through the market level. However, conducting this analysis in the civil engineering domain presents organizations with unique problems that are not found in other industries. Specifically, the fragmentation and specialization in the civil engineering industry is a unique attribute that fundamentally changes the results obtained in a business process reevaluation. To explain the difference, let us start with the expected analysis results from an outside industry. For example, using a small tool manufacturer, Pacific Tools, this organization can analyze the various levels of business operations to find several opportunities for strengthening the organization around their core competencies. Pacific Tools will find opportunities such as strengthening their product line through new tool introductions, focusing their distribution to a better-defined customer base, or building alliances with retailers to market products to a broader audience. In each of these cases, Pacific Tool has an opportunity to strengthen core competencies through a clear growth and expansion path. This opportunity is not as clear in the civil engineering domain.

The civil engineering industry is characterized by many organizations, each focusing on a small part of the engineering–construction process. Outside of a few large organizations, the majority of the industry is focused on garnering enough revenue from a small piece of the process to allow the organization to remain a viable entity. Crossing boundaries to explore opportunities in other parts of the process are generally seen as outside the scope of operations. For example, the typical structural design firm has built its strength and reputation on its ability to perform structural design functions. The firm does not desire to cross discipline boundaries to enter the construction domain. Similarly, the environmental engineering firm that focuses on environmental impact statements has very little incentive to enter the transportation design business. This specialization represents a fundamental characteristic of both academic and industry pursuits within civil engineering.

The existence of this specialization is the difficulty that faces civil engineering organizations attempting to reevaluate business practices in accordance with core competency development. The issue arises as to how an organization can pursue external market opportunities when internal strengths are so narrowly defined. This is a serious difficulty for many civil engineering organizations. The fragmentation of the industry often leads to an organization focusing its business development opportunities on obtaining the greatest number of projects that fit within the organization specialty. While this approach is sound in its basic business dynamics, it fails to provide the organization with options in the growth pattern. Put in another context, increasing revenue does not necessarily equate to building on core competencies. Any organization can increase revenues by pursuing additional projects within a narrow domain. This fact is reinforced by the annual ranking of firms based on revenues by the professional publications (ENR 1998). The rankings are not categorized under a general heading of civil engineering organizations, but rather are classified under narrow categories that convey to the reader the success of the organization in obtaining additional work within the narrow categorization. However, these same organizations will be the ones that are forced to downsize when their market niche experiences economic difficulties. In contrast, the successful organizations are the ones that have the flexibility to address multiple markets, each of which is related through a core set of strengths. The difficulty for the civil engineering organization is to move beyond traditional industry boundaries to identify these multiple market opportunities.

With an increased focus on the entire project lifecycle as a business opportunity, the definition of market niche is expanding for civil engineering organizations. However, the first step in achieving this expansion is recognizing that opportunities exist outside the narrow definitions placed upon the industry. Following closely behind this realization is the understanding that the development of core competencies is not limited to these narrow definitions. Returning to the concept of the organization as a system, the successful balancing of the system does not equate with limiting the system definition. Rather, the system needs to be balanced by ensuring that each component is operating cohesively to obtain an overall objective. The balance is achieved through a focus on core competencies, which does not necessarily equate to limiting the definition of the system.

Creating a Business Foundation

With a focus on reevaluating business practices, the organization is establishing a foundation that will remain solid beyond a short-term horizon. This foundation serves to solidify three organization components, the vision, the workforce, and the market focus (Fig. 4-3). The first of these components, the vision, is solidified by core competencies through a demonstration of commitment to the vision. Given the broad focus of a vision statement, organization leaders have the responsibility to translate that statement into demonstrable actions. At the center of these actions is demonstrating a commitment to achieving the vision. The strengthening of core competencies provides an avenue for this achievement. Although a vision provides a long-term goal, the specifics within the statement are intentionally left vague to allow for changes over time. To fill this ambiguity, concrete actions must be initiated to provide organization members with an indication that steps are being taken to fulfill the vision. The establishment of core competencies represents a strategic first step, since it addresses how the vision is going to be achieved and who is going to accomplish the goals.

To illustrate this role in solidifying the vision statement, the general civil engineering firm introduced in Chapter 3 is revisited. In this example, the firm emphasizes the preservation of environmentally sensitive watershed areas as a core element of their business. The vision established by the firm is as follows:

> The vision of this organization is to demonstrate national leadership in the preservation of environmentally sensitive watershed areas through the application of unique environmental knowledge and creativity.

Using core competencies as an avenue to implement this vision, the organization can demonstrate commitment to the vision by initiating actions that reflect the underlying intent of the statement. First, the organization can address the need for appro-

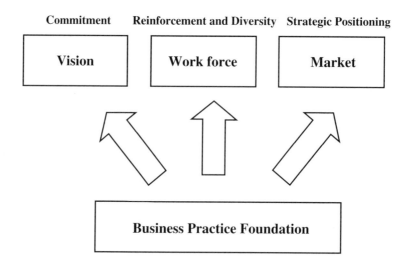

Figure 4-3 Building a focused business competency will reinforce the pursuit of a vision, strengthen the work force, and position market efforts.

priate personnel to achieve the vision by emphasizing the hiring of new employees that have strengths in the core areas. Specifically, in the example case, the organization should focus on identifying individuals who have interests in watershed preservation, environmental engineering or environmental policy, as well as traditional civil engineering knowledge. The organization should de-emphasize hiring individuals who have primary interests in areas such as structural modeling or construction engineering, since these specialties represent a deviation from the fundamental focus of the vision. Similarly, the organization should place an emphasis on actively pursuing projects that have the potential to gain publicity due to opportunities for emphasizing unique approaches to preserving environmentally sensitive areas. Although it is impractical to say definitively that the organization should refrain from accepting jobs in other areas, it is the emphasis and proportion of work that is crucial to establishing core competencies. The combination of these factors serves to generate a mutually beneficial situation. Hiring personnel with a focused set of competencies provides the company with the knowledge to pursue projects within the core focus domain. At the same time, obtaining visible projects that emphasize this strength serves to attract individuals with areas of expertise that complement the core competency. Together, the projects and personnel demonstrate a commitment to achieving a vision as well as establishing a foundation for long-term growth and expansion within the core area of expertise.

The development of a foundation that demonstrates commitment to a vision is the first step toward achieving long-term success. However, as illustrated, an organization requires individuals with the knowledge in the core area to ensure that the expertise is available to build and expand a core competency. In contrast to conventional wisdom, this expertise is not necessarily found exclusively within the civil engineering knowledge base. As illustrated in Figure 4-4, the firm emphasizing environmentally sensitive areas has interests in a broad spectrum of topics. From environmental policy to biology and computer simulation, the organization relies on inputs from numerous specialties to develop project solutions. Although civil engineering resides at the center of these specialties, it represents only one component of a larger picture. Given this larger picture, the organization has a choice to make: Should additional civil engineers with traditional backgrounds be hired, or should specialists from related fields be found, or is it possible to find civil engineers who also have expertise in the related specialty areas?

Responding to the work-force issue, an organization should adopt a structured focus for employee hiring. Specifically, an organization should pursue a path that initially ensures a strong emphasis on core knowledge, followed by an emphasis on supporting knowledge, and finally, an emphasis on expanded knowledge. Placed in the terms of the example organization, the initial emphasis must be on obtaining core expertise in civil engineering as it relates to environmentally sensitive watershed areas. Developing a staff of civil engineers with this interest is paramount in achieving long-term expansion. Similar to Honda's initial emphasis on building a work force with knowledge in small motor development, the example firm must establish its core reputation on a recognized knowledge domain. However, once this core set of person-

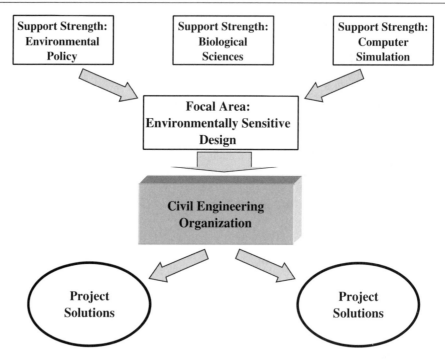

Figure 4-4 Core competencies are not limited to civil engineering expertise. Often, a combination of areas are joined to develop an organization's core competency.

nel is in the organization, the value of expanding in this direction begins to diminish. Although an expanded work force with knowledge in this area will allow the organization to pursue additional projects, the organization fails to establish a complete core competency since it continues to rely on external subcontractors or consultants to provide essential services such as habitat analysis and policy interpretation. A preferred hiring pattern would emphasize obtaining complementary personnel to strengthen the activities performed by the civil engineers. If possible, the organization should pursue members who have knowledge in both civil engineering as well as complementary disciplines. However, in the absence of these individuals, the organization should place a priority on obtaining individuals with strengths in the complementary disciplines. In terms of the example, the civil engineering firm would benefit from the addition of a chemist, biologist, or computer visualization expert to a greater extent than the addition of an additional civil engineer with knowledge similar to existing staff members. Internally, the addition of the nontraditional staff member is evidence that the organization is committed to its vision of unique solutions for existing problems. Externally, the organization is conveying a message to clients and competitors that it intends to demonstrate its expertise in its chosen domain and to acquire the resources necessary to build on its core expertise.

Given a commitment to the overall vision and a work force that reinforces this vision, the core competency concept can be used to expand into the area of strategic markets. The concept of tying core competencies to market strength is one that is rapidly gaining favor among management theorists and consultants (Dutton 1997). Once an organization establishes core competencies through a combination of vision and work force, the logical extension is to use this strength to forge a strategic direction within the marketplace. As illustrated by Hewlett Packard, a single market strength will rapidly expand into multiple opportunities as the organization establishes itself as a leader in the field. Assuming that customers are more likely to contract with a perceived leader in the field than a smaller player when it comes to specialty knowledge, an organization must place itself in a position to be perceived as a leader in the field in which it establishes core expertise. The direct path to achieving this status is through focused attention on projects that reflect organization strengths and utilize the full complement of expertise available to the customer. Put another way, an organization should focus on obtaining projects that are both visible and require a number of experts within the organization. Strategically, this focus sets the basis for long-term operations. By establishing precedence for the selection of projects, organization leaders set the direction for entering, pursuing, and eventually leading a particular market segment.

Summarizing this relationship between strategy and core competency in the context of the example civil engineering firm, the emphasis of the strategic market direction must reflect the vision and work-force directions previously established. Given the emphasis on environmental consciousness and the acquisition of employees that reflect this strength, the strategic development of a market area should leverage these strengths. Focusing on projects that emphasize a small component of this expertise will successfully sustain the operations of the organization, but these projects will fail to provide a basis for strategic advantage. Establishing strategic advantage within the marketplace requires an organization to demonstrate that it contains a set of unique core competencies that set it apart from competitors (Porter 1987). To demonstrate this advantage fully, the organization must actively pursue projects that demonstrate the need for the complete set of services provided within the organization. For the hypothetical civil engineering firm, this translates to pursuing projects that not only utilize the traditional civil engineering skills, but also provide opportunities to demonstrate the ability to integrate scientific and policy strengths in project solutions. While every project may not provide this opportunity, it only takes a few visible success stories to establish a reputation and strategic advantage within a market segment.

Setting a Course for Long-Term Expansion

The previous sections provided an overview of the role core competencies play in developing a strategic direction and advantage for the civil engineering organization. Additionally, it was pointed out that the civil engineering industry faces unique difficulties in building upon core competency concepts due to fragmentation and specialization between the civil engineering disciplines. However, the existence of traditional barriers does not inherently limit an organization from building a core competency strategy for long-term success and expansion. Rather, the opposite is true. Civil engineering organizations must view these traditional barriers as opportunities to build strategic advantage positions. While the majority of organizations consider the implications of breaking these barriers, the progressive organizations have the opportunity to develop strategies for bypassing these barriers and establishing leading market positions. As demonstrated by Honda, breaking from tradition is an opportunity for those who have the foresight to see the long-term benefits. With this as a starting point, the next step in the process of developing strategic advantage through core competencies is to identify, establish, and leverage these competencies within an individual organization. The following sections provide the basis for achieving these objectives by addressing two central issues: (1) how to identify and establish core competencies, and (2) how to leverage core expertise into expansion opportunities.

■ Identifying Core Competencies

Proponents of the core competency concept differ on the specifics of implementing the concept within an organization (Goddard 1997). The focus spans from emphasizing core competencies as a guideline for research and development efforts to emphasizing core competencies as the fundamental component of long-term strategy. Although each of these theories differs on the purpose of core competencies, each proponent agrees on the importance of identifying core competencies as the basis for further organization development. In terms of the civil engineering industry, this may seem like a trivial exercise given the specialization of the industry along discipline lines. In contrast to a manufacturer such as General Electric that develops multiple product lines, or a service provider such as McKinsey Consulting that furnishes clients with diverse services, civil engineering organizations are seen as focused entities. This perception is not only incorrect, but it is detrimental to the long-term development of civil engineering organizations. If a civil engineering organization views itself in terms of a narrow definition, then it likely will follow that perception with reinforcing actions until the organization ultimately reflects the perceived definition. Given this tendency, the identification of core competencies attains a greater importance for civil engineering organizations than it does for organizations in many other industry sectors. Specifically, without this identification, the civil engineering organization is likely to overlook long-term opportunities that may exist outside its traditional boundaries.

Guidelines for identifying core competencies vary among core competency proponents (Snyder and Ebeling 1992; Kozin and Young 1994). However, to bring the concept back to the relevance of civil engineering, this identification process exists on a spectrum (Fig. 4-5). At one end of the spectrum is a focus on the individual and the strengths each individual brings to the organization. At the other end of the spectrum is a focus on the work the organization performs. The task for organization leaders is to determine where on this spectrum the combination of focal points will pro-

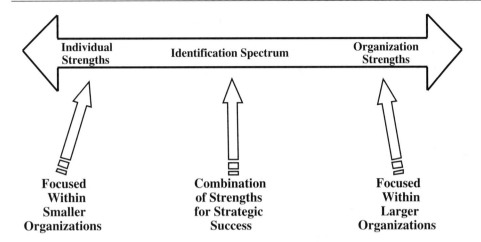

Competency Identification Spectrum

Figure 4-5 Core competencies reside in a combination of individual strengths and organization capabilities.

vide the greatest opportunity to identify core competencies within their specific organization. The following sections highlight the opposite ends of the spectrum as anchor points for this identification process.

Individual Competencies

The side of the competency spectrum anchored by individual strengths is a contrast to the traditional statement that "the whole is greater than the sum of its parts." Unfortunately, specialization within many engineering disciplines including civil engineering creates overly defined career paths (Porbahaie 1994; National Academy of Sciences 1995). In these career paths, specific knowledge is considered essential at different stages of an individual's career. While this knowledge may in fact be required for an individual to succeed within the industry, the uniformity of the career ladder also serves to minimize the opportunity to capitalize on the differences that each individual brings to the organization. The side of the spectrum emphasizing individual strengths advocates changing this procedure by identifying the unique abilities of each individual as a precursor to strengthening long-term goals. However, to achieve this emphasis, the organization must look beyond traditional indicators to analyze the subtle differences between individuals. This analysis process can be generalized to three different classification of organization members: new members being hired out of college, project managers or design engineers in the early to midstages of their careers, and senior-level personnel who are established in senior management positions.

The college graduate. The first of these classifications is the new hire out of college. Traditionally, civil engineering organizations take one of two approaches to the new-hire category. One approach is to assume that the employee knows very little about what he actually wants to do long term due to insufficient exposure to the industry. Therefore, to assist in the decision, the organization places the employee in a short-term rotation of set positions whereby at the completion of the rotation the employee is expected to decide on a predefined specialty area to enter as a career path. The other approach is to assume the employee has made a conscious decision to enter a specific career path and is ready to be placed into an entry-level position. In either approach, a fundamental flaw exists that the organization is neglecting to examine the underlying strengths of the individual beyond the selection of civil engineering as a career path. While it is true that very few entry-level civil engineers know exactly where their career path will lead, that does not imply that each should be molded into a predefined concept of a structural, construction, geotechnical, or environmental engineer. Consider the three upper-division civil engineering transcripts depicted in Figure 4-6. Transcript 1 illustrates a student who has taken the required civil engineering core courses and has supplemented them with additional courses in environmental engineering. Transcript 2 illustrates a student who has taken the required civil engineering core courses, but has elected to supplement them with electives from structural, geotechnical, and construction engineering. Finally, transcript 3 illustrates a student who has supplemented the core courses with electives in public policy, biology, and chemistry.

Taking a traditional approach, student 1 would be considered a good candidate for the directed entry-level position. The assumption in this case is that the student has consciously selected an area of interest and would like to pursue it in a professional setting. Similarly, student 2 would be considered a good candidate for a rota-

Partial Transcript 1 Environmental Focus Core Courses	Partial Transcript 2 General Focus Core Courses	Partial Transcript 3 Policy/Engineering Focus Core Courses
Intro to Geotechnical Engineering	Intro to Geotechnical Engineering	Intro to Geotechnical Engr.
Hydrology	Hydrology	Hydrology
Structures	Structures	Structures
Environmental Engineering Systems	Environmental Engr. Systems	Environmental Engr. Systems
Ethics	Ethics	Ethics
Environmental Electives	*General Electives*	*Electives*
Microbiology in Environmental Engr.	Construction	Environmental Policy
Applied Hydrology	Transportation Engineering	Negotiation Management
Environmental Fluid Mechanics	Structural Masonry Design	Land Development
Hazardous Waste Site Assessment	Concrete Structural Components	Construction Economics
Water Quality Engineering	Water Resources Development	Society and Environment
Hazardous Substance Engineering	Computer Applications	Environmental Assessment

Figure 4-6 Three example transcripts from upper-division civil engineering students illustrate the diversity of backgrounds that can enhance an organization's existing and future competencies.

tion experience in a larger civil engineering organization that performs both design and construction services. The assumption here is that the student is interested in pursuing a traditional civil engineering career, but has yet to focus on a specific discipline. Finally, student 3 would also be placed in the rotation with the assumption that the extra classes might be helpful in the long term, but the initial focus should be on the development of a strong civil engineering foundation. While this traditional approach will successfully produce civil engineering professionals, it fails to address the concept of looking beyond the surface indicators to examine the core competencies of the individuals. Each of the students is indicating underlying strengths through their selection of courses. The student who elects to take additional courses in areas related to, but are outside, the traditional civil engineering disciplines is indicating that a strong multidisciplinary competency exists. This strength should not be diminished through traditional career paths. Rather, it should be fostered through opportunities to work with senior personnel who are exploring new income opportunities. Similarly, the student who takes courses from several civil engineering disciplines should not be viewed as requiring additional exposure to decide on a specific career path. Rather, the student is indicating strength in bridging civil engineering barriers and should be considered as a good candidate for emerging trends in design–build or other alternative procurement strategies. Finally, the student who takes courses in a directed discipline is indicating not only a core competency in one of the underlying civil engineering components, but is also indicating a strong competency in the areas that comprise these disciplines such as creative design, scientific analysis, management, or earth sciences. In each case, the organization has a responsibility to explore these underlying strengths and develop outlets to use them effectively.

The message to comprehend from these examples is that traditional human resource viewpoints in the civil engineering industry only begin to touch the opportunities represented by students entering the industry. However, to extend beyond traditional hiring patterns, the organization must examine the whole picture portrayed by a student's decisions during his or her academic career. Selecting a civil engineering major indicates an interest in the profession. However, it is the selection of courses outside the required courses that provide the insight into the student's individuality and core competencies. If a civil engineering organization desires to complement its existing work force, then it has the responsibility to examine these indicators. Furthermore, once these indicators are identified, the organization must make the effort to find nontraditional opportunities to build upon these strengths.

Midcareer engineers. In contrast to the college graduate who is the focus of many organizational recruiting efforts, the evaluation of a midcareer project manager, designer, or specialist is a less defined process. Many organizations make the decision to utilize a staff member at this level solely on the basis of incoming work. When projects are identified, staff members with appropriate experience are identified to handle the new influx of work. Similarly, some organizations are constantly looking for potential hires from other organizations who may fill a need to improve current project capabilities. In either case, the emphasis is representative of the focus on assigning resources to fulfill short- or near-term project requirements. The placement of the organization member focuses on filling a slot on the organization chart or in the project responsibility matrix. Far less emphasis is placed on examining the individual strengths of these individuals and their potential for contributing to the long-term organization success. Consider the three biographical sketches illustrated in Figure 4-7. The first sketch highlights the career of an Army Corps of Engineers officer who has left the military and is preparing to enter the private sector or a public agency. The second sketch highlights a traditional career engineer who has worked her way up through the structured step of a large engineering–construction organization. Finally, the third sketch highlights the career of a civil engineer who has learned the profession within the context of a small consulting firm.

Given the need to satisfy the two fundamental components of a midcareer engineer, technical capabilities and personality fit, organizations need to go beyond these surface requirements to analyze the strengths that characterize these individuals. In a quest to fill a project or office requirement, organizations often fail to calculate the long-term benefits that an individual brings to the organization. Of greater importance, the organization often fails to analyze how these strengths fit within the long-term vision and the goals put in place to obtain that vision. Returning to the illustrated biographies, each of these individuals brings a unique set of strengths to the organization. First, the Corps of Engineers officer brings a strong background in leadership and independence. Placed in charge of engineering units from the outset of his career, the officer brings a greater degree of leadership knowledge than individuals who have spent their entire careers in the private sector. Concurrently, the individual has worked in remote areas of the world that require ingenuity and resourceful-

Career Sketch 1 Retired Military	Career Sketch 2 Traditional Career	Career Sketch 3 Small Consultant
Colonel Steve Taylor (ret.)	Ms. Sandra Keets	Mr. Ron Holiday
Colonel Taylor was commissioned into the Army Corps of Engineers in 1975 after graduating from West Point. After serving two tours in the South Pacific constructing military training facilities, Col. Taylor served as a District Engineer overseeing the construction of several bridge and dam facilities on the Mississippi River. Col. Taylor concluded his career with a faculty position in the civil engineering department at West Point.	Ms. Keets is currently a Vice President with JJ&R Engineers in St. Louis. With a civil engr. degree from MIT, Ms. Keets has specialized in Process Plant Engineering. Since joining JJ&R in 1985, Ms Keets has had the opportunity to work with Fortune 500 clients on an international basis. She is currently the head of the engineering design group with over 300 engineers in her department.	Mr. Holiday is currently a Sr. Project Manager with PM Consulting. Mr. Holiday has been with PM since graduating from Stanford in 1987. Focusing on construction management activities, Mr. Holiday has participated in field, office, and management activities within the PM organization. Currently in charge of business development, Mr. Holiday is developing new risk analysis procedures for the organization.

Figure 4-7 Three career sketches illustrate the different backgrounds and strengths that midcareer professionals can bring to an organization.

ness to accomplish given projects. An organization looking to expand into new geographic regions could gain significant benefits by placing this professional in a leadership role within a new geographic office.

Similarly, the career engineer brings unique strengths. First, succeeding within a large-organization format requires an individual to understand the diverse elements that lead to success within such an organization, including politics, small-group management, teamwork, and self-promotion. An individual who demonstrates the ability to succeed in this environment brings a strong element of commitment and drive to organization objectives. This individual may bring the commitment that is required to propel the organization to the next level of achievement.

Finally, the engineer who began his or her career in a small firm brings a third set of strengths to an organization. Of specific interest in this case is the strength related to independent thinking and problem solving. Being in a small-firm environment, projects are undertaken with fewer procedures and rules. Individuals are provided greater independence to approach and solve problems based on personal discretion. Although this independence may not fit within some environments, it can provide a significant competitive advantage for organizations developing strategies for new business ventures and opportunities.

Similar to the student examples, the midcareer professionals bring more to an organization than the experience listed on a resume. Depending on the professional environment in which they work, the individual has built upon an academic foundation to enhance individual core competencies. An organization has the choice to build upon these competencies as an additional tool to achieve long-term objectives, or continue to focus on short-term requirements and the need to fill slots on a project responsibility chart.

Senior management. Identifying core competencies in senior management differs from the previous two categories in that the activities these individuals perform are less uniform than those found at lower levels. Depending on the type of organization and the size of the organization, senior managers may have very different roles to play. However, a single emphasis should prevail at this level during a core competency analysis. Specifically, an organization should determine whether or not its senior managers are performing the jobs that make use of their greatest strengths, or are they in senior positions primarily because their seniority warrants such a title and that position happens to be available. The statistics reveal that civil engineers are promoted to leadership positions as much for their organization loyalty as for their competence and abilities (Goodman and Chinowsky 1997). Organizations need to analyze these individuals to determine their inherent abilities and strengths.

In performing this analysis, organizations must consider the basic functions that senior management must undertake. Technical management, marketing, finance, business development, and technology management are all areas that are required for a successful business enterprise. However, the majority of individuals do not contain the core competencies required to manage all these business areas. Therefore, the responsibility of the organization is objectively to match the strengths of senior managers with required business responsibilities. In performing this match, a basic guideline can be observed. First, every area is important to long-term organization success. Therefore, placing less emphasis on a given managerial responsibility will ultimately hurt the organization. Second, individuals who have narrowly defined backgrounds should not be placed in business development positions. These positions require individuals with the ability to recognize opportunities both inside and outside the current market domain. Third, leadership does not equal oversight. Leaders provide an organization with vision, motivation, and direction. Placing an individual in a leadership position who believes that the position is an invitation to impart personal preferences is a quick road to mediocre returns and status quo. Fourth, expertise within an engineering discipline does not equate to business expertise. Engineering disciplines such as computing, electronics, and automotive have demonstrated that placing individuals in charge of the organization who have the capacity to make visionary business decisions will have a greater long-term impact than those who focus primarily on technical competence. Finally, senior managers are caretakers of the organization vision. An individual who understands the vision and has the ability to translate it into current market actions will provide greater strategic advantage then the individual who places the vision solely in the context of annual objectives.

Putting it together. The previous sections highlight the differences between individuals at critical stages in their careers. The advantages that one individual may bring to an organization are very different from those brought by another. However, this is not to say that one is superior to another. Rather, the strengths combine to form the core competency of the entire organization. The key is to place the individual competencies into a proper pattern that builds the organization into a dynamic entity that is able to move beyond its perceived focus and limit. To assist in achieving this goal, competency maps are introduced as a tool for evaluating individual personnel. The focus of this tool is to identify the underlying area of competency that ties together an individual's areas of interest. The map is divided into two segments, above-the-line competencies and below-the-line interest areas. Figure 4-8 illustrates the competency map in the context of an entry-level construction engineer. Developing this competency map proceeds as follows:

1. Chart the areas of interest: Along the horizontal axis, place the individual's focus areas. These should include both primary interests and secondary interests (Fig. 4-8a).

2. Identify surface characteristics: For each focus area, identify the surface characteristic that is commonly referred to as the specialty knowledge required to complete tasks successfully within this area. For example, the entry-level construction engineer with work experience in commercial construction and classroom interests in public policy and computer-aided engineering has surface knowledge in construction engineering, government regulations, and technology, respectively (Fig. 4-8b).

3. Establish the boundary: After identifying the surface characteristics, draw a line that represents the boundary between surface knowledge and underlying competencies (Fig. 4-8c).

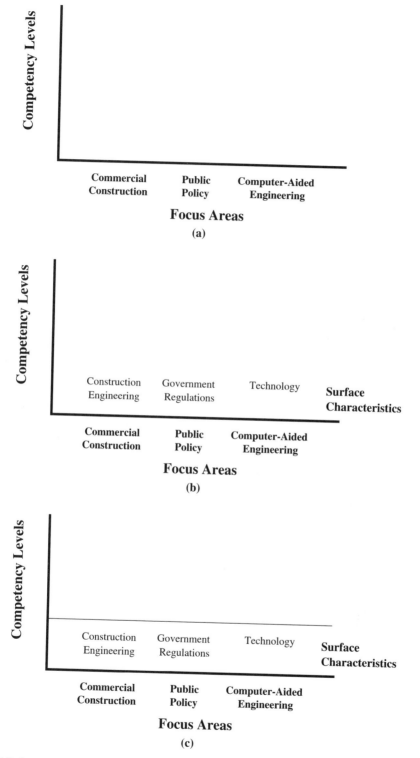

Figure 4-8 Competency maps provide a structured methodology for identfying underlying competencies of both individuals and organizations.

Competency Levels

Project Management Regulatory Analysis Analysis Tools **Competencies**

Construction Engineering **Government Regulations** Technology **Surface Characteristics**

Commercial Construction **Public Policy** **Computer-Aided Engineering**

Focus Areas

(d)

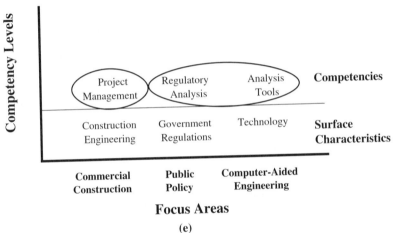

(e)

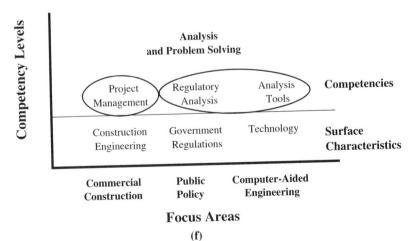

(f)

Figure 4-8 *(Continued)*

4. Identify underlying competencies: For each focus area, identify an underlying competency that enables the individual to pursue the area of interest. For example, Figure 4-8d illustrates underlying competencies in project management, regulatory analysis, and analysis tools.

5. Analyze competency linkages: Given several areas of expertise, the next step is to identify commonality between the areas. The example in Figure 4-8e illustrates that the construction engineer has a strong competency in engineering analysis, but a second competency also exists in management.

6. Repeat connections if necessary: If a clear connection has not been identified by step 5, it is necessary to repeat steps 3–5 until a common basis is identified. As illustrated in Figure 4-8f, the construction engineer has two competencies identified at the conclusion of step 5. Repeating the steps one more time, it is determined that analysis and problem solving is the common link that binds the competencies together. This is determined by a combination of decisions, including: Project management requires a strong competency in analysis and problem solving; so it is refined at the next level; and engineering analysis is a core competency by itself; so it is retained at the next level. In some cases, a link may not be found between an element and the remaining competencies. In this situation, the decision must be made whether or not the additional area is adding to the core competency, or if it represents an interest area that is outside the core competency of the individual. Figure 4-9 illustrates competency maps for several of the other individuals in the above scenarios.

Given an individual's core competencies, the organization can determine whether these competencies provide added benefits. The focus of this analysis is not to find a certain "correct" profile. There is no proven formula for obtaining a correct set of competencies. Rather, the organization should use the analysis as an opportunity to evaluate what are the preferences, individual strengths, and collective directions displayed by organization members. An organization stating that leading-edge practices and visionary leadership are key to long-term success should have individuals with strengths that reflect this statement. Similarly, organizations that emphasize project success and customer satisfaction cannot succeed with an office full of visionaries who have little interest in day-to-day operations. However, the other side of the analysis is the opportunity to discover strengths that extend beyond the stated organization direction. Finding a group of individuals with strong leadership and expansion interests could place the organization in a position to explore new revenue opportunities that were previously considered outside the scope of interest. In these discoveries will emerge the long-term organization direction. Individual core competencies do not predispose an organization to a specific direction, but they provide a foundation for pursuing long-term objectives.

Organization Competencies

If the identification of individual core competencies illustrates that the sum of the individual parts is often greater than the whole that is created, then the identification

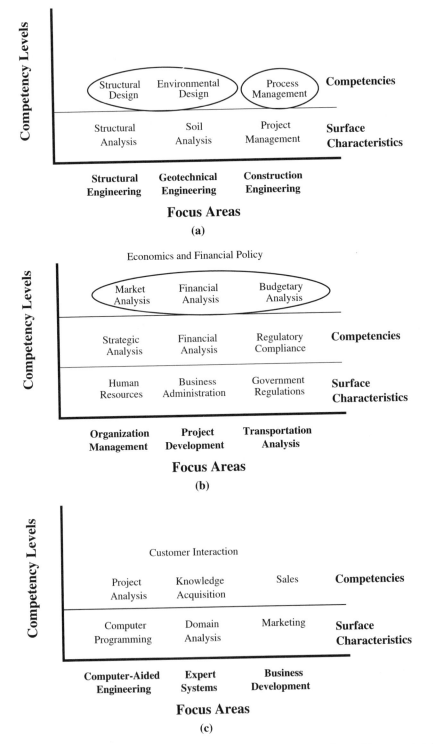

Figure 4-9 Additional examples of competency maps including (a) a college graduate with a general civil background, (b) a senior individual from a small firm with a broad industry exposure, and (c) a midlevel engineer with extensive client interaction experience.

of organization-level competencies illustrates that focused strengths can emerge from analytic numbers. In developing an understanding of an organization's strengths, the end of the spectrum that advocates projects as a reflection of strengths emphasizes examining projects an organization undertakes beyond their surface characteristics. In this analysis, the intent is to identify the core competencies that allow the organization successfully to undertake and complete these projects. The key to this identification process is looking beyond the obvious answers such as strong project management skills or advanced structural design abilities. Although these traits may be components of an organization's success, they do not reflect the core strength that lies below these elements. Rather, these surface strengths represent the knowledge that allows the organization to succeed in its narrowly defined discipline. Extending beyond this market niche requires an understanding of how the organization has achieved its current status. For example, an outside observer would probably state that Honda's success in the early 1960s was due to its superior strengths in building motorcycles. Although this would have been correct, it represents the surface layer of Honda's competency. If in fact the ability to build motorcycles was the organization's strength, then its future would have focused almost entirely on competing with manufacturers such as Suzuki and Harley-Davidson for dominance within the motorcycle marketplace. This competition might have placed Honda in a dominant market position, but it would have limited its ultimate potential to create substantial market shares in the automobile and household equipment markets. These markets become logical extensions only after the underlying core competency in small engines is identified. Similarly, civil engineering organizations are limited in their potential growth when they characterize their strengths in narrow market definitions.

With the need to expand strategic boundaries as a basis for identifying organization strengths, the process can be undertaken in one of two variations. First, the organization can elect to analyze the current income streams to determine where underlying patterns exist that may reflect core competencies. This approach emphasizes the identification of technical and business strengths that tie specific projects together in a coherent pattern while eliminating others as outside the core strength. Second, the organization can elect to analyze the clients who engage the organization to perform work. In this approach, the organization is identifying strengths by identifying the common characteristics that tie the clients together and attract them to the organization.

Projects as strength indicators. The difficulty that organizations face in using competencies is the difficulty of moving beyond surface connections. For example, general contractors often focus success on the fact that they obtain a number of projects in narrow areas such as health care, education, or transportation. The organization emphasizes their expertise in managing and completing the projects with superior quality and attention to customer requirements. Although these elements are factors in the organization's success, they do not reflect underlying strengths such as understanding market forces, ability to manage complex processes, or strength in coordination and communication. These strengths are not limited to the construction process, but rather provide opportunities for expansion into related domains. Identifying such links can range from an obvious connection to an obtuse relationship. To illustrate an easier connection, consider the graph and competency map in Figures 4-10a and 4-10b. In this example, an engineering–construction organization is focused on the power industry. Revenues are heaviest from the construction of power generation facilities, but significant percentages are also represented from management services, oil-field development, and maintenance contracts. The connection that ties these revenue streams together emerges from the identification process. As illustrated in

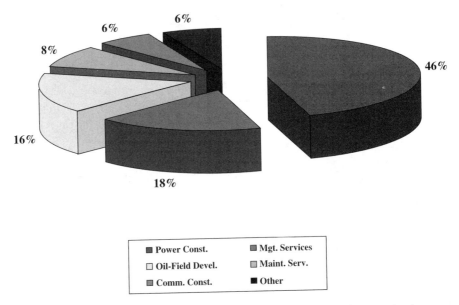

Figure 4-10a The first step in using projects as competency indicators is to map the proportional revenue from each project type.

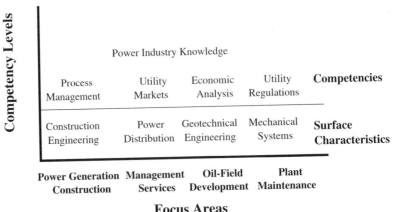

Figure 4-10b Given a division of project types, a competency map can be developed in the same manner as those drawn for individuals.

the competency map, the strengths above the competency line all relate to a strong knowledge and understanding of the power industry. It is clear that the organization has acquired expertise in the legal, technical, political, and market aspects of power generation and dispersal. Although the surface connection may indicate a strength in constructing these facilities, it is the understanding of when these facilities will be required, the customers who contract the work, and the political changes that regulate the industry that provide the competitive advantage. The engineering and construction of the facilities is the technical implementation of the knowledge; the identification of the needs and influences of the industry is the underlying strength.

The vertical competency reflected in the power example has a parallel example in horizontal competency. In the graph and map illustrated in Figures 4-11a and 4-11b, an organization has revenues originating from several diverse operating units, including infrastructure design and construction with a specialty in railroads, erection of athletic stadiums, design and construction of water treatment plants, and rehabilitation of urban structures. In contrast to the power-focused organization, this example does not provide a natural link beyond division boundaries. However, an underlying link does exist between these units beyond the engineering–construction tasks. Specifically, the knowledge related to coordinating complex assembly processes that incorporate diverse constituencies and multiple regulatory bodies underlies each of the operating areas. From the environmental concerns related to water treatment facilities to the urban redevelopment concerns related to athletic stadiums, the organization demonstrates the ability to work with multiple agencies, clients, and team members to complete projects. This management strength is a competency

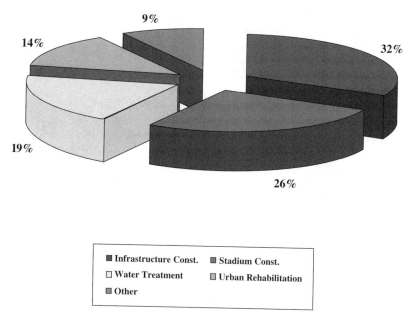

Figure 4-11a An organization example that illustrates a diverse business foundation.

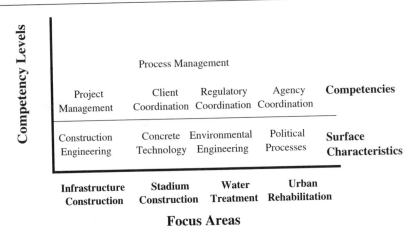

Figure 4-11b The competency map reveals a common basis underlying the diverse market sectors.

that extends beyond the design–construction process. Rather, the management of these processes is a core competency that can be transferred to services outside the design–construction process to the entire lifecycle of projects. However, this strength does not reside within the individual operating units, but rather is reflected in the combined strengths of the overall organization.

Figures 4-12a and 4-12b illustrate a final example of projects as an avenue to understanding core competencies. In this example, an organization has diversified

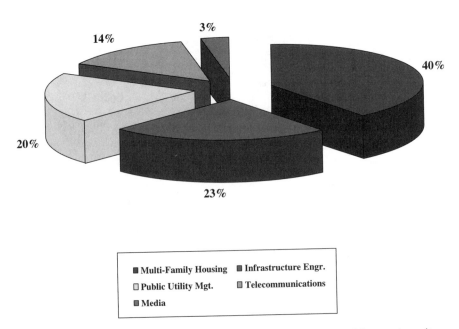

Figure 4-12a A final example illustrates an organization with multiple operating units.

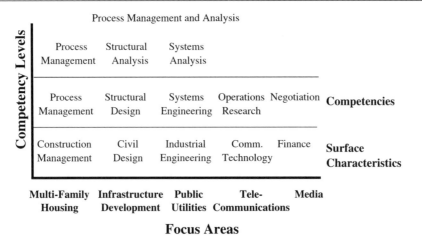

Focus Areas

Figure 4-12b The competency map illustrates how several iterations may be required to identify the underlying core competency.

into multiple operating units that cover diverse areas, including multifamily housing engineering, infrastructure engineering, management of public utilities, telecommunications, and media interests. In contrast to the previous examples, a clear connection between these areas is not apparent. In fact, as the competency map illustrates, a third level must be analyzed to find a connection that binds the core competencies. What the second level of the map illustrates is that the organization has overextended itself into areas such as media and telecommunications services that have little in common with its core strengths in design and construction. These external areas may be interesting to the organization, but the assets and knowledge required to manage these activities are outside the core strengths. This organization faces a difficult decision. It must decide whether to continue with these external interests and stretch its core competencies or to divest these external interests and return to a central focus. The identification of the core competencies alone does not provide the input for this decision. However, as the competency map illustrates, this organization is moving in multiple directions that require distinct operating skills. As illustrated by well-known organizations in prior decades, this diversification will more often than not result in eventual retrenchment and reengineering (Markides 1997).

Clients as strength indicators. A close parallel to the use of project types as indicators of core competencies is the use of clients as core competency indicators. The difference between the two categories is that projects provide an organization with an internal perspective of strengths. By analyzing project data, the organization is analyzing from an internal perspective the strengths that underlie the organization success. In contrast, the use of client data provides an organization with an external perspective of core competencies. In this perspective, the organization is determining what external constituents view as the organization's core competencies. For service organizations such as those that exist in the civil engineering industry, this external perspective is just as important as the internal perspective. While it is possible for an organization to refocus its internal strengths by refocusing its personnel decisions, it takes far longer to readjust client perspectives concerning organization strengths and capabilities.

Performing a client analysis for core competencies follows the pattern established for project analysis. As illustrated in Figures 4-13a and 4-13b, client data are consolidated into a single graph reflecting the relative percentages of income. In the ex-

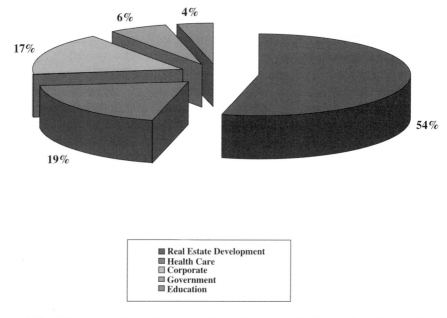

Figure 4-13a Clients can also be used as a core indicator. Once again, the first step is to determine the relative revenue from each client base.

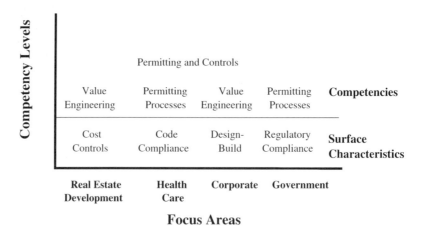

Figure 4-13b The competency map process follows the same steps as the individual and business sector illustrations.

ample, an engineering–construction organization has a significant percentage of its work originating from real estate developers, with an emerging presence in the health and corporate client sectors. A smaller presence also exists with government and education clients. From a surface analysis, these clients tell the organization little about their expertise beyond the fact that the organization understands the importance of cost controls within each of these client constituents. However, a different picture begins to emerge when examining the competency map reflecting the interests of these clients. First, government, education, and health-care clients have a common interest in working with organizations that meet the strict regulatory requirements that exist within their industry sectors. Second, corporate and real-estate development clients have common interests in working with organizations that understand the benefits of value engineering and developing design–build alliances. Very quickly, these two connections provide the organization with an indication of how the external world views their core competencies.

A second example of client analysis is illustrated in Figures 4-14a and 4-14b. In this example, a full-service engineering–consulting organization competes for projects in several sectors, including environmental remediation, petrochemicals, transportation, and microelectronics, as well as offering information and economic consulting services. The client breakdown illustrates a strong presence in the petrochemical, aviation, and environmental sectors that has been established over many years of business development efforts. From the organization's perspective, this success is founded upon their commitment to quality, customer satisfaction, and technical superiority. However, in analyzing the competency map, a slightly differ-

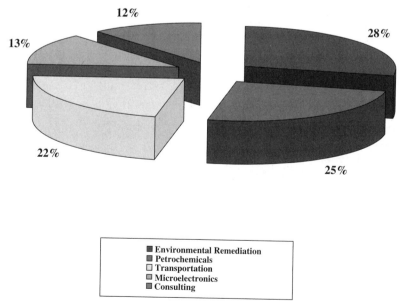

■ **Environmental Remediation**
■ **Petrochemicals**
□ **Transportation**
▨ **Microelectronics**
■ **Consulting**

Figure 4-14a An example of a full-service engineering–construction organization and the diversity of its customer base.

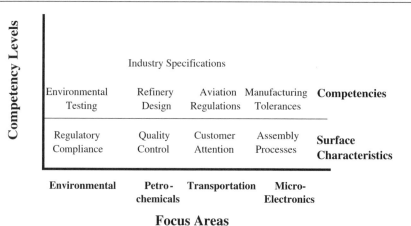

Figure 4-14b The diverse customer base ultimately results in an underlying competency that emphasizes the organization's focus on understanding and complying with complicated specifications.

ent picture emerges. Specifically, the connection that is spotlighted in this analysis is the strength of the organization rapidly to understand industry-specific concerns. In contrast to commercial properties that have many common elements between projects, petrochemical and microelectronic projects require strong industry knowledge. Similarly, aviation projects require a strong understanding of airport operations that impact the design and delivery of new additions or expansions. This external perspective represents a finer definition for success than that perceived internally. The client need for sector knowledge, responsiveness, and problem analysis is a strength that can be built upon with greater focus than the ambiguous strengths represented by quality and customer satisfaction.

Whether external perspectives are in agreement with an organization's internal analysis is not guaranteed. External clients may have a very different perspective of strengths from those identified internally. It is the responsibility of organization leaders to ensure that internal strengths are brought into alignment with external perspectives. Although internal analysis of core competencies is important for understanding organization strengths, it is the external viewpoint that brings clients back to the negotiating table for additional work.

Leveraging Core Competencies

The effort put into identifying core competencies can be significant. Organizations with multiple divisions or numerous members will quickly realize that many diverse interests and project areas exist within the organization. Analyzing these for a common set of competencies is not a trivial task. A commitment of time and resources must be made to identify these competencies and identify the links that represent the core strengths. However, once this identification process is complete, the focus needs to switch from identification to strategic leveraging. Specifically, the identification of competencies provides little benefit unless they are used to focus long-term strategic efforts. This support can take the form of setting future business development activities based on organization competencies, setting objectives to enhance specific competencies, or conducting business development activities that follow core competencies. In each example, the common component is a focus on long-term plans based on the enhancement of existing competencies. With this as a premise, an organization can choose to leverage individual competencies, organization competencies, or a combination of both to achieve a stronger focus on achieving stated objectives.

Building on Individual Competencies

The identification of individual competencies provides an organization with a greater understanding of the contributions that an individual member can make to the long-term vision. By combining complementary competencies, a team can be assembled that has the capacity to achieve enhanced results by capitalizing on unique talents. Rather than molding each individual into a predetermined career mold, the organization can identify the unique competencies of each individual. Through these unique competencies, opportunities will emerge to expand income streams from market segments that reflect previously unexplored strengths. For example, competencies in regulatory processes, environmental management, and engineering design can be combined into a new consulting division that provides engineering organizations from various markets with input on methods for reducing environmental pollutants. However, the issue that underlies this analysis is the determination of whether or not individual competencies are focused on common objectives.

The answer to this issue may be difficult for many organization leaders. The traditional pattern of building a successful career mold and placing each member into a variation of this mold is one that many leaders will be hesitant to release. Rather than threatening this tradition, leveraging core competencies can modify this tradition by focusing it into slightly expanded directions. Rather than placing each individual into a similar career pattern, the organization can acknowledge individual competencies by expanding traditional positions to accommodate expanded competencies. The level of responsibilities associated with a position does not need to change. Preferably, the organization will change the expectations of the position to coincide with an individual's competencies. To achieve this transition, the organization must diverge from its standard review and interview processes to focus greater attention on the relationship of the individual to the long-term vision and mission. As illustrated in Table 4-1, this focus can occur through a small number of directed questions that emphasize long-term plans. The questions in the example are representative of the types that an organization should ask to examine these long-term trends. The specific questions that are adopted for examining current or future employees can vary. However, the questions must remain focused on the individual's ability to enhance the long-term growth objectives.

Leveraging Organization Competencies

Whereas leveraging individual competencies emphasizes building on unique characteristics to establish new opportunities for growth, the leveraging of organization com-

TABLE 4-1
Examples of Issues that an Organization Should Consider to Build Successfully on Individual Competencies at all Levels in the Organization.

Member level	Question	Focus	Impact
Entry-level	Depth vs. breadth of background?	Individual's strength in analysis of problems or synthesis of solution opportunities.	Enhancement of business diversity or strengthening of central market focus.
Entry-level	Long-term vs. short-term vision	Individual's career emphasis in terms of visionary strengths or objective strengths.	Enhancement of long-term vision or strengthening of annual objective attainment.
Entry-level	Independence as opportunity vs. structure as stability	Individual's comfort level with working independently in new business ventures vs. working in structured domains to strengthen traditional business practices.	Development of diverse income streams or strengthening of existing market domains.
Mid-level	Developer of vision or implementer of plans	Individual's strength in developing visions and missions or ensuring the successful implementation of plans.	Creating new options and directions for organization development or ensuring the successful completion of current objectives.
Mid-level	Leadership vs. management	Individual's strength in motivating project personnel vs. managing project requirements.	Development of senior executives or enhancing project completion practices.
Mid-level	Engineering focus vs. market focus	Individual's interest in furthering knowledge in technical field or expanding knowledge of market factors	Reinforcing technical reputation or expanding business plans.
Senior-level	Expansion vs. focus	Individual's strength in analyzing market opportunities.	Where is the organization direction going to develop?
Senior-level	Innovation vs. standards	Individual's comfort level with innovative procedures and technologies.	Is the organization going to be on the cutting edge or safely in the group?
Senior-level	Performance vs. experience	Individual's interest in performance or experience as a key to organization success.	Is the organization going to reward loyalty or performance as a metric for advancement?

petencies emphasizes building upon a small number of overall strengths. For example, the construction management organization that has built a strong reputation for managing complex projects such as hospitals and microelectronics manufacturing facilities is in a position to build upon this experience. After developing a competency map that identifies an underlying competency in process management, the organization is in a position to explore additional areas that require similar expertise. In this case, a number of opportunities are possible, including municipal management, utility management, and corporate facility consulting. The common thread throughout these opportunities is an underlying reliance on process management, the same competency that exists within the construction management organization. This is not to imply that the organization can walk into an opportunity in these areas and instantly succeed. Domain knowledge is required to succeed in an expanded market sector. Of greater interest is the potential to transfer the management processes that serve as the core competency of the organization. By modifying the processes and the practices that established the organization's reputation in construction management, the organization has a proven basis for making the transition to a related domain.

A decision such as this to expand beyond existing market boundaries can be a daunting one due to issues of risk, unknown barriers, and established comfort zones. These issues are understandable and real. Expanding upon established strengths is a significant decision. However, to place the need for this decision in perspective, a few basic market issues can be analyzed. First, an organization must examine what is the ultimate benefit that can be achieved by becoming the dominant player in the market. Specifically, is the market sector large enough that becoming the dominant player ensures consistent profits, growth, and economic stability? In fact, very few market sectors provide such benefits. Second, an organization needs to examine the stability of the market sector. Is the market sector immune to economic cycles and external influences? Third, an organization needs to analyze the level of competition in the market sector. Is the competition at such a level that expanding profit margins would be difficult for an individual organization to put in place? For most civil engineering market sectors and organizations, independently increasing profit margins is not a viable long-term alternative. Finally, an organization must realistically examine the difficulty that exists for a new organization to challenge its position in the market sector. Is the expertise that the organization has established unique enough to have established barriers for other organizations to compete effectively? Without these barriers, the organization will face increased competition that will in turn require the organization either to compete more effectively in the limited market sector or to examine new business opportunities.

In each of these alternatives, the organization is facing a similar question, "Should the initiative be taken to expand market interests while the option exists and the time is available to thoroughly analyze options, or should it wait until competitive forces make it necessary to expand in a shorter time frame?" The answer to this question is unique to each organization. However, as the core competency analysis demonstrated, each organization has strengths that go beyond its existing market focus. Once these strengths are identified, the opportunity exists to move outside the established comfort zone. The size of the organization is irrelevant to this expansion. Size is only relevant to the degree of expansion that should occur. Smaller organizations can focus on increasing services along a greater extent of the project lifecycle, while larger organizations can examine this option as well as the option to move into related disciplines. In either case, the central point is to focus outward on opportunities. The organization that continues constantly to focus inwards on a single market sector will eventually succeed in having the best people to perform a single task. This same organization will find itself attempting to respond to external market forces with no established alternative capabilities. In contrast, the organization that focuses outward on the possibilities of building on core competencies will have a competitive advantage through a combination of increased attention to expanded capabilities and increased attention to long-term goals.

■ Taking the Next Step

The identification of core competencies and the analysis of growth opportunities are a logical step forward from the establishment of long-term visions, missions, and goals. The identification process forces organizations to return to the realities imposed by existing conditions rather than remain in the realm of visionary objectives. Similarly, the opportunities for expanding market segments place the organization with a need to obtain or reassign resources. Chapter 5 responds to this need by focusing on the human and technological resources required to effectively compete as a twenty-first-century civil engineering organization.

DISCUSSION QUESTIONS

1. A central question in building upon core competencies within an organization is, "Are current objectives focusing on what the organization does best?" How does an organization ensure that objectives correlate with core competencies?

2. Chapter 4 introduces a hypothetical organization, MRB, and its potential expansion opportunities. What are other examples of actual civil engineering organizations and their expansion opportunities?

3. Hewlett-Packard is an excellent example of building on core competencies. Starting from a core technology and slowly expanding into areas such as calculators and computers, HP continues to demonstrate market awareness while remaining focused on its strengths. How does the fact that HP is in the manufacturing sector rather that the service sector change the way it approaches core competencies from a civil engineering organization?

4. Does the concept of core competencies conflict with the civil engineering organization emphasis on projects as a central management focus?

5. Can a focus on core competencies restrict the development of innovative market sector initiatives for organization expansion?

6. The core competency map methodology is introduced as an opportunity to identify individual and organization competencies. How can this same methodology be used as a tool to evaluate market opportunities? Competitive positions? Employee development?

7. What should organization leaders do if they find that individuals currently in the organization do not have core competencies that match the overall core competencies identified for the organization?

Shifting Core Competencies

Building on core competencies to establish a sound business foundation is a cornerstone of modern business process. Core competencies provide direction, boundaries, and focal points for organization investment. Of greater importance, core competencies define the strengths on which to achieve long-term visions. However, at times, fundamental business practices must be questioned in terms of their applicability to specific situations. Core competencies are no exception to this rule. Although core competencies are an essential component of organization success, organizations have long sought to diversify as a hedge against economic swings. In this case study, we examine one such diversification and its impact on the underlying organization culture.

Heery International is a full-service architecture, engineering, and construction management firm with over 800 employees located in offices throughout the United States, Canada, and Europe. A subsidiary of the London-based BICC Group, Heery International is consistently ranked as one of the top design firms in the United States. Founded in 1952, Heery has grown to achieve design revenues approaching $100 million. Heery currently ranks in the top ten of several design categories including education, general building, and sports facilities. Working with global customers such as Motorola and Home Depot, Heery has built a reputation of providing excellent service to each of its clients while delivering innovative design solutions. An example of this innovation is reflected in the recently completed Motorola Mexico facility in Chihuahua, Mexico. Located on a rural desert site, the project includes offices, training facilities, wastewater treatment, an electrical substation, and recreational amenities. A design award winner, the facility represents Heery's commitment to satisfying unique customer needs throughout the world.

The Issue

The quality of the architecture and engineering (AE) services provided by Heery are unquestioned both within the organization and by its clients. Considered as the core competency of the firm by its founder George Heery, the AE component of Heery has defined the image, business, and marketing for most of the organization's existence.

When combined with its significant presence in its core business areas, the focus on core design competencies has been a strong presence throughout all of Heery's activities. However, as with all other components of the business world, change is the only consistent element in the operating environment. Consistent with that theme, Heery is experiencing forces of change within its organization. Specifically, the long-held belief that Heery is defined solely as an AE services firm is being challenged as the organization increases revenue in the construction management (CM) arena. Originally established as a secondary business element to support customer requirements, the construction management group has grown significantly over the past decade.

Today, the construction management group accounts for over $70 million of revenue for Heery International. Although praised by all segments of the organization and the parent company, the success of the construction management group is challenging traditional beliefs within the Heery organization. Specifically, the belief that design is the core competency of the organization is being questioned. With CM revenues approaching that of the design group, and projected to exceed design revenues within the next decade, the success of the CM efforts is causing Heery traditionalists to examine their view of the organization. The days of comments such as "We will never be a construction organization" are over, as CM revenues account for an ever increasing component of the financial bottom line.

In contrast to an expansion from one design or construction sector to a related sector, the Heery move is an expansion into a new business area. In this expansion, Heery is addressing the integration of two fundamentally different operating and experience factors. In contrast to the architectural emphasis on concepts and aesthetics, the construction operation is bringing a primary emphasis on budgets and schedules. These differences are reflected in the approaches that each group brings to client projects. For the designers, the development of aesthetically innovative and pleasing concepts is the top priority. The time required to develop the design is important, but is seen as secondary by many designers to the need to develop a successful design concept. In contrast, the construction managers have the responsibility to the client to monitor budgets and schedules. As such, each element of the design–engineering–construction process is viewed as a deadline and budget requirement that Heery has an obligation to meet.

With these differences, Heery International must address a fundamental business decision. The traditional emphasis on design as a core competency is being challenged as construction management revenue meets and exceeds design revenue. Of greater importance, the increase in construction specialists is challenging the traditional culture within the organization. A greater emphasis on business practices and budgets is emerging as the original design partners retire from the firm. Accompanying this change in leadership is a change in market focus. No longer is construction management viewed as a secondary element in the business portfolio. Rather, construction management is being marketed as a strength of the organization and a central element of Heery's full-service options for clients. As reflected in the sentiments of an increasing number of Heery employees, construction management is emerging as a second core competency within the organization.

Looking Ahead

The success of the construction management group within the Heery organization cannot be overlooked. The increase in revenues, number of employees, and client recognition is bringing the CM group to the forefront of the organization. However, with this increased visibility comes corporate friction. Any organization founded and built on a single vision develops a corporate culture to support that vision and make it thrive. The CM developments at Heery are challenging this concept within the Heery organization. The design emphasis established by George Heery in 1952 has decreased, and will continue to decrease, if the CM growth continues on its present course. However, accompanying this change in emphasis will be an increase in corporate friction. Visions and core competencies provide employees with a comfort zone in which to operate. Every individual understands the emphasis of the organization and its direction for future business initiatives. Changing this foundation with the introduction of a business venture that promises equal or greater long-term activities can create instability and unease.

As Heery approaches its sixth decade of operations, it is facing this prospect of instability and unease. The continued success of the CM group will pose a challenge to the designers within the organization. The design culture that lies at the heart of the Heery organization will be challenged as clients approach Heery exclusively for CM services. In summary, the construction management group is an emerging force within the Heery organization and the issues associated with developing a second core competency will emerge together with this development.

QUESTIONS

1. Placed in the role of a senior AE manager, how would you address the issue of a changing corporate image to the AE staff who came to Heery based on its reputation as a leading design firm?

2. Placed in the role of a senior CM manager, how aggressively would you pursue the development of your business unit when faced with the knowledge that corporate friction will accompany this growth?

3. Placed in the role of the Heery President, would you continue to advocate the development of a second core competency, and how would you justify your decision?

4. Is it possible for Heery to avoid the corporate friction associated with business expansion?

5 KNOWLEDGE AND INFORMATION RESOURCES

What processes have been put in place to enhance knowledge and information resources within the organization?

The third question in the question–answer process focuses on the need for civil engineering organizations to reexamine the human and information technology resources available to the management team. The emergence of new competitors and the need to explore new income opportunities challenges organizations to reexamine these traditional resources from new perspectives. This reexamination emphasizes the need to look beyond traditional departmental boundaries. The strength and creativity of an organization is not found within individual departments, but rather in the combination of nontraditional alliances. One such alliance that lies at the core of this new perspective is the link between human resource management and information technology. Traditionally, human resource and information technology departments have had little connection beyond supporting functions such as hiring employees and keeping the management information system operating. However, the knowledge and information that links these two departments together is critical to the long-term success of civil engineering organizations.

Knowledge and information are the keys to success in the twenty-first century. The world is emerging from an industrial basis of power to an information basis of power. The size of an organization no longer dictates the level of service that can be provided to clients. Information access and remote communication capabilities are breaking physical boundaries that traditionally have limited smaller organizations from competing with larger organizations (Ross 1997). The ability to provide clients with solutions to complex problems is no longer limited to the domain of a few dominant industry players. Rather, organizations that leverage knowledge and information to the greatest benefit will emerge as the new leaders within this information age. Placing this in the perspective of the civil engineering industry, the organizations that leverage the combined strengths of employee knowledge and information technology will have the greatest opportunity to break out of traditional industry niches to capture a segment of emerging client markets.

In this chapter, the converging roles of human resources and information technology are examined in the context of the resource enhancement question posed above. This chapter emphasizes the need to combine human and information technology resources into a combined knowledge-based resource that supports twenty-first-century business requirements, including coordinated problem solving, new service identification, rapid communications and coordination, and world-wide information access. This knowledge-based resource combines the creativity of human resources with the processing capability of information technologies to create a work force that utilizes expanded knowledge to achieve results that were not previously possible through individual efforts. Achieving this level of resource integration sets the challenge for this chapter—*to examine the integration of information technologies and human resources to determine if the work force is maximizing the knowledge-based opportunities presented by emerging technologies.*

■ The Changing Resources of Organizations

The union of information and human resources is not unique to the civil engineering industry. This union was originally identified in 1967 by Peter Drucker, considered to be one of the founders of management theory, in the context of a knowledge economy (Drucker 1967). As defined by Drucker, "the knowledge economy is one in which knowledge is the central resource, the major factor of production." The significance of this statement for the civil engineering industry is broad and far-reaching. Although the introduction of knowledge workers is creating a work environment where both traditional engineering activities and emerging knowledge-based activities will be valued, the concept of the knowledge economy focuses on a larger-scale transition. Specifically, the knowledge economy is placing greater emphasis on the ability of organizations to meet customer demands in the context of a broad and global perspective. In the traditional model, a project such as an office building or chemical plant would be proposed based on local factors, and an AEC team would combine their skills to develop a series of viable solutions. The greatest asset for this team was the relative continuity that existed in terms of local economic and business conditions that could be relied upon to base solutions. In contrast, today's rapidly evolving technological, economic, and political environment demands these same teams find solutions that not only meet the local conditions, but also have the flexibility to respond to changing global conditions (Nonaka 1991). Accomplishing this task requires individuals that are not only skilled in their industry segment, but also have the ability to identify, analyze, and use diverse resources to address global issues. These individuals do not reflect the traditional labor force that worked in predominantly hierarchical environments. Rather, these individuals reflect the emphasis on broad knowledge as an asset and the key to economic success (Boudreaux 1984; Drucker 1993).

The model for this transition is being spotlighted by national consulting firms such as Anderson Consulting and Ernst & Young. In these national consulting firms, traditional strengths in business consulting are being leveraged into new revenue opportunities from both existing and new clients. With the recognition that clients are searching for organizations that can provide comprehensive services, the national consulting firms are expanding their core business consulting services into related areas such as construction management and real estate development. In these cases, the organizations are developing employees with broad business and industry perspectives to introduce clients to the service solutions that the consulting firms can provide. Combining tax, strategic, product, and facility perspectives into a comprehensive solution, the new generation of knowledge workers in these consulting organizations are setting a new standard for competing organizations. Specifically, the challenge has been set for organizations such as civil engineering firms to either respond with similarly broad knowledge perspectives or risk losing clients to nontraditional competition.

This challenge represents a transition for civil engineering organizations from narrowly focused professional knowledge such as computer-aided drafting to broad, global tasks such as computer-integrated planning. These knowledge-based tasks diverge from the traditional emphasis on well-developed industry knowledge to an emphasis on manipulating global knowledge resources. Within the civil engineering industry, the knowledge economy is creating the need for a new generation of knowledge workers. These knowledge workers will augment traditional industry knowledge with diverse influences to enhance work processes by infusing new problem-solving perspectives. To illustrate, a traditional worker entering the construction management sector would begin the career ladder as either an assistant field engineer or assistant office engineer. After several years of learning the construction management process, the employee would progress to a project management position, where refined knowledge would be used to manage the completion of construction projects. The focal point of this promotion process is the learning and refinement of narrow fields of knowledge such as scheduling and budget controls that are traditionally required to manage projects to completion. Similarly, entry-level workers in the transportation, structural, and environmental fields enter the profession by performing fundamental tasks to acquire specialized expertise that is combined at higher levels in the career ladder. This progression from entry-level, narrowly focused positions to incrementally higher professional levels has remained stable over generations of industry workers.

The transition to a knowledge worker industry will change these traditional scenarios based on catalysts from outside clients and entering professionals (Webber 1993). As a knowledge worker, the entering employee has been exposed to a broader spectrum of knowledge domains. The worker has witnessed the interrelationships between global economic and political forces in both education and professional settings. When combined with the expectations of the new client base, the organization must acknowledge the need for these broader knowledge perspectives. With this broader knowledge perspective, the knowledge worker enters a transformed career ladder. At the entrance stage, the knowledge worker uses broad technological, economic, and client knowledge to identify and access data on project partners and client organizations. Additional focus may be placed on enhancing project communications through the development of advanced computing technologies. The common thread between these duties is the focus on broad perspectives as the avenue to enhance traditional project solutions. Progression up the career ladder focuses on the ability of the knowledge worker to enhance operations within the organization rather than refine traditional knowledge. For example, the ability to establish new information transfer processes, introduce global economic influences, or develop new market niches is emerging as a valuable commodity in the knowledge economy. This is not to say that traditional civil engineering knowledge will no longer be required. Rather, the emergence of a supplementary set of knowledge, or a second career path, is required to reflect the changing environment in which knowledge workers and civil engineering organizations operate.

■ Knowledge Workers as an Organization Investment

The introduction of the knowledge worker can be viewed by organizations as either an upheaval of traditional methods or an opportunity to establish an enhanced position for competing in the knowledge economy. In the former, the organization resists the transition due to long-held traditions and beliefs concerning the manner in which business should be done and the manner in which professionals should be developed. Examples of this approach can be found in many industries. Recognized organizations such as General Motors, IBM, and Sears each resisted the change to the knowledge economy (Webber 1993). As a result, each of the organizations suffered repercussions at a later date. For example, Sears was required to redefine itself and its market as competitors such as Wal-Mart adopted technology-based solutions to rapidly adjust to changing economic conditions and consumer preferences. The civil engineering industry is not immune to such results. The failure to grasp opportunities to enhance the organization knowledge base will result in significantly longer periods of transition. In contrast, organizations that view the knowledge worker as an opportunity for enhanced competitive positions have the opportunity to enhance operational practices with broader perspectives. Recognized organizations such as Nucor Steel, Xerox, and Motorola demonstrate the potential benefits offered by the knowledge economy transition. In each of these cases, the organization seized the opportunity to broaden operational perspectives to gain market share or price advantages over competitors. The same opportunity is available to civil engineering organizations. However, to grasp this opportunity, organizations should examine two fundamental factors, the cost of employee mobility and leveraging knowledge versus skills.

Costs of Employee Mobility

Traditionally, the conformity of education found in civil engineering graduates provided organizations with a common baseline from which to establish requirements for entry-level positions. Given the relative uniformity of education, the career pyramid found within civil engineering organizations is similar in design. Furthermore, the uniformity provides an avenue for both employers and employees to examine a number of potential employment alternatives. If the fit between an organization and an employee is not satisfactory for either party, then the uniformity of knowledge ensures a supply of potential applicants who have commensurate skills to fill the position within the pyramid. Similarly, the employee, while not as resilient as the organization, has the opportunity to take his services elsewhere, assured that the knowledge obtained during a college education is transferable to another organization. This combination of uniformity and replacement ability has provided civil engineering organizations with the confidence that employee replacement could be achieved with a minimum of disruption to daily operations.

With the introduction of the knowledge worker, this view of employee mobility must be altered. In contrast to industry knowledge that is uniform among traditional entering employees, the global knowledge perspectives that knowledge workers bring to an organization will be a distinct and highly variable commodity. Each employee will enter the organization with a unique set of knowledge perspectives. Some employees may emphasize global communications, while others may emphasize market analysis or client analysis. While it is true that traditional employees will continue to bring varied benefits to an organization, the scope of the knowledge that these workers bring to the position is restricted within a tradition-oriented professional perspective. The difference with a knowledge worker is the scope and breadth of knowledge that these professionals bring to the work process. In response to the globalization of the world's business environment, the new knowledge workers will bring new perspectives to traditional problems. Whereas the traditional worker will initially focus on a common set of professional knowledge and later introduce broader perspectives, the knowledge worker may immediately focus on unique and diverse knowledge areas, foregoing traditional knowledge boundaries to concentrate on global organization influences. The ramifications of this differing focus are seen as the employee progresses through an organization.

Rather than following the traditional career path, the knowledge worker challenges the organization to create an environment where he or she can provide the greatest service (Mandt 1978). The development of this environment permits the knowledge worker to provide the organization with broad, interdisciplinary perspectives that are potentially unique within a market niche. Although this global perspective may be transferred to additional employees, the original knowledge worker remains the source of the innovation. The issue to the organization becomes whether the knowledge worker can be replaced based on the traditional civil engineering paradigm, or whether the employee brings a unique set of perspectives that cannot be replaced based on traditional measurements. For example, the senior-level knowledge worker who incorporates global economic indicators through remote communication facilities cannot be replaced by a senior-level project manager using traditional civil engineering perspectives.

What are the ramifications of these unique capabilities to a civil engineering firm? Of greatest importance is the impact of replacement. Although civil engineers have historically remained stable within their work environments, organizations have felt comfortable during down cycles to lay off employees with the understanding that replacements could be readily obtained when necessary. The knowledge worker changes this equation. In contrast to the traditional viewpoint, the replacement of knowledge workers must be viewed from a value-added perspective. First, knowledge workers bring breadth and global perspectives to an organization. Given the nontraditional career paths that these workers may follow, their contributions are not limited to traditional job descriptions. Some may provide new perspectives on client interaction, others may provide new work process innovations, while others may provide new management perspectives. In each case, the unique backgrounds of these employees will make it difficult to find a competent replacement. More important, the organization faces the loss of the creativity of the departing knowledge worker. Although the loss of experienced professional knowledge is important, that loss is temporary until a new employee attains a commensurate professional level. In contrast, the loss of a knowledge worker is a permanent loss of creative perspectives and global awareness that a course or experience cannot change. In this manner, the emergence of the knowledge worker requires civil engineering organizations to reevaluate their commitment to employees. Employees can no longer be thought of as easily replaceable components within an overall organization. Rather, knowledge workers are unique contributors to a unique service provided by the organization.

Leveraging Global Knowledge Perspectives

The second focus of employees as an organization investment, the leveraging of global knowledge perspectives, is on the ability of organizations to leverage the diverse perspectives of knowledge workers. In this concept, organizations must identify the areas where broad perspectives can be synthesized into enhanced project solutions. To contrast this with the traditional civil engineering perspective, the context of the traditional structural engineering design firm is examined. In this traditional context, individual knowledge is collected into a comprehensive whole through a pyramid process (Fig. 5-1). In this pyramid, the lowest levels represent the design calculations developed by entry-level engineers. The knowledge developed to calculate design requirements is leveraged at the next level of the pyramid by senior designers, who place the individual calculations into a system-level solution. This design knowledge is subsequently leveraged by division managers into project deliverables, including designs, invoices, reports, and proposals for additional tasks. This pyramidal approach to civil engineering operations is replicated in each of the disciplines after being refined over decades of organizational development. The benefits of such a process are

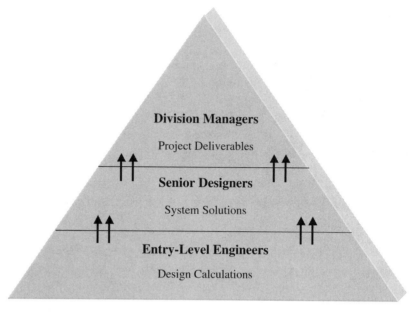

Figure 5-1 Traditional career pyramids focus on the incremental development of solutions through the transformation of individual components developed by staff members into final deliverables by senior organization members.

obvious. By building upon the industry knowledge of lower-level employees, the organization reduces the cost to the client, reserves the advanced knowledge of senior employees for system-level tasks, and provides an avenue of advancement for employees based on the development of traditional civil engineering knowledge.

The development of a comparable pyramid based on global knowledge perspectives is the challenge for the next generation of civil engineering organizations. How should this challenge be addressed? Foremost, organizations need to reevaluate the levels of the pyramid. Specifically, what are the basic elements that a knowledge worker brings to the organization? In some cases, the answer is the ability to access and process client and project information. In others, it may be the ability to analyze global environmental trends on future revenue streams. In either case, the abilities brought to the table by the knowledge worker are going to diverge from traditional civil engineering knowledge. Similarly, higher levels of the pyramid need to be reevaluated in terms of aggregating the knowledge developed at the base levels. Rather than developing system solutions, senior knowledge workers may develop comprehensive business solutions for clients that include traditional civil engineering solutions as well as illustrations of installation, cost, future expansion, availability of suppliers, and similarity to existing client facilities. The focus of these new solutions recognizes that the client has interests that go beyond individual project components, to issues that center on the value of the facility to the overall client organization. This recognition diverges from the traditional role of the civil engineer. In this revised role, an organization builds upon the ability of its staff to provide comprehensive solutions that recognize overall client needs, rather than solely providing technical solutions that are limited to the needs of an individual project.

Similar to the change in solution generation, additional avenues exist for the introduction of knowledge-based workers into the civil engineering organization: for example, the development of budgets and schedules that address the purpose of the facility rather than just the construction of the facility, or the development of environmental reports that demonstrate the regulatory, social, and physical impacts of projects on the surrounding community. In these examples, the addition of a knowledge-worker perspective establishes opportunities for success by creating value-added bonuses for clients. In addition to the information processing capabilities that the client is expecting, the organization places itself in the position of providing knowledge that is unanticipated by the client. Similar to the introduction of technologies such as three-dimensional modeling, the introduction of knowledge-based activities can be leveraged into greater revenues if the organization can present the knowledge in a manner that proves its worth in terms of long-term profits for both the client and the civil engineering organization.

■ Technology Resources: Enhancing Knowledge Exchange

The move to a knowledge-based industry emphasizes the need to support knowledge processing with increasing levels of technology resources. From executive management to engineering departments, a vision of knowledge and information integration must be established as part of the strategic direction adopted by the organization (Boynton, Victor, and Pine 1993). However, it is unrealistic to assume that an organization can establish this direction in a single day, week, or even a month. Rather, it is a slow, gradual process that begins with an internal champion carrying the appropriate message to as many decision makers as possible. The focus of this message must emphasize the new role of technology in enhancing knowledge resources rather than the traditional arguments of technology lifecycles and useful life scenarios. In this context, the message must emphasize a focus on the positive movement of the organization to a direction where technology is integrated with knowledge development. However, a positive-sounding statement by itself does not create a strategic position for technology. Rather, a structured approach is required to build an argument for strategic technology integration. One such approach is to structure the technology integration process into the following steps:

1. Understand the internal and external catalysts for technology integration
2. Clearly define the new organization role for technology and its relationship to emerging technology opportunities
3. Evaluate the technology needs of the organization
4. Establish an integrated technology strategy

The Technology Catalysts

Developing a plan requires knowledge of the situation that the participants are preparing to enter. The greater the knowledge, the greater the advantage the planners have to develop a strategy that works within the context. In practical terms, the development of a technology strategy must respond to the realities of the business environment while setting a foundation for future growth and opportunities for business development. Identifying this full range of catalysts presents a daunting task to an organization that is maximizing its use of resources to remain just ahead of project deadlines. As repeatedly stated by organization leaders to the authors, while the intent to accomplish this task is clearly evident, the time to implement the task is rarely available. Breaking from this dilemma requires a motivating factor that places technology integration on an even level with project concerns. As illustrated in Figure 5-2, this motivation now exists from a combination of internal organization and external business catalysts. Developed from management of technology models, field analysis, and emerging trends in the civil engineering industry, the catalysts provide a synthesis of the emerging forces of change (Brisley and Fielder 1983; Sviokla 1996; Price 1996).

Internal

The individual work processes adopted by an organization reflect the culture and perspective of the individuals leading the organization. Conservative or progressive,

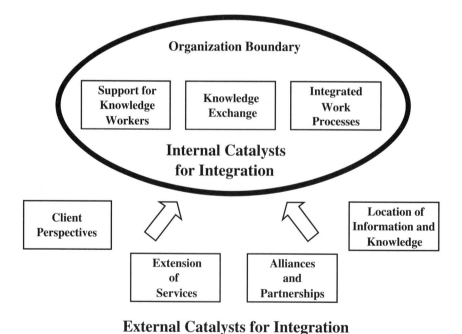

External Catalysts for Integration

Figure 5-2 A combination of internal and external catalysts is forcing civil engineering organizations to plan for the strategic integration of technology into business operations and objectives.

processes are established reflecting the beliefs, knowledge, and perspectives of the organization leaders. Challenging these processes is a difficult task, since it is equivalent to challenging the culture of the organization. From this perspective, a powerful set of catalysts must be in place to encourage the internal analysis of operational standards and practices. Focused inward toward the role that technology will play in supporting knowledge transfer, the catalysts analyze fundamental work processes and traditional procedures used to complete projects.

Support for Knowledge Workers The first catalyst for internal change is the support for knowledge workers. The previous section outlined the need to push the organization forward in respect to its focus on employees as manipulators of global knowledge rather than as narrowly focused professionals. Underlying this push is the need for knowledge workers to have a technology infrastructure that supports the sharing, access, and dissemination of global knowledge. In response to the evolution of multimedia technologies, this infrastructure can no longer be limited to the support of text-based documents and communications. Rather, a technology infrastructure must be established that supports the full range of traditional, emerging, and future knowledge representations that exist or may exist within the framework of the problem-solving process (Fig. 5-3). Achieving this flexibility requires an examination of operational processes with a perspective on future growth rather than past practices. For example, an organization that focuses its services on the generation of environmental impact statements may traditionally have focused on the use of technology to exchange individual report sections eventually to compile a single client document. In the context of future knowledge exchange, technology infrastructure should be put in place that allows this same team to communicate three-dimensional site models, video site studies, virtual-reality development studies, and collected information links to agencies and repositories around the world (Chinowsky 1998). The team members should even consider the possibility of future technologies that may institute communication methods such as distributed virtual worlds that enable designers, consultants, and owners to examine site characteristics

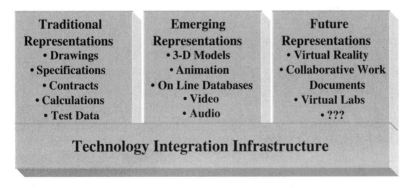

Figure 5-3 Technology integration strategies must address the needs of traditional, emerging, and future knowledge representations.

collaboratively through a virtual experience from their individual offices (Fruchter 1996).

The underlying message from this internal catalyst is that the effective organization will leverage knowledge-worker capabilities by successfully providing the infrastructure for these workers to maximize their interdisciplinary perspective (Grant 1996). Without this infrastructure, knowledge remains as isolated islands within the organization. The organizations that are best able to bring these islands together into a coherent, integrated structure will have the greatest advantage in the knowledge-based business environment. Concurrently, it is necessary to acknowledge that some members of the organization may use this catalyst to focus on technological control of knowledge rather than technological exchange of knowledge. The function of the organization leader in responding to this catalyst is the establishment of an organization mission that clearly states that knowledge is a corporate asset, and as such belongs to the benefit of the organization and not to the benefit of any single individual.

Knowledge Integration The transition from a narrowly focused profession to a globally focused profession requires organizations to combine specialized islands of knowledge into comprehensive knowledge networks. In the traditional firm, technology provides organization members with access to individual knowledge islands. Technology provides the ability to find knowledge islands, query members within these islands for solution components, and receive information generated by the individual islands (Fig. 5-4). The strength of the organization resides in the few members within each island who control the knowledge. Peripheral members who query these islands remain replaceable until they in turn build enough knowledge that they create a new knowledge island. This model is found in civil engineering organizations where a senior designer, project manager, or cost estimator is considered the primary expert on a given topic and the remainder of the staff serves as an entry point to this centralized knowledge. In contrast, the emerging focus on integrated knowledge is changing this model as clients, new employees, and related industry constituents expect every member to reflect the total knowledge accumulated by the organization. This emerging technology catalyst is changing the focus of technology from a conduit of questions and answers between peripheral and central employees to a support framework for hyperarchical knowledge exchange (Evans and Wurster 1997).

As the concept of hyperarchical knowledge exchange emphasizes, every member of an organization has access to, and can exchange knowledge with, every other member of the organization (Fig. 5-5). Traditional knowledge island barriers are removed in favor of free access and exchange. The advantage of this scenario is that every member of the organization is seen as an equal part in the integrated knowledge base. Through rapid knowledge exchange, the overall experience of each member is enhanced as access to any knowledge source is enabled through technological connections. However, on the negative side, the hyperarchical model creates an environment where the weakest link in the organization is the member who does not have access to the knowledge network. In this case, the organization loses its com-

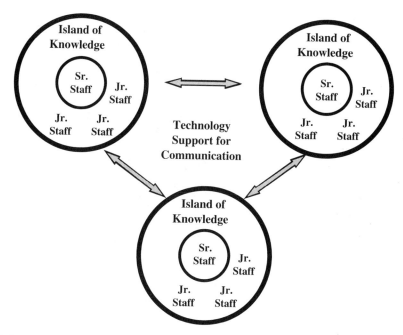

Figure 5-4 Traditional islands of knowledge isolate organization members into separate areas that develop solution components independently of the overall project perspective. Technology serves as the communication component between these islands, reinforcing the separate nature of the island personnel.

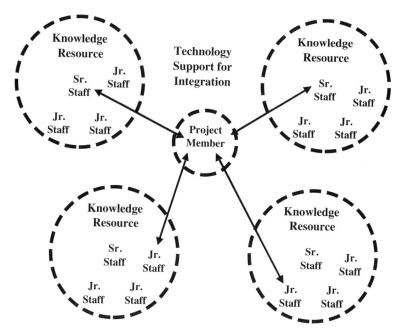

Figure 5-5 Knowledge integration reduces knowledge islands by using communication technologies to break barriers and provide access to the complete organization knowledge base. With this broader perspective, a broader project perspective can be obtained for project solutions.

petitive advantage since the client's perception of knowledge within the organization is restricted to the limited knowledge of the individual. In the case of a design organization, this isolation could result in a designer not being aware of a new material introduced through work on a previous project. The lack of electronic integration diminishes the overall image of the organization as the weak link gets exposed to the outside operating environment.

Integrated Work Processes The drive to create integrated knowledge fabrics within organizations creates a complementary catalyst that examines existing work processes and the role of technology in executing these processes. Although current statistics indicate that almost all civil engineering organizations use computers within their work processes, the role of these computers is focused notably on office automation tasks rather than enhancing knowledge-based tasks. Although this has been an acceptable vision in the past, it falls short of meeting the requirements of the emerging knowledge-based organization. As demonstrated by the aerospace and automotive industries, an integrated knowledge framework requires a redesign of work processes to leverage the new access to organization knowledge. For example, the design of the Boeing 777, a complex structure equal to any civil engineering design task, has been a focus since it achieved a groundbreaking result by being designed entirely with three-dimensional modeling (Binder 1997). Although this fact in itself is an interesting design development, it is the strategic integration and redesign of the work processes implemented by Boeing that makes this an interesting technology integration example. Boeing focused on the use of electronic models to allow each member of the design team to access and collaboratively develop design solutions. The focus of the task moved from individual department or employee ownership to collective development of a complex entity. By providing each member of the team, from design to manufacturing, with the ability to access, review, and query the models, the teams obtained a greater understanding of the context in which design decisions were made. Finally, by adding the ability to communicate electronically with all members of the design team at any time, the need to make assumptions concerning design intent was significantly reduced. In the end, the time, cost, and quality factors of the project were all enhanced since the process integrated functions that previously were separated by organization barriers.

The Boeing example represents one extreme of the work process spectrum. However, the integration concept is applicable to organizations of every size. It is not the size of the project that dictates the need for process integration, but rather the opportunity to enhance team communication and coordination. However, for an organization to leverage knowledge, the focus on tasks must move from individual employees to integrated teams. Examples include examining design tasks to determine opportunities for reducing errors and omissions through central design models, examining field operations to determine if electronic links to the office can bring greater participation, and bringing project managers together electronically to exchange project status and control information. In these examples, work processes are enhanced by electronically connecting isolated individuals to leverage collective knowledge.

The resulting impact requires organizations to emphasize that autonomy and self-direction is not being abandoned, but, rather, individual input is being increased as the opportunity to provide input to a greater set of projects is realized.

External

Although inward analysis is important for an organization to understand its capabilities and liabilities, it is rarely sufficient by itself to prompt change within an organization. As documented in cases such as BP Oil and Encyclopedia Britannica, the need to integrate technology may be evident based on the inefficiencies caused by existing procedures, but it is not until an external force such as a client, government requirement, or project partner provides a stimulus that an organization undertakes technology integration (Prokesch 1997). Previously, these external catalysts were rarely present in the civil engineering industry except for cases such as the client requirement to produce drawings electronically and the introduction of computer-based analysis tools that created new design possibilities. However, this situation is rapidly changing as the civil engineering industry begins to operate within the context of a public that assumes technological proficiency and expertise.

Client Perspective of Industry When reviewing publications and professional reports, the civil engineering industry is often characterized as a conservative industry that moves slowly to adopt emerging technologies. Is this characterization a fair generalization to an entire industry? It depends on the perspective from which the viewpoint is being generated. From the industry viewpoint, the argument that is often put forth is that civil engineers have a responsibility to the public to be conservative since public safety is in question. Mistakes caused by irresponsible adoption of new technologies are not acceptable from a social or legal vantage point. However, from a public viewpoint, the pervasive presence of the Internet, computers, video cameras, and hand-held digital organizers is creating a perception that technology should be an integral component of any professional practice. The perceived reluctance of the civil engineering industry to adopt these technologies on a widespread scale could create a negative backlash that impacts owner relationships.

Addressing this backlash is not something that can be accomplished by new organization brochures or subtle changes in organization marketing. Rather, client perspectives compel organizations to examine internal work processes from external perspectives. Specifically, organizations must examine work processes from the perspectives of potential clients and emerging markets. The perceived importance of technology integration creates an environment where this integration can no longer be relegated to specialized areas such as structural calculations. Rather, technology integration is evolving as a visual representation of the capabilities encompassed within an organization (Magnan 1997; McKinney and Fischer 1997). For example, civil engineering organizations can learn from the finance industry, where company representatives never arrive at a meeting without a laptop computer that has the capability to demonstrate instantly the ramifications of investment decisions (Schwartz

1997). While it is true that civil engineers do not interact with the general public on a daily basis in the same way that financial consultants do, the emerging impact of technology integration during interaction with design, regulatory, and financial professionals as well as the general public cannot be underestimated. The climate that previously accommodated the conservative nature of the civil engineering industry is changing. The demand for equal technology capabilities by all project constituents will soon override the traditional viewpoint of not changing a process that has served the industry faithfully for decades.

Extension of Services The need to satisfy client expectations, whether it is a private, professional, or regulatory client, is a constant that exists for every organization in every industry. However, it is the breadth of these expectations that varies from industry to industry. Within industry sectors such as retail merchandising, the client expectation is one of full service to the customer's needs. Companies such as Nordstrom have carved a strong customer niche by providing a comprehensive service environment, including personal shoppers, sales, and tailoring. Similarly, in home building, customers now expect national builders such as Ryland Homes to provide architectural, construction, and the option of interior design services. Traditionally, fragmentation within the civil engineering industry has negated this issue for the majority of civil engineering organizations. Only the largest organizations such as Bechtel and Fluor-Daniel had the capacity and diversity of personnel to offer comprehensive design, engineering, and construction services (Kostem and Shephard 1984). Although an argument can be made for either viewpoint concerning the rationale of retaining this status quo, external client demands are making it difficult for organizations to retain a narrow specialty as a complete client offering. Specifically, the opportunity to extend existing professional services through the use of technology integration is emerging as the next wave of organization expansion.

The drive to expand existing services is one that serves as a challenge to civil engineering organizations. Organizations that view this catalyst as an opportunity to use existing and emerging technologies to reinterpret professional services will emerge stronger in the perspective of prospective clients. For example, as illustrated in Figure 5-6, traditional civil engineering services tend to focus on a single component of the design–construction process. The next generation of civil engineering organizations have the opportunity to break this mold and focus on the entire project lifecycle as a series of service opportunities (Goodman and Chinowsky 1997). The underlying basis for this opportunity is the ability to store, process, and disseminate knowledge within a single integrated framework. By examining internal work processes and placing the processes in the context of an integrated knowledge repository, organizations have the opportunity and capability to extend existing services to any segment of the design–construction process. Analysis opportunities include the capability to extend existing two-dimensional drawings into three-dimensional virtual-reality design models, to extend design models into construction control tools, and to take construction control tools to the realm of operations and maintenance. It is the organizations that are willing to explore these opportunities through the inte-

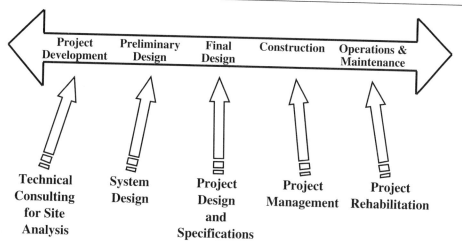

Typical Single Focal Point Services

Figure 5-6 Traditional civil engineering business practices emphasize single source specialties and the use of technology to support the focused work processes.

gration of knowledge that will emerge as the comprehensive service providers of the next decades.

Alliances and Partnerships The introduction of remote computing technologies such as the World Wide Web and videoconferencing is revolutionizing the concept of the engineering office. The physical boundaries of office walls and jobsite distance are crumbling as technologies create the possibility for instant communications from any location at any time. This communication capability is coinciding with the trend to establish partnerships and alliances effectively to leverage individual organization assets into integrated team efforts. The trend to establish partnerships is steadily increasing within the civil engineering industry. Issues including risk reduction, client service, delivery options, and liability protection are primary drivers for this evolving project delivery option. Projecting the current trend for alliances, this project delivery option may become the predominant approach to project development. However, to ensure success within these partnerships, knowledge exchange must become transparent between members from each organization within the partnership.

Currently referred to as the virtual office concept, the use of technology provides the ability for structural designers, construction engineers, consulting engineers, and other team members to eliminate the co-location requirement (Handy 1995). Each member of the team remains connected to every other member via the use of knowledge-transfer technologies. These may be synchronous technologies such as videoconferencing, or asynchronous technologies such as electronic mail and electronic work centers (Ross 1997b). Either method provides the foundation for civil engineers to expand their internal capabilities by leveraging the strengths of allied organizations. To the client, this alliance represents greater strength, service, and knowledge about a particular project under consideration. However, to achieve the full potential of these engineering partnerships, each member of the team must have the opportunity freely to exchange knowledge with every other member of the team. This is the essence of the alliances and partnership catalyst in terms of technology integration. As internal integration of technology becomes prevalent within the industry, organizations and clients will expect that members of the partnership will have the technological infrastructure to develop a virtual office rapidly. The failure to have this infrastructure will create external perceptions that the organization is either unwilling to invest in technology, or does not have an interest in entering the era of knowledge exchange.

Location of Information and Knowledge It has been reported that traffic over the Internet is doubling every 100 days and over 100 million people are now on line (Weise 1998). Given this international exposure to information technology, public- and private-sector clients are increasingly setting expectations that civil engineering organizations provide data availability to all project team members throughout the project. For example, in the case of construction, clients are raising the minimum expectation for information and reporting. The knowledgeable client understands that technology integrated into the work processes provides the ability transparently to connect a field office with a central location. With this understanding, the client can expect that information on any project can be accessed from a virtual repository. As illustrated in Figure 5-7, the virtual repository may include construction videos on a site-based server, budget and schedule information from the main office, or material delivery information from a supplier warehouse (Khedro et al. 1995). It is the responsibility of the construction organization to recognize that clients will begin to request electronic access to this information. Establishing the capability to respond to these requests requires a long-term, strategic vision that transfers control of information from individual planning, operations, and business development departments to virtual organization repositories.

The compromise in this catalyst between traditional departmental control and unlimited access is the virtual repository concept. As repeatedly emphasized during interviews with organization leaders from all segments of the civil engineering domain, control of knowledge is a fundamental part of human nature and lies at the center of departmental divisions within a civil engineering organization. Technology integration provides one solution to this control argument. Rather than physically moving knowledge out of individual locations and departments, the establishment of virtual connections to knowledge creates a nonthreatening atmosphere for individual employees. Specifically, the individual locations and departments retain complete control over their internally generated designs, schedules, and operational data, but provide virtual access to other members of the organization (Pawar and Sharifi 1997; Anumba and Duke 1997; Mowshowitz 1997). From an internal organization perspective, the team is integrated within a single project, but retains individual identi-

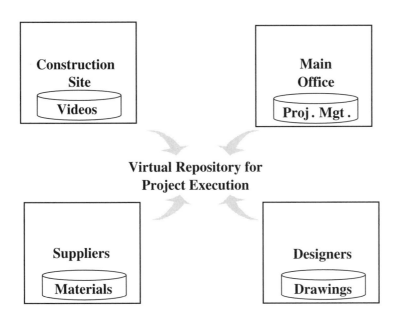

Figure 5-7 Virtual repositories will include links to a diverse set of information and knowledge located in many constituent organizations.

ties in terms of knowledge responsibility. From an external client perspective, the team appears as a single, integrated unit with unlimited, on-demand access to project knowledge at any time that the client requests a project update. Although the virtual repository approach does not resolve the fragmentation issues that reside within the civil engineering industry, the focus on technology integration as a foundation for reducing department barriers is a basis for examining the long-term effects of external demand for greater knowledge access.

■ A New Role

Once an organization identifies, acknowledges, and understands the internal and external technology catalysts, the focus of strategic integration changes to defining the new role for knowledge within organization objectives. At the center of this process is the new emphasis on technology as a corporate asset. Technology changes from a support role to a primary role, where it is featured as a central part of the problem-solving process. For example, in the context of a structural design firm, the role of technology changes from a computer-aided design and visualization tool to a lifecycle analysis tool that enables team members to integrate structural design decisions into the construction process. In this example, the computer is elevated from design assistant to integrated project facilitator by focusing on the extended use of the design information beyond the localized use required by the designer. However, to achieve this level of enhancement, the traditional concepts of technology, work processes, and knowledge must undergo a fundamental transformation.

The first of these concepts is the concept of technology itself. Specifically, the technology perspective must be altered to focus on the integration of technology into the overall knowledge strategy (Stalk, Evans, and Shulman 1992). Rather than developing a knowledge strategy first and then identifying the technology that is required to support that knowledge as a secondary issue, the two components should be developed in a single, integrated plan. This focus is a direct extension of the core competencies topic addressed earlier. Since the success of an organization is highly dependent on its ability to focus and build upon core competencies, it follows that the organization should enhance these competencies through greater use and leverage of knowledge. The organization that integrates technology into a knowledge enhancement role will further leverage core competencies by making the organization knowledge base available to every member. This integration process can be viewed as a series of steps (Fig. 5-8). The first level is the identification of core competencies that establish a focus for organization operations. The second level is the refocusing of knowledge to leverage core competencies into enhanced organization opportunities. Finally, technology integration is incorporated to bind organization knowledge together into a foundation for pursuing expanded business opportunities.

As the binding element, technology must be viewed from the perspective of leveraging and supporting knowledge transfer. Every task that occurs within an organization is a component of a larger project. As such, technology integration serves as the medium through which task assignments are integrated into the overall project.

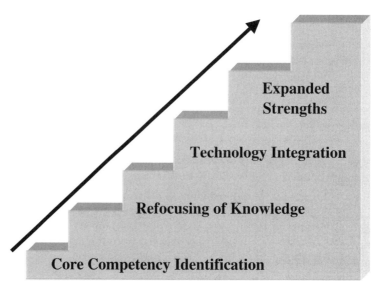

Figure 5-8 Technology integration can be achieved through a series of steps that build upon core competencies and extend into business opportunities.

Once again the Boeing 777 example provides a guideline for this integration. By using electronic models as a replacement for traditional aircraft mockups, the organization focused on the use of visualization technologies as an electronic connection between design team members. The central models transferred the design focus from individual parts to the collective development and enhancement of an overall project entity. The same transformation can occur in the civil engineering organization. By integrating emerging technologies with the organization knowledge base, conduits are created through which knowledge can be exchanged and strengthened. In this role, technology serves to emphasize centralized knowledge enhancement rather than individual task improvement. However, this role as knowledge conduit is significantly removed from traditional office automation tasks. This does not mean that technology is no longer required for these traditional tasks. Rather technology undertakes a second role that elevates the perspective from a service position to a strategic position.

■ The Strategic Transition

The refocusing of the technology perspective from a support role to a strategic role provides a starting point for the strategic integration of technology. Successfully proceeding with this transformation requires a second change in perspective that moves beyond organization goals to focus on project work processes. In this next level of refinement, self-analysis is required to examine critically how projects are undertaken. Traditionally, the integration of a work breakdown structure has been a fundamental component of the project management task (Oberlender 1993). Individual task assignments are developed from the structure and provide the core elements for schedules and estimates. Additionally, the structure provides the basis for separating individuals into independent islands of knowledge to solve very specific problems. Solutions emanating from these islands are configured into coherent entities at project meetings or by senior team members independently of the individual task processes. In these traditional scenarios, the introduction of technology follows the independent island concept to emphasize the enhancement of individual task assignments rather than the enhancement of overall work processes.

The new role of technology within civil engineering organizations diverges from this knowledge island approach to emphasize unified knowledge exchange and problem solving. The introduction of technology creates an environment that supports the collective problem-solving process. Returning to the Boeing 777 example, the company instituted a central repository that allowed each member of the team to contribute to the overall solution. The concept of the central repository was established by the company prior to the implementation of technology enhancements that assisted individual designers. In this manner, the design process received the initial attention and the focus of the overall team. By putting in place an overall design strategy that emphasized a technological component as the centerpiece, all contributory elements of the work process were required to be redesigned to conform to a central problem-solving paradigm. As a consequence, each department associated with the design task was required to establish procedures and methodologies that would tie traditionally independent processes into the centralized design models. This process could not have succeeded from a bottom-up approach that emphasized the independent development of task-oriented automation. Success was achieved by starting with an overall concept and pushing that concept out to the redesign of all work processes associated with the project.

Understandably, designing an airplane has some fundamental differences and goals from a civil engineering project. However, the concept of redesigning a central element of the design process and then filtering this change to individual task processes is not limited to the realm of aerospace design. The same impact can be achieved in the civil engineering domain. Existing work in both the academic and professional environment attests to this possibility. In the professional realm, the use of three-dimensional modeling with attached design attributes is being introduced in several large civil engineering firms, including Stone & Webster, Bechtel, and Fluor-Daniel. In these organizations, the move to central design models has already begun. Although full technology integration may not be realized at this time, the basic process of examining the overall work process to determine the greatest opportunities for technology integration is successfully being implemented. Similarly, in the academic domain, several civil engineering research centers, including the Center for Integrated Facility Engineering at Stanford, the Intelligent Engineering Systems Lab at MIT, and the Engineering Design Research Center at Carnegie-Mellon University, are beginning to break through the traditional civil engineering barriers to develop projects through integrated technology frameworks. At the core of these frameworks is the concept that an integrated approach to civil engineering problem solving will reduce project errors by providing each member of the team the knowledge required to make informed decisions. This approach will succeed if organizations demonstrate the will to reexamine overall work processes, but is destined to fail if the concept is broken into pieces and implemented in the traditional bottom-up approach to problem solving.

Technology Needs Assessment

The identification of the new role that technology will play in the civil engineering organization provides the starting point for the next component of the technology integration process—evaluating the technology needs of the organization. Civil engineering organizations are similar to organizations in the manufacturing, financial, and other engineering specialties in that significant budgetary mistakes occur in the area of technology. When conducting interviews with civil engineering executives, one of the most common responses in the area of technology investment is related to an incident in which the organization was economically damaged from a technology investment. Whether it was an investment in computing technology, field technologies, or communications technology, each organization had at least one story related to excessive hype, unfulfilled promise, or failure of adoption. However, in contrast to other industry sectors that traditionally have been forced to innovate or die, technology mistakes in the civil engineering industry have not been viewed as the cost of doing business. Rather, these mistakes have reinforced the conservative nature of many civil engineering organizations to the point where it is difficult to embark on a new technology integration effort. Breaking this barrier and establishing a new vision is the focus of the technology assessment process.

Evaluating the strategic technology needs of any organization can be a daunting task. Determining appropriate technologies, level of investment, and long-term requirements are all issues that must be addressed during the planning process. However, in the civil engineering industry an additional factor must be taken into consideration, the ability of technology to assist in multiple, independent projects. In contrast to a manufacturing company that designs for large-scale manufacturing runs, the civil engineering industry develops solutions for one-time projects. Although many projects may be similar, they are not duplicates like automobiles or dishwashers. Subsequently, the investment in new technology must focus on flexibility rather than optimization of a specific task such as the manufacture of a part casing. Achieving this flexibility while retaining a focus on fiscal responsibility and relevance to the civil engineering task is the balance that is required during the needs analysis process.

The Technology Pursuit

The constant pursuit of leading-edge technology often lies at the core of budgetary disasters. The pursuit of the leading edge in technology is a losing game in the strategic development of a civil engineering organization. While it is intellectually logical for an executive to dictate that an organization will refrain from excessive technology purchases, translating this refrain from intellectual reason to fiscal responsibility is often a difficult task. The increasing pace of technology improvement creates enormous pressure to keep pace with the technology curve. Organizations are often led to believe by technology manufacturers that falling behind the technology curve is equivalent to losing competitive advantage. Although this thought pattern is easier to identify than to break, the necessity of breaking the pattern cannot be overstated. However, the key to eliminating any negative process is to increase knowledge of the targeted offense. In this case, the target offense is the futility of staying on the leading edge of technology development.

Figure 5-9 tracks the progression of the technology curve over a 25-year period from 1972 to 1998 based on the evolution of the personal computer ("Intel"). The evolution illustrates two important facts: (1) Any technology considered on the leading edge today will be obsolete in an increasingly shorter period of time, and (2) performance capability continues to accelerate at a pace that is considered unforeseen at every technology release. To use a common analogy, the first of these facts is the carrot that instigates technology pursuit, while the second is the stick that punishes organizations for attempting to capture the carrot. Given this analogy, a focus to eliminate the punishment from the stick is a reasonable place to begin. Consistently, in-

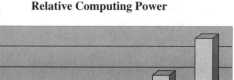

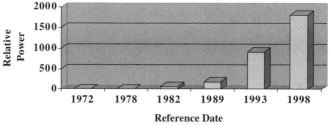

Figure 5-9 The rapid increase of computing power over the indicated 25-year period illustrates the difficulty in remaining at the edge of technology development ("Intel").

formation technology advances are accompanied by claims that the new technology will remain at the leading edge for the foreseeable future. Inevitably, these claims are proven incorrect as the next technology advance appears close behind. When viewed independently, it is reasonable that claims related to increases in productivity, enhanced communications, and competitive advantage should be assumed to be correct by organization leaders. However, this is exactly what lies behind the stick that punishes the organization. While there is no guarantee that a new technology release will meet its claims, the rational approach in the near term is to assume that the technology curve will continue to follow the established pattern. Specifically, technology releases will continue to occur at an ever-increasing rate of speed. This assumption should result in one specific outcome—caution on the part of strategic planners when viewing the possibilities that are presented by the technology carrots.

Although a segment of technology claims are rooted in accuracy, the vast majority of claims are irrelevant to the majority of firms. However, the issue of release claims are secondary to the underlying issue of obsolescence. The length of time that any product release retains its position as a cutting-edge product is fleeting at best. For an organization to remain at this cutting edge, technology upgrades must occur every 18–24 months. To illustrate this in the context of mainstream civil engineering hardware and software, the evolution of the Intel microprocessor and the progressive development of CAD software can be examined. In the case of hardware, Intel has released a major chip upgrade on the average of every 2–3 years and a minor upgrade every 5–6 (Table 5-1). While the price versus performance curve of these products (Fig. 5-10) has consistently reduced the cost of the hardware in relation to the power provided by the processing chip, the cost of a new computer that includes the chip at the time of its release has not demonstrated the same drop in price. Rather, computing companies elect to provide additional features to retain an entry-level price point that remains consistent at approximately $2500–3500 for a typical business computer. Given only these facts, a civil engineering organization is faced with the choice of remaining with old technology or reinvesting the same amount every 18–24 months. However, an additional set of facts is relevant to the civil engineering firm: specifically, the question of needing cutting-edge technology when it is first released versus waiting a period of time for the price of the technology to be reduced. Con-

TABLE 5-1
The Evolution of the Microprocessor Illustrates the Increase in Computing Capability and the Reduction in Associated Costs Since the Introduction of the Personal Computer ("Intel").

Date	Processor	Cost Per Chip	Speed (in millions of instructions per second)
June 1978	8086	$360	0.75
February 1982	80286 (12 MHz)	$360	3
October 1985	80386 (16 MHz)	$299	5
February 1987	80386 (20 MHz)	$219	6
April 1988	80386 (25 MHz)	$219	9
April 1989	80486 (25 MHz)	$950	20
May 1990	80486 (33 MHz)	$1056	27
June 1991	80486 (50 MHz)	$665	41
August 1992	80486DX2 (66 MHz)	$682	54
March 1993	Pentium (66 MHz)	$965	112
March 1994	Pentium (100 MHz)	$995	166
March 1995	Pentium (120 MHz)	$935	203
June 1995	Pentium (133 MHz)	$935	219
November 1995	Pentium Pro (150 MHz)*	$974	332
November 1995	Pentium Pro (200 MHz)	$1325	448
May 1997	Pentium II (300 MHz)	N/A	639
April 1998	Pentium II (400 MHz)	$959	699

*In 1995, the processor testing was changed to reflect the introduction of multimedia capabilities. The speeds from this point forward are approximated to reflect previous testing procedures.

sistently, about 9–12 months after a new processor is released, the cost of comparable computers decreases by an average of 20%. The impact of this reduction in cost can be seen when comparing a technology upgrade for a company of 10 employees over a 5-year period of time.

In the first scenario, the organization elects to upgrade their computers each time a major chip release is available. Assuming that the company purchases an average system capable of performing graphic, scheduling, accounting, and office automation tasks, the cost of the 5-year technology pursuit would be as follows:

> Over a 5-year time period, three purchases are made to correspond with the introduction of major chip releases. For example, between March 1993 and April 1998, the company upgrades from Pentium 66-MHz machines to Pentium Pro 150-MHz machines and finally to Pentium II 400-MHz machines. Buying a $3000 machine for each of the ten employees results in a $30,000 purchase at each upgrade. Given three upgrades, the company will expend $90,000 in computing equipment over the 5-year period.

In contrast, if the company elects to wait 9 months after the release of the new technology, the cost of the 5-year pursuit would reflect the 20% reduction in computing costs and result in a savings of almost $20,000. However, another scenario is also appropriate to examine. In this scenario, the organization elects to upgrade technology on a slower schedule, deciding to upgrade only at major chip releases. In this case, the following scenario occurs.

Price–Performance Curve

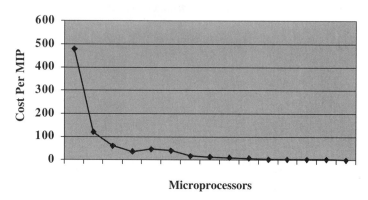

Figure 5-10 The price–performance curve of computing capabilities illustrates the continuous decline of costs for equivalent computing power.

Over the same 5-year time period, two purchases are made to correspond with the introduction of every other major chip release. For example, between March 1993 and April 1998, the company upgrades from Pentium 66-MHz machines to Pentium II 400-MHz machines after deciding to skip the introduction of the Pentium Pro chip. Buying a $3000 machine for each of the ten employees results in a $30,000 purchase at each upgrade. Given two upgrades, the company expends $60,000 in base computing equipment over the 5-year period.

The decision to skip one chip release results in a direct savings of $30,000 for the organization as well as expenditures on installation, training, and procurement. When combined with the 9-month delay savings, the organization could save another $12,000. For a small organization, this savings is significant. For a large organization, the scaled-up costs can have an even greater impact on the overall operating budget.

While not as significant as the hardware example, the pursuit of the latest software upgrades can also prove costly for a civil engineering organization. As illustrated by the CAD market, if an organization desired to remain at the cutting edge of current CAD software, a company upgrade would be required on the average of every 12 months for some component of the program. Using an average upgrade price of $150 per seat, the cost over a 5-year period would be $7500. This does not include the additional costs of training, procurement, and installation.

Similar to the hardware scenario, the pursuit of updated software proves to be a costly struggle as the releases appear in a more rapid manner. More important, the pursuit of this technology curve requires greater focus from the organization as the need to set technology upgrade plans becomes increasingly frequent. Eventually, the pursuit of the technology infringes upon the focus of the tasks the technology is supposed to benefit. Although the determination of when a technology upgrade should be made is not as simple as setting a specific time frame, it is also not as complicated as some vendors would like strategic planners to believe. Specifically, the decision to upgrade should rest on the evaluated needs of the organization and not on the perceived benefits projected by others.

Evaluating Needs—Not Marketing Hype

The evaluation of strategic technology needs requires an awareness of both short-term and long-term priorities. Although long-term benefits are the ultimate focus of the integration objectives, analysis must take into account current computing situations realistically to set priorities. In addition to this dual time frame, strategic integration must incorporate business goals as a central component of the integration plan. When these goals are neglected (as often happens), technology integration proceeds in a vacuum, resulting in an integration plan that requires organizations to purchase components that are significantly costlier than those actually required for daily activities. However, integrating these diverse elements into a comprehensive technology plan requires a structured planning approach. Although multiple planning examples exist for organizations to follow, these examples exist almost exclusively for industries outside civil engineering (Szakonyi 1990; Wysocki and DeMichiell 1997). As such, the burden of transferring the plan into a context that works for individual civil engineering organizations is placed on the organizations themselves. While no plan can alleviate this problem by providing a generic solution that works for every organization, several issues should be addressed by every civil engineering organization when undertaking the development of a strategic technology integration plan:

- Avoiding the technology trap
- An analysis of work process requirements
- An analysis of distributed computing requirements.

Avoiding the Technology Trap The first issue that needs to be addressed by every civil engineering organization is to determine if the organization is falling into the technology trap. As discussed previously, the pursuit of the technology curve is an endeavor that afflicts many civil engineering organizations. Based on interviews and the experience of the authors, extricating an organization from that trap, or preventing one from entering the pursuit, can be traced to three critical issues: image, perceived needs, and perceived competitive advantage.

Image Image lures organizations into the technology trap by creating a technology mystique. Specifically, the image trap can be defined as the pursuit of technology based on the belief that others outside the organization will respect the organization due to its ownership of advanced technology. On the surface, this appears to be a mundane statement that has reasonable elements of truth. However, the image trap becomes an issue when the consideration of the image begins to overtake fundamental business decisions. When purchasing decisions begin to focus more on outside individuals than the internal organization needs, then the image trap claims another victim. To illustrate this trap, a small firm was studied that provides both architectural and civil engineering services. The firm became successful because it obtained a reputation of producing innovative work using CAD technology. After several years of this success, income began to decline as the local demand for commercial buildings began to decline and competition increased from new competitors. Rather than focus on the services that the organization was providing, the partners focused on their original image as technology innovators, believing that they had fallen behind the technology curve. This thought transformed very rapidly into the basis for a constant technology pursuit. Since success was equated with image, and image was equated with new technology, the firm became solidly entrenched in the technology image trap.

It can be very difficult to recognize the image trap from within an organization. Once a belief is established that clients are choosing the organization due to its image as a technology innovator, this belief can quickly permeate numerous decisions. In the extreme, the organizational culture believes that the success of the firm is completely grounded on its ability to portray itself as a leader in the technology arena. At this stage, emerging from the image trap requires a complete change in organizational philosophy that equates to rebuilding the foundation on which the organization operates. Given this potential consequence, the need to avoid the image trap becomes quite real. To accomplish this, an organization must examine its technology strategy from the perspective of the work processes. If an organization focuses less on what is required to produce work product, and more on the need to have the technology in the firm, then a flag of concern should be raised. At the onset of the problem, these decisions may be difficult to discern since they are usually couched in terms of upgrading the firm to the next generation of technology tools. However, as the trap progresses, the organization increasingly focuses on the fact that new technology exists and not on the role technology can play as a tool to enhance work processes. Eventually, the need for technology to enhance the work process becomes lost in the pursuit of the technology itself. Therefore, the key to avoiding this technology trap is to keep the focus on work processes. If a technology purchase can legitimately enhance a work process, then a case can be made for its purchase.

Perceived Needs The second technology trap is the mistake of perceived needs. The perceived needs trap establishes an environment where technology decisions are based on artificially inflated needs. Within this trap, individual technology decisions are influenced less by actual work requirements, and more by the perception of the work processes that occur. In common terms, this is the sledge hammer approach to solving problems that require a tack hammer. A common task such as contract development can be used to illustrate the perceived needs trap. At one end of the technology spectrum, a typewriter can be used to insert data into standardized forms. While redundant and time consuming in its approach, it remains a reliable approach to finalizing documents. One step higher on the technology spectrum is the use of a word processor to customize boilerplate contracts. Minimally intensive from a computational viewpoint, the basics of word processing have changed little since the initial introduction of the personal computer. However, the technology spectrum does not stop at this point. An organization can elect to pursue a highly automated approach to contract development that incorporates artificial intelligence technologies, on-line libraries, and customized software. Given the commitment to this automated approach, the next step along the technology spectrum is to obtain equally high-tech hardware. As the spiral continues, the organization becomes convinced that the application of faster technology or further customized software will lead to an optimized application solution. Since the technology curve continues to move at an increasing rate of change, the builtin barrier set by technology limitations fails to appear, and thus the organization can continue to pursue solutions through constant updates of hardware and software.

In the context of contract development, the technology spiral is an exaggerated example that appears to be easy for any organization to avoid. However, the appearance of the spiral is not quite as obvious when caught in the perceived needs trap. When viewing the needs from inside the organization, each step into the technology spectrum appears to be a logical progression from the current state. Since the organization believes that the key to enhancing a given work process is the application of greater technology capabilities, any work process can serve as an entry point to the trap. The work process does not have to be a low-tech process such as the contract development example. The same technology spiral can occur with the most complicated three-dimensional modeling tasks. In either case, once the work process has been identified as eligible for enhancement, the potential for changing from real needs to perceived needs can occur. Navigating around this possibility requires an organization to have discipline and restraint in its analysis of internal work processes. It will always be possible to identify a technology that can enhance a given work process in some incremental degree. However, at some point, the cost benefit from this enhancement no longer justifies the enhancement. This is where the organization must draw the line on the new technology. If the cost of the technology cannot be justified by real savings, then the technology purchase must be questioned for its relevance. This does not mean that it cannot be purchased, but a sanity check should be put in place to ensure that all members involved in the decision have analyzed the purchase from a pragmatic perspective.

Perceived Advantage The final technology trap is the misperceived competitive advantage. Of all the traps, this one is the easiest for a civil engineering firm to enter. The perceived competitive advantage trap creates a belief that the organization is successful because technology within the organization creates an advantage over other firms in the same market niche. In this trap, technology decisions focus on obtaining the latest technology for the exclusive purpose of acquiring technology before the competition. Once again, on the surface, the trap appears minimally threatening due to the obvious absurdity of making decisions based on this premise. However, the trap is seldom this obvious. In fact, pressure to be technologically advanced creates the perfect environment for this trap to emerge in any organization. Civil engineering marketing often builds a picture of competitive advantage through speed, functionality, and innovation. Given this picture, it is easy to adopt a philosophy that staying technologically ahead of the competition is a central key to success. This viewpoint is constantly reinforced by print and electronic media that spotlight manufacturing organizations that have obtained competitive advantages by adopting the latest robotic and automation technologies. With this reinforcement, a civil engineering organization can transfer this success into the terms of its own work processes. If technology provides an avenue for completing the task more efficiently than the competition, then the advantage and need are clearly identified.

This philosophy of utilizing technology to increase productivity and competitive advantage is a valid strategic approach for organizations to adopt. However, the philosophy can rapidly change from competitive positioning to technology trap when the organization begins to lose external perspective on technology investment. In this scenario, an organization evaluates technology decisions from a perceived position created by noncompetitive forces rather than actual market forces. These perceived forces may be media forces, computer sales forces, or internal staff members. However, each of these forces has a stake in promoting the technology rather than the competitive advantage that it may provide. For example, while the purchase of a state-of-the-art computer may provide the fastest processing power available, the actual competitive advantage of the machine may be minimal, as the software being used on the machine fails to support the latest hardware advances. In cases such as this, the organization transfers the focus of technology integration from actual benefits to perceived benefits. Once this pattern begins to emerge, it becomes difficult to break as a pattern is set for future technology purchases. Given this pattern, it becomes easy to pursue technology decisions from the perspective of obtaining the fastest, biggest, and newest advance rather than the advance that provides the greatest return on investment for competitive advantage.

Work Process Requirements The evaluation of technology traps is a positive step for every organization that expends the time and effort to undertake the process. For those organizations that identify internal patterns that resemble the identified traps, the opportunity should be seized to eliminate the pattern as soon as possible. Conversely, for those organizations that are fortunate enough to avoid the traps, the opportunity exists to establish an awareness for future technology missteps. In either situation, the completion of the analysis process establishes a foundation to move forward in the needs analysis process. Given this foundation, the second component of the needs analysis process is the determination of where to focus the technology integration effort. In a break from current technology acquisition procedures, the next generation of strategic technology integration requires organizations to progress beyond the blanket development of computer specifications that meet the needs of every member. The advances in specialized modeling, communications, and design technologies require the strategic plan to address the divergent technology requirements of work processes as well as emerging communications and business operations.

Computational Intensity Although technology planning must provide for both current and long-term needs, organizations can misinterpret this intent by assuming that the goal is to obtain the most advanced computing environment with the anticipation that the organization may require it in the future. Instead, the organization should create the best environment that meets the current and foreseeable requirements of the work processes. Specifically, the computational intensity of the work process is the primary driver for technology planning. Computational intensity is the demand of the work process for mathematical solutions versus information delivery. The former encompasses operations such as three-dimensional modeling, soil stabilization calculations, or structural analysis. In each of these cases, the focus of the computing task is on rapid calculations to execute the task effectively. In contrast, operations such as word processing, construction scheduling, and database manipulation are data intensive. In these cases, the computing focus is on the storage and transfer of large volumes of data to and from local and remote locations; the need for rapid calculations is a secondary requirement.

The difference in these tasks is significant for strategic management. If the work processes are determined to fall into the computationally intensive category, then it is important to provide employees with technology that can perform these calculations with the greatest efficiency. This translates to obtaining equipment with high-speed processors and the capability to upgrade as new technologies are introduced. Although the initial investment may be higher than other technology alternatives, the return on investment can be justified based on the fact that organization members will be more efficient in their daily tasks since the amount of down time related to calculation time will be reduced. In contrast, if the work process emphasizes information transfer, then technology integration should focus on obtaining equipment that provides for large amounts of data storage and the capability to transfer those data to as many organization members as required. In this scenario, emphasis is placed on obtaining advanced storage devices such as disk arrays and central data servers that rapidly distribute data. The need for the newest processor is secondary to the purchasing of data storage and transfer devices. The associated return-on-investment concerns are alleviated based on the capability to provide organization members with information that was previously unavailable during daily work processes. The data transfer devices significantly eliminate the down time required to identify, locate, and contact organization members that have required work process information.

In these technology scenarios, the emphasis remains on properly identifying the needs of the work process. This central tenet is the greatest guarantee to reduce inadvisable technology decisions. The trends in technology development are creating greater opportunities to distinguish between computation and information advancements. The capability to upgrade computing systems and invest in targeted components will become increasingly important as organizations properly transfer technology decisions to emphasize the needs of individual processes. Successful organizations will set in place a strategic plan that emphasizes the differences between computing requirements rather than the need to identify a common machine that can be purchased for the lowest price and serve the greatest number of individuals.

Communications Focus The introduction of the hyperarchical computing philosophy provides a technology integration focus within an organization. Building upon this focus, the organization can expand its vision outside the physical boundaries of the office to a distributed communications environment that characterizes the next generation of civil engineering projects. Achieving this vision raises the next issue that must be examined during the needs analysis procedure. As previously introduced, the focus on communications is changing the civil engineering industry in ways that could not have been predicted even 10 years ago. The concepts of field office management, project team coordination, and electronic information exchange are rapidly changing as communication capabilities link project teams in virtual teams. However, as this transformation occurs, organizations will be required to evaluate their communications requirements in terms of the need to serve as an information provider or act as an information consumer. Loosely defined, the role of information provider is established when an organization elects to place itself in a position at the center of information exchange within a project. As illustrated in Figure 5-11, this position indicates that the organization serves as a central transfer point for design, engineering, and construction information throughout a design project. In contrast, the role of information consumer is adopted when the organization elects to serve as an information receiver within the team structure, accessing information to develop project solutions and sending results to the team as they are developed (Fig. 5-12).

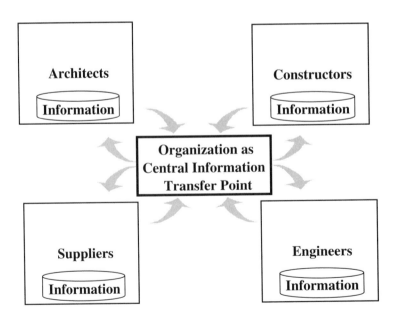

Figure 5-11 The model of the organization as a central information transfer point requires technology integration to address long-term information bottleneck possibilities.

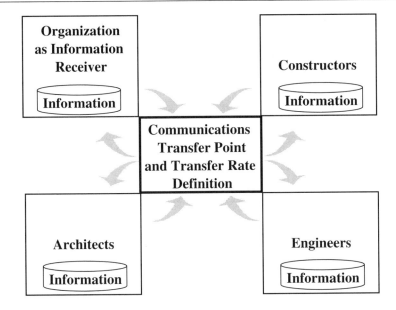

Figure 5-12 The model of the organization as an information recipient relinquishes the responsibility to plan for information bottlenecks, but requires the organization to refocus its technology requirements.

The impact of these roles on strategic technology planning is significant. If an organization intends to place itself in the position of information provider, then commensurate actions must be taken to support this role. Specifically, technology planning must focus on obtaining communications equipment that reduces the potential for the organization to become an information bottleneck and optimizes the flow of information into and out of the organization. Providing the greatest communications throughput requires equipment such as high-speed modems, dedicated transmission lines, and fiber-optic connections. Placed in the context of the civil engineering project, communications equipment must be put in place to allow complex project data such as virtual-reality models and video documentation to be exchanged as close to instantaneously as possible. Since the organization sits at the center of the information flow, insufficient emphasis on advanced technology will create an information bottleneck at the hub of the virtual team. The failure to put in place communications equipment that eliminates this bottleneck limits the communications capability of every team participant. This bottleneck eventually reflects poorly on the organization and slowly defeats the purpose of establishing a virtual engineering team.

Similarly, if the organization adopts the position of an information receiver, then the organization should avoid expenditures that are incompatible with this position. In the position of information receiver, the organization does not set the communications standard for the virtual team. Rather, the organization elects to follow the standard set by the communications hub. The resulting impact of this role is directly applied to the strategic technology plan. In the role of information receiver, the organization is limited in its ability to transfer information by the communications equipment put in place by information providers. While the information receiver may have the capability to transfer information at very high rates of speed, if the central transfer point is incapable of receiving information at that rate, then the high-speed equipment is rendered useless. As such, the decision to adopt the information receiver role cannot be overlooked when the technology plan is put in place. In this role, the organization should refrain from making purchases that are inconsistent with the role of information receiver. Purchasing communication technology that is superior to anything that the organization will encounter in a virtual team environment must be acknowledged as a gamble. The gamble will pay dividends if the information providers continue to upgrade their environment to the point where the receivers can fully exploit the capabilities of their purchases. However, if the information providers do not make this commitment, then the receivers are left with expensive equipment that can do nothing to relieve the information bottleneck generated by the central transfer point.

Centralized versus Decentralized Computing In contrast to previous information revolutions, the current focus on knowledge support and virtual teams is breaking the physical boundaries of the office. The introduction of cost-effective and reliable communications technologies is rendering the physical office barrier obsolete. As such, the needs analysis process requires organizations to analyze the role of distributed computing within their future operations. With the understanding that the following statement may instigate an argument with some information system mangers, one of the greatest reasons for lost opportunity in the integration of remote computing technologies is the lack of vision to identify the role of these technologies in civil engineering work processes. However, to mollify the same information system mangers, this lack of vision is understandable when placed in the context of traditional civil engineering computing strategies. From the era of mainframes to the introduction the personal workstations, the industry has placed considerable resources into the development of in-house technologies that could provide competitive advantages. Initially, this focus was established by the civil engineering firms that could afford mainframe technology. Companies established competitive advantages by creating large information system staffs that could control and guide computing operations within the organization. By providing clients with unique services such as computerized cost controls, the organizations had an in-house product that could differentiate themselves from competitors while enjoying the luxury of complete technology control. Similarly, the introduction of minicomputers such as the DEC PDP-11 distributed technology to a broader range of organizations, but retained the fundamental concepts of in-house control and processing. This second generation of computing technologies proliferated in the 1970s as corresponding market changes placed large-scale projects such as nuclear powerplants into the civil engineering business cycle. With the introduction of these incomparably profitable projects, organizations rushed to adopt computing advantages with customized software such as CADAM, STRUDL, and COGO (Spindel 1971). Similar to the first generation of mainframe programs, these second-generation applications adopted the philosophy of centralized processing with access through terminals that provided interaction with no independent processing. Information system departments continued to serve as the focus of computing strategies as technology investments in the high six- to seven-figure range required centralized control and policy making.

Although the era of the minicomputer introduced computing to a greater audience than the initial mainframe era, it was not until the introduction of the PC by IBM in 1982 that civil engineering organizations were provided with the possibility of breaking from the centralized computing model. With the introduction of individual processors on each desk, organizations had the opportunity to transfer control of computing operations from a central staff to individual organization members. However, in the majority of organizations, this transfer of control did not occur as quickly as it did in other industries. Three years after the introduction of the PC, only 10–15 percent of CAD systems were being sold to the A/E/C community (Teicholz 1985). Rather, the PC was seen as an opportunity to remove office automation tasks such as word processing and spreadsheet analysis from the minicomputers to allow for greater amounts of computing time dedicated to complex engineering tasks. Justifiably, at that point in time, civil engineering software still required the processing power of a centralized computer. It is only recently that civil engineering organizations were presented with the opportunity to abandon the central processing model and transition to desktop UNIX-based or high-end Windows-based workstations with the processing power to handle complex design and visualization tasks. However, with the momentum developed over a 20–25-year time frame, it is understandable that the transition to desktop workstations has not broken the mindset of computing as a centralized organization function. Specifically, the introduction of the desktop workstation has not created the democratization of computing that many had predicted. As repeated industry surveys demonstrate, increasingly powerful computers, and more recently computer networks, are still controlled by information specialists as supplements to the core design, project management, or construction tasks conducted by highly trained professional staff.

It is this mindset that must be challenged by organizations analyzing the strategic role of computing in the next-generation professional office. The core of this challenge is the role computers will play in relationship to the physical boundary of the office walls. The introduction of mobile computing technologies, including laptop computers, personal information devices, and high-speed networks, provides the opportunity for organizations to conduct operations beyond the limitation of a desktop workstation. The possibility exists to place mobile computing technologies into field offices for long-term projects, give members access to these technologies for site or client visits, or create a virtual network of operations with no boundaries. The only limitation is in the vision that the organization wants to place on itself in terms of ambition or scope of operations. However, to transition from vision to reality takes foresight and planning. The first step in this process is evaluating the organization goals.

Rather than investing haphazardly in mobile computing technologies, an organization needs to focus on the need to expand beyond the physical office boundary. If the organization determines that market forces, clients, operational effectiveness, or regulatory bodies require the introduction of mobile computing, then distributed computing can be added to the technology integration plan. The range of these forces can cover the entire lifecycle of the building or infrastructure process. In each situation, the emphasis is on the use of a core technology by field personnel to enhance the work process. For example, mobile computing technologies in the infrastructure rehabilitation process could enhance the testing process by bringing nondestructive testing capabilities directly to the site. Augmenting this testing capability with remote network capabilities, the organization could inform local transportation agencies of the test results immediately upon completion. By reducing the time required to take the test as well as introducing new testing techniques, the organization is strategically integrating computing technology into the work process. In the analysis of remote computing needs, organization personnel must determine if equivalent opportunities exist that may be realistically explored for technology integration.

Who Needs Control of Information? Historically, the issue of information control has been inconsequential to civil engineering organizations as centralized computing solutions dominated the industry. Through these centralized solutions, all information generated, processed, and requested within the organization resided in a central location. Control of this information has traditionally resided with the individuals or department deemed as information specialists. Given the responsibility for information control, these specialists developed the procedures, manuals, and controls for the organization to access centralized data repositories. This gatekeeper responsibility extended to every project as well as to long-term technology purchasing decisions.

The introduction of remote computing technologies is changing this traditional focus of the information gatekeeper (Wilson and Shi 1996; Jin and Levitt 1996). Specifically, the break from centralized computing solutions to a combination of in-house and remote computing capabilities is opening the door for multiple organization members to lay claim to information control. For example, the introduction of a virtual team that includes regional offices, field offices, the home office, and individuals with laptops who travel to client sites creates an environment that is far less clear in terms of information control. First, the traditional boundary of information creation is broken as information creation is no longer limited to in-house software. Second, the introduction of hyperlink technology eliminates the need for a central repository. Through these links, information can be stored at any location, transparent to the users requiring access. Third, the refinement of data exchange standards reduces the need for standardization of data and software as users begin to have the flexibility to exchange data between programs. Finally, the introduction of remote computing technology allows field and client personnel to generate solutions at remote sites on demand. Collectively, these issues establish a legitimate position for professional staff to claim individual control over project data.

From an opposing perspective, this diversity of computing components and play-ers introduces the opportunity for information chaos. The elimination of centralized control opens the door for every project member to establish communications and data centers on their personal workstations. Taken to an extreme, this open informa-tion policy presents the organization with an almost impossible task of tracing in-formation flow and existence (Edwards et al. 1996). In the event an individual leaves the organization, the possibility exists that a data center will be lost together with its accompanying information repository. Rather than achieving the democratization of computing, the complete removal of controls can result in information anarchy.

Reasonably, an organization has the right to wonder where the logical balance exists for information control. Although a single answer to this issue does not exist for every organization, similarity exists for every organization seeking to position computing technologies strategically. In the emerging generation of remote comput-ing technologies, the ability for professional staff to generate information, commu-nicate from remote locations, and rapidly produce project solutions cannot be overem-phasized. For an organization that has determined that remote computing will play a strategic operational role, supporting these work processes is an essential element of the strategic plan. Creating an information control balance therefore needs to strike an accord between supporting distributed work processes and retaining a degree of order through information specialists. A key to this accord exists in the division be-tween information in progress and finalized solution elements. This division may be defined differently for each organization, but it retains an underlying philosophy that can be used universally for strategic planning purposes. For example, a regulatory agency is charged with overseeing the allocation of environmental permits for a state or region. Given the large geographic area that needs to be regulated, a series of re-gional offices exist to assign field personnel closer to their area of coverage. Using a distributed computing model, the regional offices are interconnected with the dis-trict office as well as to other regional offices with the capability of exchanging and reviewing information. This example provides the perfect environment for informa-tion control problems. Rather than requiring the regional offices to send all infor-mation back to the centralized district, the regional offices can maintain their infor-mation locally while the current project is considered "in progress." By providing external links to the information for regional and district offices, the local individu-als in charge of the project can retain logistical control over information components. However, as components of the project are considered past the "in progress" stage, the information can be transferred to a central "electronic vault," where it falls un-der the supervision of central staff for long-term protection. Through this model, a compromise exists between all parties in the distributed network. Each party retains the control that is required for the information that justifiably should stay within their control during project execution. However, flexibility lies in the ability of the orga-nization to define the dividing line between "in progress" and complete. Although this line may differ from project to project by establishing the model as a central tenet of distributed information management, the organization sets in place a flexi-ble control framework for long-term strategic operations.

A Technology Integration Checkpoint

Distributed computing is the final component of the technology needs analysis. At this stage, an organization should be able to identify the drivers, requirements, and technology needs impacting strategic technology integration. A complete survey of technology integration possibilities can be completed that covers both traditional and nontraditional technologies. Of greatest importance, the organization can identify the technology fallacies that could lead to inadvisable technology expenditures. With this identification complete, the last step of the technology integration process can proceed, establishing a formal strategy for technology integration.

■ Establishing a Technology Integration Strategy

Completing a needs analysis often leads to a common conclusion, implementation paralysis. The needs analysis often infuses a sense of optimism and excitement as organization members identify opportunities for technology integration. This optimism is a short-lived phenomenon. Unless organization members see tangible results directly emanating from the needs analysis, enthusiasm can diminish, leaving thoughts of status quo and cynicism to replace the enthusiasm. Thus, in the age of fast-moving technology advances, the traditional conservatism of civil engineering organizations is facing an unprecedented challenge.

The decision to create a technology strategy must be made quickly, but thoughtfully, to demonstrate to internal and external stakeholders that the organization is serious about strategic integration. The following sections provide an introduction to establishing this strategy. The three-phase analysis process incorporates the needs analysis previously completed and expands them into an overall emphasis on long-term organization planning. However, the issues do not, by themselves, establish a technology plan. Purchasing decisions, hardware and software selection, and other specific operations, are processes that must be tailored to each organization. In response to this individuality, the following sections provide a foundation on which an implementation plan can be constructed.

Concentric Circle Integration

As discussed several times in this book, the fundamental key to successful strategic management is to build on existing organization strengths as the foundation for growth and expansion. The development of a strategic technology integration plan does not stray from this formula. The integration of technology should initially occur in locations where the greatest potential for acceptance exists, and then proceed to other areas with decreasing potential, until finally, the area with the greatest potential difficulty is addressed. Two questions immediately arise: (1) Why not attack the whole organization at once, and (2) why not use an organization center with reduced capabilities as an illustration of the advances that can be made through successful technology integration? In response to the first question, the conservative nature of civil engineers results in pockets of resistance to change in almost every organization. Rather than delaying departments that are primed for success while the remaining areas are convinced, it is far more effective to establish initial success stories through quick action. Similarly, in response to the second question, the time required to establish an initial success with a reluctant department versus the time required to establish a guaranteed success with a primed department is vastly different.

With this focus on initial success as an introduction to long-term technology integration, a concept of concentric circles of integration is provided as a model for analyzing the integration plan. As illustrated in Figure 5-13, the concentric circles concept emphasizes a technology integration plan that builds upon priorities together with the internal strength of the organization and then pushes outward to address areas less likely to have technology awareness prior to the integration process. At the core of these circles is the home office. Encompassing the champions of the integration process and the senior policy makers, the home office provides a natural base from which to expand. Focusing on this circle, the technology plan should initially address the needs of the primary work processes characterizing the organization. If the organization has a primary responsibility to generate environmental impact statements, then the initial technology integration concerns should focus on enhancing the home office support for generating these statements, such as creating accessible boiler

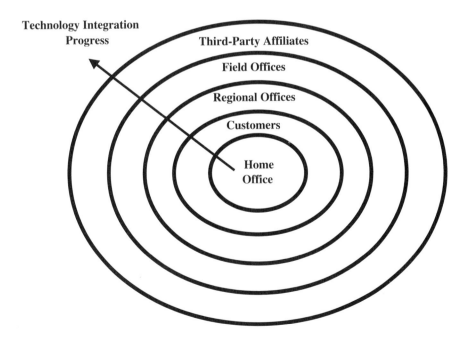

Figure 5-13 The concentric circles approach to technology integration progressively builds upon organization strengths to ensure both short-term and long-term success stories.

plate materials, accessible examples to a variety of projects, and infrastructure for storing multimedia project documents. Creating an environment that assists organization members to work smarter establishes the tangible results sought after by executive and entry-level position alike. Additionally, creating success within this central circle provides an intangible asset, the generation of positive word-of-mouth reviews by home office employees to stakeholders located in the outer rings of the planning process.

The first level of these outer rings is the customer level. Viewed as clients, permit applicants, or industry observers such as financial analysts and bond guarantors, the customer is the life blood of a service organization. As such, the customer focus must change with the introduction of the information era. In traditional settings, choreographed interactions such as lunches, golf outings, and cocktails formed the center of the organization–customer relationship. However, a second relationship is emerging with the technology age, asynchronous interaction. In contrast to the planned and highly choreographed interactions, asynchronous interactions occur as unplanned interaction between the organization and the customer through electronic communications. These communications may take the form of e-mail, Web browsing, or any number of emerging electronic technologies. In these interactions, the customer determines when an electronic visit will occur, where it will be initiated, and what will be learned. The organization is placed in the position of accommodating these visits through the establishment of an electronic infrastructure that can best illustrate the capabilities of the organization without the benefit of a personal visit. While intimidating to many civil engineering organizations, the era of asynchronous communications has arrived and will form the basis of many interactions for years to come. Thus, to ensure a proper appearance to the customer, the second circle of integration emphasizes developing an electronic presence that appropriately depicts the organization and provides access to information that previously may have been considered off-limits. Achieving this presence requires the organization to determine several elements including what the customer should see, how this information is presented, what are the limits of the information access, and who should be allowed access to what components of the organization. While daunting, this electronic presence provides an equalization factor in the industry, allowing small, medium, and large civil engineering organizations an equal opportunity to establish an electronic presence that creates a level of interactivity on a world-wide scale.

The third concentric circle once again turns toward the needs of the work processes. In the third concentric circle, the work process focus extends beyond the boundaries of the home office to the regional offices. Given the importance of networks on the ability of the organization to function as a single virtual unit, the demand to incorporate regional offices into the network is evident. Supporting the concept of a virtual office, the organization must analyze and specify requirements to bring the regional offices into an easy-to-use communications connection with the home office. Of particular importance is the ability of regional staff members to transmit and receive project data on demand. Without the ability independently to transmit and receive project data, the virtual link fails to achieve the potential it advertises. To require approval for this data exchange is the equivalent of telling the regional offices that they cannot be trusted with main office data. The strategic technology plan must break these barriers by emphasizing transparent connections that disregard individual member locations. However, this demand for transparent interconnections can create a difficult situation for the organization. While technology demand is a positive move in that it establishes a greater potential for success, technology demand also can create a negative potential. Specifically, technology demand can lead the organization to develop plans that exceed the level necessary to implement the virtual team. For example, although the installation of high-speed communication links are desirable by all employees, the organization must evaluate these communication links on a case-by-case basis. Establishing an incremental advancement policy where communication links are based on return on investment is a legitimate response for an organization that is concerned about potential cash traps. The communications solution can focus on sufficient rather than optimal solutions as long as the regional offices retain a capability to communicate in a manner that does not impede with work process effectiveness.

Closely related to the regional office expansion is the expansion of strategic computing to the fourth concentric circle, the field offices. For the purposes of strategic planning, these offices include temporary field trailers associated with construction projects or temporary testing sites associated with preconstruction or rehabilitation projects. In either situation, the focus of technology integration is on the ability for field personnel to integrate daily work processes and information flow into the virtual web centered at the home office. This integration effort is placed on an outer circle for two primary reasons: (1) Field personnel are schedule restrained, and (2) standard operating procedures are essential elements of success. The first of these issues recognizes the fact that field personnel are under strict time and cost constraints. Given these constraints, creating additional tasks for these individuals will be viewed as intrusions into an already overloaded daily schedule. Similarly, introducing technology into tradition-oriented procedures can be construed as an attempt to alter time-tested methods for successfully completing engineering tasks. Therefore, to expand the strategic integration plan into field operations, a clear objective must be set that outlines the necessity for the integration of technology into traditional operations.

Creating this necessity is the focus of the field office expansion. The method used to establish this necessity can be varied based on the preference of the organization. However, one path to this establishment is a four-step process as follows. First, the development of the strategic integration plan must bring forth all the analysis conducted during the needs analysis process to establish a clear statement of goals and objectives for field personnel. Second, the benefits of such an introduction should be clear and concise to assist home office personnel in cooperatively developing the integration plan with field personnel. Third, the successful integration of the previous circles is used to present a scenario of planned integration within the firm. Previous success is critical to demonstrate to field personnel that an established pattern of integration has been undertaken and the field application is the next step in this progression. Finally, organization leaders must deflect potential resistance by establishing a strong statement that technology integration will be the revised methodology for performing field operations. Note the theme here is not technology augmentation, but the replacement of traditional manual operations with enhanced technological solutions. Given this statement, combined with clear objectives and patterns of previous integration, the organization can aggressively pursue the extension of the integration plan into the domain of the field office.

The final circle in the concentric circle concept builds upon the complete integration plan for the organization by extending the integration concept to third-party affiliates. Affiliates including suppliers, partners, consulting engineers, and owner representatives are each embarking on technology integration plans to various extents, as illustrated by efforts by 84 Lumber and North Pacific Lumber Company. In these examples, suppliers are placing electronic versions of their catalogs on-line to allow engineers to specify, order, and track materials electronically. This progression provides an entry point for civil engineering organizations to bridge the gap between outside organizations and internal work processes. Through the definition of electronic linking procedures, the organization has the opportunity to build upon advances taking place in each of the affiliate industries. In this manner, advances within the organization are multiplied by building upon advances instituted by outside affiliates. The extent of these advances may not be known at the time the technology integration plan is developed. However, the technology integration plan should be as broad as possible in terms of external affiliates to provide the organization with the opportunity to incorporate external links as they become available.

Technology Requirements to Support New Business Ventures

The development of a concentric circle plan is the primary element in the development of a strategic integration plan. However, transforming this plan from an abstract document that sits on a shelf to one that can be used actively to move the organization forward is the last stage of strategic integration. Within this stage, strategic integration returns to the primary emphasis of strategic management, expanding the organization. This expansion may focus on projects, increased interaction with the general public, or a number of other growth horizons. In each scenario, the focus of technology integration should enhance external opportunities as well as enhance internal work processes.

Successfully retaining this focus requires the organization to focus on the integration of strategic objectives, not technology as an isolated corporate expense. As introduced in the strategic knowledge pyramid in Chapter 1, the successful integration of technology into long-term strategic management is based on removing technology from the traditional role of isolated departmental control. The placement of these decisions in a central strategic position requires the management team to recognize the role of technology in assisting with strategic objectives. Foremost within this transition is the recognition that information and knowledge resources are no longer limited to large computers in the back room. Rather, the emergence of knowledge workers, rapid communications, mobile computing, and distributed technology centers is opening a window of opportunity for organizations of all sizes to redefine their approach to civil engineering. From enhanced work processes to transformed field operations, knowledge resources serve as an equalizer within the industry. The organizations that recognize this equalizer have the opportunity to stake a position that capitalizes on emerging technologies to establish long-term strategic advantages. Initializing this process through sound planning that sidesteps traditional technology traps, incrementally integrates technology throughout internal and external operations, and focuses upon strategic objectives is no longer an option for traditionally conservative organizations. The window of technology opportunity is limited; taking advantage of the window takes initiative, planning, and strategic foresight.

DISCUSSION QUESTIONS

1. The evolution of the knowledge worker has been developing steadily over a number of generations. These knowledge workers are becoming pervasive in many industries. Does the knowledge worker concept conflict with traditional civil engineering career paths?

2. The knowledge worker concept emphasizes information manipulation as the focus in the new world economy. How does global information access influence local projects during design and construction?

3. The catalysts for technology integration include both internal and external catalysts. Each of these catalysts is driving civil engineering organizations to redefine the role of technology in the organization. How is this new role going to impact the hiring patterns within civil engineering organizations?

4. The evolution of the personal computer has repeatedly demonstrated that organizations cannot cost effectively chase the technology curve. Given this as a premise, how can a civil engineering organization develop a technology plan that balances technology competence with fiscal responsibility?

5. One component of technology integration is the analysis of work process requirements. Given that the definition of civil engineering work processes is changing with each introduction of a technology advancement, what timeline into the future should a civil engineering organization adopt to determine technology requirements?

6. The control of information has traditionally been a divisive issue within the AEC industry. Who controls information such as design concepts, calculations, schedules, and budgets is a constant subject of contracts, arbitration, and court cases. How does the introduction of new computing technologies influence this traditional issue?

Knowledge Resources for Market Competition

Case Study

Can a small organization compete with national and international organizations? If a small organization has set strategic objectives to compete for the same customers as a large organization, how can it present a case that it has the capacity to produce similar or superior results? Can knowledge resources serve as an equalizer in the increasingly technology-based market of the twenty-first century? As senior managers of small organizations and divisions understand, these questions are at the forefront of today's evolving civil engineering marketplace. With the introduction of technology has come the changing of the industry definitions that previously characterized the competitive and market boundaries. Today, the size of an organization may have less impact on the capabilities to provide services than the commitment of an organization to incorporate knowledge resources into operational procedures and strategic objectives. However, the question of how far to rely on these technologies is an unresolved issue that must be addressed by each organization based on its individual preferences, culture, and objectives. Is one approach better than another? The following profiles represent two approaches to this question based on similar objectives, but different philosophies.

Profile One—EDI, Ltd.

EDI, Limited, is an Atlanta-based consulting engineering firm that specializes in developing quality indoor environments. Founded 12 years ago as a partnership of four civil and mechanical engineers, EDI has grown to 40 employees and broadened its focus to include electrical, security, and telecommunications consulting. Providing innovative solutions that exceed customer expectations, EDI emphasizes the development of long-lasting relationships with its customers. With a customer base that includes developers, building managers, brokers, and large corporations, EDI is focused on expanding to 100 employees and $10 million of revenue over the next 5 years. As a start to this expansion, EDI has adopted a national focus for its services, resulting in the organization developing projects in every region of the United States. Focusing on the future directions of the AEC industry, EDI consistently modifies its operations to remain at the forefront of the AEC industry and its evolving customer needs. Today, EDI's services include communications and network technologies, indoor air quality evaluation and problem solving, and security and special systems design. With full-service engineering capabilities in each of these areas, EDI offers its customers project management, planning, design, and installation management services.

At the center of this focus on the future is an emphasis on technology and its capacity to assist EDI internally as well as its capacity to meet customer requirements. EDI founder and President, Donald Curry, epitomizes this focus on technology. Since the inception of EDI, Curry has espoused the use of technology as an equalizing factor in the competition with larger competitors. Understanding that EDI is a fraction of the size of its competitors, Curry emphasizes the hiring of people who have multiple talents. To leverage these talents, EDI provides these individuals with the latest computer technologies. Curry believes the end result of this investment in both human and technology resources is an organization that is better prepared to meet the changing needs of the AEC industry than his much larger competitors who require significantly longer periods of time to adjust operating and marketing procedures. With business sectors incorporating the latest in building and communications technologies, these competitors are significant players. Organizations such as AT&T, IBM, and MCI each have consulting divisions that provide similar services as that offered by EDI. With this level of competition, EDI cannot afford to make mistakes in its resource investments.

As a current example of EDI's commitment to leveraging knowledge resources, EDI has made a full commitment to using evolving World Wide Web technologies to extend the capabilities of each employee. Giving each employee access to the Web at their desk, EDI has developed Intranets and corporate knowledge bases to allow each individual to access both internal information and global data whenever needed to assist them in developing customer solutions. Understanding that EDI does not have the research staffs of its larger competitors, EDI knows that it must find ways to assist its employees to work more efficiently. Spending 10 times the amount that an average AE firm spends on education, EDI acknowledges the importance of leveraging knowledge resources. Committed not to make 70- and 80-hour work weeks a standard in the EDI culture, EDI views the World Wide Web and education as equalizers in the development of customer solutions. Curry encourages each employee to use the Internet capabilities to learn about prospective customers, new products being developed, and evolving technologies. In essence, the World Wide Web is viewed as a personal research staff for each EDI employee.

The results of this commitment are difficult to dispute. With a customer base that now includes such internationally recognized organizations as IBM (also one of its biggest customers), UPS, and Hewlett Packard, EDI is successfully defeating its larger competitors in the pursuit of customer awards. Additionally, Curry believes that this commitment to technology and knowledge resources is at the core of employee commitment to EDI. With a turnover rate of less than 10%, EDI has a solid foundation on which to build for long-term strategic objectives.

Profile Two—Boyken International

A second approach to incorporating technology resources into the civil engineering organization is that taken by a second small firm, Boyken International. With offices in London, Orlando, and Atlanta, Boyken specializes in cost management services throughout the project lifecycle. Working with large municipal and private owners on recognized projects such as Disney World and airport expansions, Boyken has established itself as growing leader in the field of cost management. Today, Boyken provides services from project concept to completion, with an emphasis providing customers a price benefit of up to 5% of the total construction amount. Building on both in-house resources and outside experts, Boyken serves as an outside member of the owner's team, overseeing operations to ensure the owner receives the best product for the best price from the design and contracting team.

Similar to EDI, Boyken also emphasizes a focus on the future and the changing needs of the AEC industry. Developing new project management and cost analysis processes, Boyken continually emphasizes the need to incorporate evolving business developments into the AEC industry. Led by the co-founder of Boyken International, Sandy Arnold, the company strongly emphasizes the incorporation of economic theories and analysis procedures into the cost management process. Subsequently, Boyken customers receive the latest in cost management procedures founded on principles that are proven in a number of global industries. In this manner, Sandy Arnold sets the organization apart from its competitors who rely heavily on traditional construction management techniques and procedures. For example, Boyken has introduced the concept of the Development Program Manager to assist customers throughout the project, delivering value engineering, bidding, control, and auditing advice in a comprehensive package.

Similar to EDI, Boyken also recognizes the need to leverage human and technological resources. Focusing exclusively on experienced personnel who understand customer issues, Boyken utilizes experience to respond quickly to project requirements. To build on this experience, the organization has also incorporated electronic communications and intranets to keep every member of the organization connected wherever they are currently located. However, in contrast to EDI, Boyken has not incorporated the World Wide Web as a central focus of employee operations. Believing that the tool should be viewed as an additional reference material, Arnold has restricted the use of the Web to the organization library where it is one of several traditional and emerging tools available to organization personnel. Through this philosophy, Boyken has taken the position that the Web should be used by employees whenever they believe it will be of benefit to them in their development of customer solutions. However, by removing access to a library location, Boyken is also stating that the Web can be a tool that consumes significant amounts of time without a direct purpose. Since Boyken is working on agreed-upon project cost percentages, efficient use of time is critical to a small organization attempting to compete with larger organizations. Boyken managers believe that although the Web is a useful tool, it should remain just that, a tool that is accessed in the library while project solutions are being developed.

Once again, as with EDI, the Boyken approach is difficult to dispute. Awarded contracts by some of the largest airports, municipalities, and corporations in the United States and Great Britain, Boyken is successfully competing against larger, and more established, civil engineering consulting firms.

QUESTIONS

1. Placed in the position of starting a new firm, which of the previous two approaches would you adopt for your organization and why?

2. Placed in the position of an employee looking for a new employment opportunity, which of the two organizations would you prefer?

3. Placed in Donald Curry's position, how do you avoid allowing the emphasis on technology to turn into a technology trap?

4. Placed in Sandy Arnold's position, how do you respond to new employees who have become accustomed to working with the Web as a central part of their job at either other organizations or at a university?

6

ORGANIZATION EDUCATION

What formal and informal procedures are in place for lifelong learning within the organization?

Continuing the emphasis on resource development, the fourth strategic management question builds upon knowledge resources through a focus on organization education. With the half-life of civil engineering knowledge becoming increasingly shorter, the days of education culminating in an undergraduate degree are correspondingly coming to an end. The era of ending formal education pursuits when an entry-level position is accepted can no longer be supported. Rather, the successful organization is increasingly facing the need to establish lifelong learning as a centerpiece of strategic management. Whether the focus is on updating core engineering knowledge or expanding skills to areas such as management and technology, education is the emerging avenue to organization improvement.

In contrast to the structured education that has previously been the rule in the civil engineering industry, the current education focus includes both formal and informal dissemination mechanisms. Following the education trend that is progressing throughout the international business community, the civil engineering industry is slowly adopting the concept of learning organizations (Cayes 1998). By placing the entire organization in the position of dynamic learner, the organization retains a focus on emerging competitive forces. Specifically, by providing employees with the opportunity to gain insights into both cutting-edge and established engineering and business practices, the organization leverages the education investment with returns of innovative and forward-looking ideas. It is the potential uniqueness and barrier-breaking thoughts within these ideas that represents the opportunity to establish the organization as a leader within a given market segment or industry sector.

In this chapter, the development of education strategies, both formal and informal, is examined in the context of lifelong learning. The chapter emphasizes the need to develop these educational structures as building blocks for establishing a learning organization. An examination of the formal and informal education opportunities available to civil engineering organizations are explored in conjunction with the costs associated with both accepting and declining these opportunities. The chapter emphasizes the multiple opportunities available for organizations to establish lifelong learning practices. Concurrently, the chapter provides the financial and pragmatic justifications for adopting this strategic position, including technological and personnel advancements. Establishing the impetus to implement this education focus sets the challenge for this chapter—*to examine the educational needs and opportunities within the organization and set the education objectives required to create a civil engineering learning organization.*

■ Why Education as a Focus

The topic of education is guaranteed to raise a debate between members of the civil engineering industry. Whether the audience is practitioners, academics, or public service officials, education and its role in civil engineering is a hotly debated topic. Continuing a debate that has steadily gained attention over recent years, issues including where education belongs in a professional organization, what subjects are important at a professional level, and what methods of learning are effective aim directly at the purpose of professional education. Furthering this debate, the entry of organizations such as the National Science Foundation, the National Academy of Engineering, and the American Society of Engineering Educators has politicized the concept of engineering education. With all these constituents and opinions, the civil engineering industry has not come any closer to arriving at a consensus on education requirements.

With this disagreement as a backdrop, this discussion presents a basis for building an emphasis on professional education within civil engineering organizations. At the center of this presentation is a response to the pointed question of why lifelong education should be a focus of civil engineering professionals. Although many reasons have been put forth by individuals and organizations, the authors group the justification for education into four categories (Fig. 6-1):

- Bridging the management knowledge gap
- Enhancing career advancement opportunities
- Enhancing personal and professional development
- Ensuring competency in discipline knowledge

As illustrated in Figure 6-1, the categories divide into quadrants along two axes, position responsibilities and knowledge base development. The position responsibilities axis emphasizes the need for education in response to existing or future responsibilities of individuals. The knowledge base enhancement axis emphasizes the need for education to either build on existing knowledge or enhance the individual's knowledge base with new knowledge areas.

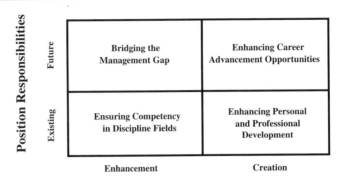

Figure 6-1 The justification for expending resources on education can be divided into four quadrants corresponding with the needs to enhance an individual's knowledge and prepare the individual for future organization responsibilities.

Bridging the Management Knowledge Gap

Research conducted by the authors, together with research from other bodies (National Academy of Sciences 1995; National Science Foundation 1996), has identified management education as a critical issue for the long-term success of the civil engineering industry. The project management tradition has created an atmosphere that downplays continued management education, since the prevalent belief emphasizes projects as the basis for organization success. Taken to the extreme, the belief equates organization management with project management by equating the management of the organization with the management of projects. In this line of reasoning, the higher the position in the organization, the more projects the individual must oversee. Although the level of detail that the individual is responsible for is reduced, the same premise exists throughout the management ladder.

While this belief is stated or believed by a segment of the industry population, the authors believe that this argument cannot stand up to the challenges of the twenty-first-century business environment. As described by the management knowledge gap principle, fundamental differences exist between the concerns of project managers and organization executives. Emerging trends in technology, global economics, knowledge workers, and leadership will force organizations to make a fundamental choice. Either the organization obtains the knowledge required to address these issues effectively, or elects to use traditional business practices. As illustrated in Figure 6-1, addressing these issues requires an enhancement of the organization knowledge base with a focus on preparing individuals for future responsibilities. Unfortunately, this enhancement does not adequately occur during the course of daily operations. Even individuals with the best intentions to acquire additional knowledge through professional journals must face the reality of daily operational pressures. In the context of running an organization, deadlines, client demands, and project fires will consume the free time of an individual that was set aside to acquire enhanced knowledge independently.

Given the reality of daily operating pressures, the need for an educational structure is critical. The individual with the desire and aptitude for executive management requires education opportunities to prepare adequately for the challenges of a strategic management position. The absence of these opportunities will result in one of two outcomes. Either the individual will obtain a senior management position and rely on the project management model of management, or the individual will attempt to acquire the needed knowledge rapidly. In either case, the competitiveness of the organization will suffer because the individual is not adequately prepared to meet the strategic challenges. Given this reality, civil engineering organizations must acknowledge that the management knowledge gap is not going to disappear. Rather, the gap will increase unless the individuals charged with bridging the gap are given the opportunity to build the bridge on a solid educational foundation.

Enhancing Career Advancement Opportunities

The existence of a well-defined career ladder was a frequent comment heard during research for this text. Given the long history of the civil engineering profession, career paths have emerged within the individual disciplines. From entry-level positions to senior project managers, design, construction, and consulting firms have built well-developed career paths for civil engineering professionals. Concurrent with these career paths is an accepted set of knowledge areas that an individual must acquire to proceed successfully. With the relative stability of the profession, this knowledge has traditionally been acquired through on-the-job training. As an individual progresses along the career path, the individual is given the responsibility commensurate with the position. Through this increase in responsibility, the individual absorbs the knowledge required to execute successfully the responsibilities of the position. Although some organizations such as the Army Corps of Engineers vary in this process by giving organization members formal education at each stage of the career progression, the common denominator remains that the civil engineering profession is very regimented in its career progression.

This traditional perspective is being challenged by changes in the business environment. With the need to expand revenue streams through diversification and market analysis, the need for individuals with diverse knowledge and strengths is emerging as a strategic priority. Business development, economic analysis, technology development, and strategic planning are only a few of the new areas of knowledge that organizations must foster. As illustrated in Figure 6-1, fostering knowledge enhancement requires the anticipation of future responsibilities and career path alternatives. For example, an individual with strength in developing alternative procurement relationships will be underutilized if he is put in a traditional project management or design position. This individual represents new opportunities to the organization. Pursuing these alternative venture relationships, the client base may be expanded to include projects that were previously considered outside the accepted market sector.

Changing business conditions require the implementation of new business practices. Although organizations may not feel comfortable acknowledging that new business practices are becoming a part of operational requirements, ignoring the requirements will not address the challenge. Rather, breaking out of the comfort zone by introducing new career goals, education opportunities, and enhanced knowledge requirements is the first step towards developing an enhanced career development environment.

Enhancing Personal and Professional Development

The third justification for education acknowledges the continuing changes that face organization members in their daily responsibilities. Personal and professional development is a broad category that encompasses everything from developing communication skills to enhancing technical competencies. However, the underlying commonality between these diverse elements is an understanding that individuals must continue to be challenged to enhance personal satisfaction. In a national study focusing on why individuals elect to leave organizations, a significant percentage (28%) note the lack of challenge or fulfillment provided by the position (McSulskis 1998). Managers at any level should not overlook this evidence. From a strategic position, the development of a loyal, cohesive team is critical to long-term organization success. Identifying with the organization and its objectives is central to developing an organization culture.

Providing opportunities for members to expand this culture must be a central component of any strategic plan. Specifically, as illustrated in Figure 6-1, enhancing personal and professional development provides an opportunity for an individual to address the challenges of current responsibilities through the creation of new knowledge resources. Examples of these opportunities encompass a wide spectrum.

- Providing foreign language courses for employees who require enhancement of communication capabilities with field personnel from foreign countries
- Providing public speaking training to individuals with business development responsibilities
- Providing basic financial accounting classes to project managers to enhance their understanding of profit and loss calculations
- Providing computer-based management and analysis education to individuals whom have responsibilities to plan and control operations
- Providing business plan development knowledge to individuals responsible for presenting new revenue stream opportunities to directors

Each of these examples demonstrates a commitment to enhancing the effectiveness of individuals in performing their specified responsibilities. The focus on these professional development opportunities not only provides the organization with the opportunity to enhance productivity and competitiveness; it also signals to members that the organization is committed to enhancing their professional development. Although civil engineering organizations have traditionally emphasized long career tenures as one basis for advancement (Goodman and Chinowsky 1997), the prospect of advancement alone is no longer enough to retain employees. Increasingly, individuals are focusing on the intangible components of a position such as satisfaction, challenge, and fulfillment. Correspondingly, the need to implement education mechanisms that enhance these elements lies at the core of long-term planning to meet these needs.

Ensuring Competency in Discipline Fields

The final area of education justification is the emphasis on enhancing discipline knowledge. Although underlying math, physics, and engineering principles remain stable over time, discipline specifics are in a constant state of change. For example, research in earthquake engineering research continues to have a profound impact on the civil engineering profession. Every instance of an earthquake such as those experienced in Northridge and San Francisco, California, and Kobe, Japan, serve to rewrite existing knowledge in design, construction, and material development. Similar knowledge advancements occur in areas such as flood control, beach erosion, and hurricane protection. In each scenario, the civil engineer is legally, ethically, and professionally obligated to remain knowledgeable about the current developments within an individual discipline.

As illustrated in Figure 6-1, this type of education is the least threatening to the majority of civil engineering organizations. Organizations are aware of their responsibility to provide employees with the education necessary to remain current in their selected fields. However, this awareness changes from professional responsibility to strategic limitation when organizations forget that discipline knowledge is only the first quadrant of a graph that encompasses three additional sectors. Discipline education should represent a foundation for building broader knowledge bases within individuals and organizations as a whole. While it is true that individuals require discipline knowledge to succeed within their fields, the existence of this knowledge does not equate to a complete foundation for career success and options for career advancement.

Placed in context, the existence of education opportunities to enhance discipline competencies should be viewed for what it actually represents, a continuation of the lifelong learning process that accompanies any given profession. It does not represent an enhancement of business opportunities for an overall organization. With this conflict, discipline education should remain as a central feature of any civil engineering organization. However, it should not represent the boundary of the education activities that an organization elects to pursue.

■ What to Study

Once an organization establishes the need for a professional education program, the next issue that arises is what topics should be studied. Organizations cannot provide the flexibility to allow professionals to dabble in a number of topics without a sound focus. Organizations are responsible for showing a profit to owners and shareholders. It is very difficult to justify a course in nineteenth-century romantic music when a direct analysis of profit and loss is summarized on a balance sheet. However, this example should not be confused with an argument against liberal arts as a basic element in undergraduate education. An employee with a well-rounded education will be a critical asset to civil engineering organizations attempting to broaden revenue streams and business opportunities.

Rather than entering the debate concerning the validity of organization and academic studies, or endorsing the private individuals who publish articles and comments in professional magazines and journals, we return to the knowledge pyramid introduced in Chapter 2 (Fig. 6-2). The basis of the knowledge pyramid is the idea that civil engineering knowledge is one component of a continuous set of steps that reinforce each other as they proceed up the pyramid. Eventually, the top of the pyramid reflects the need to understand strategic management functions and responsibilities. Based on this progression of knowledge, the knowledge pyramid provides a basis for implementing a professional education initiative. Placed in the context of the four ed-

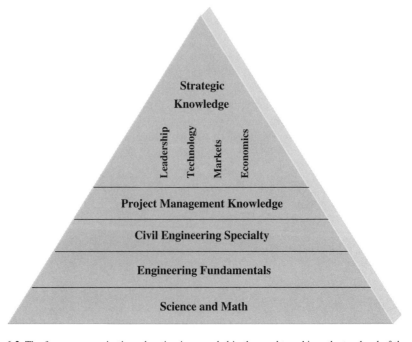

Figure 6-2 The focus on organization education is grounded in the need to achieve the top level of the strategic knowledge pyramid.

ucation quadrants identified earlier, the knowledge pyramid can be translated into the four quadrants to form the basis of a long-term education initiative (Fig. 6-3).

Discipline Advances

Long-term education begins with an emphasis on the present level of responsibilities and the extension of existing knowledge. The need to stay current within the civil engineering profession as a whole and within a particular discipline is the first priority for an organization. Organizations no longer have the luxury to retain "old" ways of conducting business. Regulatory, technical, and discipline advances are changing the industry on almost a daily basis. The organization that elects to ignore these changes may face a number of problems, including legal action, if a new regulation is not incorporated into a design, a reduction in work due to lost productivity from not adopting new technological advances, or finally, the organization may find itself unable to meet a client's request to use new materials in a project since the expertise is not available. Each of these issues represents an educational concern that is at the foundation of strategic success. If an organization is to succeed over an extended period of time, then its members must be knowledgeable of the current state of their discipline.

To illustrate this need, consider a fictional geotechnical consulting firm that provides recommendations for soil stabilization to agencies such as Departments of Transportation and Departments of Public Works. This type of work regularly includes the design of solutions encompassing slopes and embankments where roads, drainage swales, and riverbeds traverse urban and rural areas. The introduction of geotextiles and geosynthetics has fundamentally changed the manner in which geotechnical engineering is approached. The new materials offer design options that were unavailable a decade ago. However, the organization that fails to educate its members on these advances immediately finds itself at a disadvantage because it no longer can

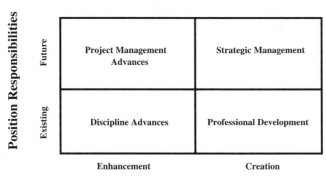

Figure 6-3 Developing an education focus returns to the foundational components of long-term knowledge development and job preparation.

provide the best service available. Although customers may be loyal to an organization for a number of reasons, lack of current knowledge is an almost guaranteed path to losing the most loyal customer.

Additional examples of discipline-specific knowledge can be found throughout the civil engineering industry. Alternative procurement methods is a popular topic that challenges traditional contracting beliefs. However, these new contract relationships incorporate legal liabilities that were nonexistent in traditional business relationships. Failing to understand these relationships will eventually lead to costly legal proceedings that cannot be defended on the grounds that the organization was unaware of the additional liability. Similarly, advances in regulatory requirements such as wetland restoration or historic preservation can impact the operational, as well as technical, components of an engineering project. For example, the construction organization that fails to follow new regulations for operating vehicles in the vicinity of a wetland may be facing Federal fines as well as potential civil actions. In each of these cases, and many other similar examples, the common thread is an understanding of the current developments in a given discipline. As a basis for what to study, the required starting point is the education of the organization in the area that directly impacts existing responsibilities and existing knowledge.

Management Advances

The second area of study progresses up the pyramid to focus on management advances. The management advances category emphasizes providing organization members with management knowledge that will assist them in both near-term and long-term advances. The management advances category addresses the need to justify education expenses by advocating knowledge that not only assists employees in doing a better job in their current position, but also prepares them for the managerial responsibilities that lie in the future. Specifically, the need for education in the area of current management practices transcends individual job responsibilities. Every civil engineering organization, every discipline within the organizations, every department within the disciplines, and every group within the departments is dependent on effective management. Whether it is project management or organization management, the effectiveness and profitability of organizations can be traced to the ability of management structures to respond effectively to current business conditions. Providing organization members with the tools to compete in this environment is a central responsibility of every organization.

To illustrate the need for this exposure, consider the contrasts between current management practices and those considered to be cutting edge only a short time ago. For example, the total quality management initiatives of the early 1980s have turned into standard operating practices today. Similar influences from manufacturing and other engineering disciplines are changing the vocabulary of today's managers. The concepts of supply chain management and enterprise management are no longer restricted to the domain of multi-billion-dollar manufacturers. Rather, management concepts that were considered to have limited value to the low-volume production of civil engineering are now appearing as influential advances. The introduction of computer-based scheduling and estimating is no longer relegated to the large-scale defense industry. Today, many owners require schedule and budget information to be delivered in program formats that are compatible with the systems used within the owner's organization. The concepts of just-in-time delivery and continuous quality improvement have become a staple in the civil engineering management vocabulary. However, these advances have not appeared as an accident. Rather, it is the concentrated effort to educate civil engineering organizations on the opportunities available in the management arena that has changed the manner in which civil engineering projects are managed.

Continuing the introduction of management advances into individual civil engineering organizations is essential to the health of organizations. Taken one step further, it can be argued that management advances will be as important to remaining profitable as productivity advances. For example, the organization that builds upon supply-chain management concepts to develop efficient supplier relationships and delivery schedules will have a strategic advantage over competitors who continue to rely on traditional bid procedures. Additional examples exist throughout civil engineering organizations. The organizations that adopt computer-based technologies to assist in managing projects will have distinct advantages over organizations that hesitate. The adoption of technology into the management process provides opportunities to increase efficiencies between geographically dispersed divisions, enhance communications between field and home offices, enhance productivity, schedule, and budget reporting, and expand awareness of project milestones. However, these advances will not appear spontaneously. Adopting new management practices that assist in increasing profitability and productivity is not an automatic process. A focus has to be placed on obtaining the appropriate knowledge, disseminating that knowledge, and ensuring the knowledge is put in place as part of an overall organization education plan.

Professional Development

The third area of education emphasizes the development of employees to assume responsibilities at all levels of the knowledge pyramid. The professional development of organization members is often overlooked because it is considered part of on-the-job experience. The concept that formal education plans should be developed to provide professional development is often hard for organizations to accept. Although reasons for this hesitancy are varied, the primary response is lack of understanding as to what is meant by professional development and how it relates to the profitability of an organization. Professional development is defined here as any activity that enhances the existing knowledge base. This is a broad definition on which to build an education plan. The enhancement of a personal knowledge base can include almost any topic. In fact, that is what makes professional development such a vital component of organization learning. By providing organization members with the opportunity to pursue diverse topics such as foreign languages, creative problem solving, computer networking, and public communications, the organization is placing itself in a position to compete effectively in the global economy.

The era of every employee following a set career path with set knowledge requirements is over. The quicker an organization recognizes this fact, the quicker it can proceed to break from the outdated tradition. In the emerging global economy, organizations must bring diverse talents together to respond to new issues with new solutions. For example, in one project undertaken by a large U.S.-based engineering–construction organization, a $700 million project was undertaken to build a chemical refinery project in a remote location in Asia. However, the location had no infrastructure to support the construction process. The organization had to build temporary ports, roads, housing, and sanitary facilities to support the multiyear project. Completing this effort required the skills of individuals with knowledge in foreign languages, monetary exchange, infrastructure planning, international legal systems, and education programs for local labor. Additional knowledge in logistics, long-range planning, communications, and foreign protocols was also considered critical elements of project success. These are not components of traditional university curricula or on-the-job training programs. Rather, these are extensions of core civil engineering knowledge that are obtained through outside educational exposure.

Although the previous example may be extreme for the majority of civil engineering organizations, the scale may be reduced to address the emerging issues in all civil engineering organizations. The need for knowledge in public communications is expanding as greater numbers of projects are moving to negotiated rather than closed-bid arrangements. The ability to present project ideas effectively is becoming increasingly important as clients evaluate proposals between competing organizations. The ability to communicate with foreign workers is becoming increasingly important as we approach the reality of the global workforce. Understanding the role of computer technologies is increasing as technology moves from the domain of the information technology department to the central engineering groups. Bringing new problem-solving perspectives into the project development phase may provide the new approach to an old problem that is the difference between competitive advantage and thin profits. Moving further from the core topic of engineering, it can even be argued that courses in travel, music, or literature can be considered essential elements for organizations as clients from foreign countries are courted during their exploratory visits to expand into new geographic regions. The ability to communicate with these individuals on social topics related to their native country displays a level of interest that may separate one organization from the competition. However, this opportunity may never occur unless the individuals are given the opportunity to explore their interests through professional development education.

Strategic Management

The final area of education focus is the strategic management category. Emphasizing both the extension of existing knowledge and a focus on long-term business success, the strategic management category forms the top of the knowledge pyramid. While professionals may argue that topics such as project management, technology, and communications can be acquired through professional experience, strategic management does not follow the same pattern. The strategic management focal points of leadership, technology, economics, and markets are outside the central civil engineering domain. In contrast to discipline-focused topics such as project management and materials, strategic management topics emphasize an outward focus on the business environment in which the organization operates.

The emphasis on upper-level management knowledge is understandably a challenge for many civil engineering organizations. Admitting that the organization requires knowledge in this area may be considered by some individuals to be an acknowledgment that the organization is not reaching its potential. To others, the concept of "staying with corporate tradition" can have great influence in organizations that tout century-long traditions. Others may feel threatened by individuals who are rising in the organization and desire extended exposure to strategic management concepts. And finally, the classic human emotion of insecurity cannot be overlooked when advocating changes in advanced management practices. Many individuals have waited their entire careers to reach upper levels of management. Changing the rules of the game after years of preparing to enter the game can be daunting to those caught in the middle. When combined with the myriad other challenges facing senior managers in their efforts to develop strategic directions, the introduction of new education focal points may be too much for an organization to undertake.

While these concerns are valid and deserve considerable attention, we must not forget why these topics were introduced at the beginning of this text. The civil engineering industry is facing new issues and managerial challenges that will change the manner in which organizations conduct business. Organizations have one fundamental choice: either adapt to change or risk being left behind as a noncompetitive player. While this is a harsh statement, it represents the future of the business environment. Business is a contact sport, and organizations that intend to play the game must assume an appropriate posture. The first step in building this posture is providing the players with the knowledge of the rules and the tactics employed within the game. In the case of global business, these rules and tactics are the focal points within strategic management.

How to Educate

The decision to implement an education plan is an important step forward, but brings organization leaders to the next strategic question: how to implement the plan. In previous years, the choices in education implementation were limited to formal options such as university graduate programs and professional society courses, and informal options such as self-study guides and professional journals. Today, the palette of opportunities is much broader. In the formal category, university programs have expanded to include traditional graduate programs, executive graduate programs, video programs, and Internet-based programs. Additionally, professional education courses have become a profit center for many professional societies as well as individual providers. Finally, an increasing number of organizations are opting to establish corporate universities where graduate-level education is provided by internal or external instructors.

Complementing the array of formal education options is the array of informal education opportunities that are evolving from both technology-based and traditional sources. In the case of technology, the use of satellite communications, Internet communications, personal computers, and video is allowing individuals to obtain professional education beyond the boundaries of the traditional classroom or conference room. Similarly, the introduction of continuing education requirements by an increasing number of jurisdictions has witnessed a commensurate increase in self-study aids such as books, videos, and computer programs. In each case, the tools provide individuals with education opportunities independent of the formal setting of a university or private seminar.

With this array of education options, organizations must balance the needs of individuals against the needs of the organization to advance in the constantly changing business environment. In terms of individuals, the organization must consider the educational objectives of the person, including the desire to advance, the commitment to the organization, the commitment to the education process, the interest in the education topic, and the appropriateness of the topic for both near-term and long-term responsibilities. For the organization, concerns include the strategic direction in which the organization is moving, the return on the education investment, the ability to disseminate knowledge beyond a single individual, and the need to provide education opportunities as both recruiting and retention benefits. In many cases, the balance between the individual's intentions and the organization's desires are going to conflict at some level. However, these conflicts can be managed and balanced if the organization retains the underlying vision and mission statements as the guiding forces for education decisions.

With the vision and mission statements as the guiding forces for education initiatives, the organization is prepared to enter the education process. From the outset, it is important to note that while professional education is a strategic advantage that is essential to the long-term success of an organization, the selection of formal versus informal methods is a decision that must be made by organizations based on individual objectives.

Formal Education Opportunities

The formal education options available to organizations are increasing at a pace that is unparalleled by previous years. The reasons behind this expansion are varied. First, increasing attention by states on continuing education requirements is prompting both private and public institutions to fill these needs. Second, the introduction of new technologies has caught many organizations unprepared effectively to integrate technology into central business operations. Third, the globalization of business has introduced numerous operational issues that most organization members never addressed during their undergraduate or initial professional careers. Fourth, the introduction of new technologies has provided educational institutions with dissemination opportunities that were previously considered outside the realm of economic viability. Finally, the pressure to remain competitive has been played upon by education providers to entice organizations to pursue professional education.

University Programs

The first of the formal education options that deserves strong consideration is the university classroom. Traditionally thought of as a commitment to full- or part-time graduate programs, universities have expanded their professional-level education to include, among others, traditional graduate education, executive programs, individual continuing education courses, certificate programs, and virtual classrooms. Acknowledging that each organization has different needs and different financial capabilities, universities are providing these expanded options in an attempt to retain and expand their graduate student population. Although each of the options is quite different in its delivery, some common elements exist throughout the university offerings:

- Courses have the backing of the university to provide legitimacy to the product
- Courses are taught in a group atmosphere at a main campus or satellite location to provide a structured learning environment
- Instructors are usually full-time faculty or adjuncts with classroom experience to ensure the quality of the product

These common elements set the tone for the education atmosphere of the formal university programs. The issue for an organization is whether or not these elements meet strategic objectives. As illustrated in Table 6-1, formal university programs have

TABLE 6-1
Representative Advantages and Disadvantages of Traditional University-Based Education Programs.

Advantages	Disadvantages
Reputation of university	Time commitment
Stability of programs	Cost
Financial resources	Distance from office
Outside accreditation	Set program
Facilities	Lack of control over topics
Prestige for organization	Isolation from industry concerns
Investment in technology	Long-term vs. immediate impact
Quality of faculty	Difficult dissemination

distinct advantages and disadvantages. In terms of advantages, the stability and reputation of a university can be used by an organization to enhance its reputation. Potential clients may base part of their decision to conduct business with an organization based on the overall qualifications of the individuals. A component of these qualifications is the reputation of the university that individuals within the organization attended. A second advantage of university programs is the ability to plan for long-term education initiatives. Education initiatives can be enhanced when the education goals are linked to formal university program that have a greater likelihood of remaining in place over an extended period of time. Finally, the investment by universities into advanced delivery mechanisms such as videoconferencing, the Internet, and computing facilities provides an infrastructure that private or informal education options may not be able to match. These alternative delivery options provide an organization with the flexibility to remain with a single education provider while matching the needs of individuals with appropriate delivery options.

Although the advantages of selecting a university as the strategic education mechanism are appealing to many organizations, Table 6-1 also highlights some of the disadvantages of the university option. Notable are the commitment and cost. In terms of commitment, many university programs emphasize extended commitments for full-time graduate study, part-time graduate programs, or executive programs. With the pressure of daily operations, individuals must examine their ability to make these commitments. Even the best strategic intentions must be balanced with the realities of business operations and the commitments of individuals to projects that may interfere with a set classroom schedule. In terms of cost, emerging executive programs or full-time graduate programs at prestigious universities may exceed organization resources. In addition to tuition costs, organizations must acknowledge the time an individual must devote to the endeavor. Enrolling in a lesser known school may lower the tuition costs, but it does not respond to the time issue nor does it assist in the strategic objective of enhancing the overall image of the organization through name association.

In summary, universities are continuing to introduce new options for civil engineering organizations to select this option as a formal education alternative. However, these alternatives do not eliminate the time and cost commitment that these programs require. Therefore, the development of a plan that details how this education will benefit long-term objectives is essential prior to sending individuals to these programs.

Private Education Providers

The commitment associated with university programs is one factor that is leading to the increase in professional education alternatives being delivered by private companies or professional societies. Combining shorter commitments with an emphasis on the working professional as the primary target audience, these formal education alternatives are rapidly gaining attention from engineering organizations (Coppula 1998). With over $60 billion spent by organizations on training in the United States in 1996, alternative education programs are being investigated at an increasing rate

(Davy 1998). Similar to university programs, these private education alternatives have common foundations:

· A focus on short courses that last from one to four days
· A focus on industry professionals, or individuals with extensive industry experience, as lecturers
· A focus on "hot" issues as topics to attract organization leaders
· An emphasis on "immediate" solutions to current industry issues

Once again, organizations must determine whether the advantages and disadvantages of these private options coincide with strategic objectives. Similar to the university programs, Table 6-2 highlights some of these common advantages and disadvantages. Notable among the advantages are three items: reduced commitment, a focus on current needs, and knowledgeable instructors. The first of these items: reduced commitment, is a direct response to the concerns associated with university programs. By providing courses that extend over one to four days, private education providers recognize the difficulty an organization has in committing individuals to long-term programs. The second item, a focus on current needs, is the backbone of private education alternatives. With an understanding that organizations are in a constant state of operational focus, these alternatives attract business by focusing on topics that provide immediate assistance to organization members. Topics such as increasing quality control, improving productivity, and responding to seismic design code changes all focus on immediate organization impact. The return on investment is easy to develop, and the results are immediately apparent. Finally, the credibility factor enters into the private education domain through the selection of instructors who have credibility through organization, project, or experience factors. The instructors selected for private education courses are chosen for their ability to attract an audience. In contrast to the university environment, where the university name is attracting clients, private providers focus on attracting clients through experienced instruction.

TABLE 6-2
Representative Advantages and Disadvantages of Private Education Programs.

Advantages	Disadvantages
Focus on immediate needs	Focus on immediate needs
Short-term commitment	Instructor training
Quality of instructors	Lack of customer recognition
Flexibility of programs	Set curriculum
One-time cost	Lack of control over topics
Increased participation options	Reduced resources
Recognized providers	Teaching facilities
Focus on professionals	Access to faculty

Although the previous advantages are attractive for organizations, it is important for organizations to understand the downside of private education. Notable among the disadvantages listed in Table 6-2 is the focus on near-term issues and the training of the instructors. The former issue may be confusing since this was also touted as an advantage. While the focus on near-term issues assists in addressing current operational concerns, it may do very little to assist organizations in planning for strategic objectives. Strategic management emphasizes building foundations for long-term growth and success. A constant focus on professional education courses that address current issues is in direct contrast to these strategic objectives. The second concern, instructor training, is one that organizations may not consider when electing to choose private education options. Although an individual may have significant industry experience, this does not equate to an ability to educate. Education is not equivalent to listening to "war stories." Organizations should not confuse the instructor's reputation with his ability to educate. Examining the teaching qualifications of the individual is just as important as expertise in the topic.

In summary, the proliferation of private education alternatives makes it difficult to generalize a statement about the category. However, in any circumstance where proliferation occurs in a profit center such as professional education, warnings should be heeded in terms of value, quality, and long-term benefits. Many short courses offer extremely beneficial options for civil engineering organizations. The challenge to organization leaders is determining which courses coincide with strategic objectives, which courses represent essential operational information, and which courses represent an attempt by providers to capitalize on business fears and "hot topics."

Corporate Universities

In a trend that is continuing at an ever-increasing pace, a notable number of organizations are turning away from outside education alternatives and crafting their own internal education alternatives. These corporate universities have traditionally been the domains of large organizations such as Motorola, Intel, and Hartford Insurance, where resources are extensive and the number of employees warrants such programs (Watson 1995). However, given the growing dissatisfaction with university offerings and the unknowns associated with private courses, organizations are starting to examine the benefits of implementing in-house education programs (Peak 1997). In-house programs are not new to the civil engineering industry. Senior organization members have long taught safety, plan reading, quality control, and field-oriented topics to junior personnel. However, the focus on in-house programs as an avenue for advanced education alternatives is a new endeavor for the majority of organizations. In this internal arrangement, instructors may be selected from within the organization, contracted privately from universities or other organizations, or coordinated through internal education coordinators. In either case, the focus is on bringing the classroom to the organization. As Table 6-3 illustrates, these programs have advan-

TABLE 6-3
Representative Advantages and Disadvantages the Emerging Corporate University Approach to Education.

Advantages	Disadvantages
Flexibility of programs	Program overhead
Focus on organization issues	Internal competition
Widespread dissemination	Customer credibility
Customization	Initial investment
Convenience	Employee commitment
Reduced course costs	Long-term retention
Continuous updates	Topic selection
Competitive advantage	Instructor qualifications

tages and disadvantages that are both applauded and criticized by industry writers (Roesener and Walesh 1998).

The advantages of internal programs focus on three principal elements: flexibility, focus, and dissemination. The first element refers to the ability to introduce courses at times that are convenient to the greatest number of employees. Rather than sending one individual at a time to a course, courses can be introduced at times when multiple individuals can attend on a regular basis or in a short-course format during operation cycles that match the organization's schedule. Second, by developing corporate universities, the organization ensures that the topics presented are appropriate to the education objectives set forth in the strategic plan. The organization ensures the appropriateness of the courses by taking an active role in determining the content, presentation, and attendance of the course. Additionally, by allowing multiple individuals to attend a course, the rate at which the information is transferred throughout the organization is greatly increased. Rather than relying on a small group to disseminate knowledge, the corporate university presents a viable approach to ensuring rapid dissemination to a large number of individuals.

While the advantages of internal universities are gaining appeal, the disadvantages associated with these endeavors are real and must be examined thoroughly. Of primary concern among these disadvantages is the overhead associated with successful corporate university ventures. Although the individual cost per employee may be significantly less than sending those employees to traditional education ventures, significant costs are associated with the organization of a corporate university. First, a person must serve as the coordinator for these ventures. This person will rapidly find that the venture is not a part-time task, but will evolve into a full-time responsibility. Second, the success of these ventures can also have an unintended negative impact due to the increased demand for new topics. As more individuals become interested in pursuing topics, the question of which topics will be addressed can become a political issue. Third, the credibility of the education can become an issue. To outside clients, the existence of courses taught within an organization may not be equivalent to working with an individual who has received outside education.

Informal Education Opportunities

The proliferation of formal education options has been closely paralleled by a proliferation of informal education options. Fueled by a combination of technology advancements, changes in the business environment, and continuing education requirements, the developers of informal education options have changed the education landscape. Whereas previous generations were limited to self-study guides and book-based materials as informal education alternatives, today's civil engineer has a wide array of options. Books, videotapes, Internet-based courses, and satellite transmissions are all available as alternatives for encouraging employee education. The common thread in each of these options is the ability for any individual to use these materials at any time without the limitations imposed by formal registration procedures, classrooms, programs, or delivery organizations. The materials are designed to allow an individual to pursue education objectives at a self-directed pace in an environment that is conducive to their work and lifestyle requirements.

Self-Study Books

A broad category of options ranging from paperback books to in-depth study guides, self-study books have long been a staple of the informal education category. Unfortunately, this category has been unfairly maligned in recent times due to the rapid increase in "soft" books designed for the business traveler. With titles such as "Dominating the Industry in 5 Days" and "Reinvent Your Business in 4 Easy Steps," these books have been designed for traveling professionals with limited time and the need to feel informed about current business trends. Realistically, the condensing of a semester of information into a two-hour read is not an equivalent education experience. However, this should not be a reflection of the entire self-study book category. The increase in continuing education requirements has resulted in the publication of well-developed educational materials that address core business and engineering topics. These study aids generally fall into one of two categories, directed study guides for continuing education units and individual reading for personal development. The former is characterized by publications endorsed by professional societies such as the American Society of Civil Engineers, the American Management Association, and the National Society of Professional Engineers. These publications are focused on specific topics such as Construction Project Administration, Structural Design, and Water Management and contain review questions designed to reinforce the materials introduced in the text ("ASCE Continuing Education"). The individual receives development credit from the appropriate granting agency by submitting samples of the completed review sections. Similarly, books in the individual reading sector emphasize focused topics, but the presentation is designed for individuals to read the material at a convenient time without the requirement to submit quizzes and review questions. Rapidly developing into a multi-million dollar industry, the publication of professional development books represents an industry influence that cannot be easily dismissed.

In contrast to formal education alternatives, enumerating advantages and disadvantages for informal methods is far less specific. The reason for this difficulty is the variability that exists between the informal options. However, at a general level, informal methods have elements in common that can be evaluated from a practical perspective. Focusing on this general level, Table 6-4 summarizes the advantages and disadvantages of self-study guides. In terms of advantages, one element moves to the forefront in terms of practical considerations: cost. The ability to purchase a self-study guide for a minimum cost provides a strong argument in favor of these education materials. With a constant emphasis on overhead and budgets, saving money by providing employees with self-study guides rather than formal education programs will reflect positively on the financial statements. In this manner, the strength of self-study books is focused less on the knowledge that the books provide and more on the financial benefits they bring the organization.

While the advantages of self-study guides focus on practical elements, the disadvantages of these guides are very tangible. Of particular concern are the unknown qualifications of the authors and the circumstances under which the books were written. In an effort to capitalize on current concerns and topics, publishers will sometimes rush a book to market without adequately focusing on the educational value of the guide. When combined with the unknown qualifications of many study guide authors, the quality of individual study guides must be questioned for their value as a fundamental education component. This is not to state that all study guides are a waste of effort and time, but it is a warning that all study guides are not equal in terms of delivering advanced knowledge.

Internet-Based Education

The second major category of informal education is the use of Internet-based communications to access education materials. What was once considered out of the eco-

TABLE 6-4
Representative Advantages and Disadvantages of the Long-Standing Use of Self-Study Guides for Informal Education.

Advantages	Disadvantages
Demonstration of initiative	Quality
Cost benefits	Author qualifications
Flexible study capability	Depth of topic
Rapid cont. education credit	Employee retention
Numerous topics	Employee commitment
Professional enhancement	Lack of dissemination
Customer familiarity w/authors	Customer credibility

nomic range of individuals and small offices has now become economically viable for every professional office. The continuing mergers between telecommunications, cable, and technology companies are setting the stage for the long-term continuation of this trend. Responding to this trend, a growing number of universities, professional societies, and private companies are making education materials available electronically. In forms such as digitized video, hypertext, animation, and real-time audio, the variations available on an on-demand basis is increasing at a rate that witnesses new innovations on an almost daily basis. However, the advances that these technologies are making raise as many issues as there are electronic options.

- Are individuals selecting Internet education options for the educational value or the entertainment value?
- Is the retention from computer-based modules equal to or better than traditional education methods?
- Is the lack of direct interaction with instructors and other students a detriment to effective learning?
- Is the investment in technology returned through notable improvements in organization development?

As illustrated in Table 6-5, these questions raise only a few of the issues associated with the advantages and disadvantages of Internet-based education. In evaluating these advantages and disadvantages, difficulty exists due to the lack of history associated with these education options. However, this lack of history does not dilute the potential impact of the Internet as an education delivery option. Specifically, the innovation, technology, and uniqueness of the Internet option are the driver that is attracting developers and professionals to investigate this new education alternative. The ability to access education modules from any location and at any time is a distinct advantage over formal education options. Additionally, the ability to distribute live video over Internet connections permits professionals to access lectures without ever leaving the office. Finally, the ability to access modules that can be constantly

TABLE 6-5

Representative Advantages and Disadvantages of the New Emphasis on Internet-Based Education Delivery.

Advantages	Disadvantages
Access to programs	Quality of materials
Cost advantages	Lack of pedagogical history
Numerous employee potential	Questionable providers
Access to distant universities	Entertainment vs. education
Flexible program options	Employee retention
Immediate impact	Instructor interaction
Build on technology investment	Lack of commitment

updated significantly reduces the questions of relevance and timeliness that often accompany traditional textbooks. Combined, these advantages set the stage for continued expansion of the Internet into the professional education domain.

While providing great promise, a major concern of Internet-based education is quality control. In contrast to an emerging belief among many individuals, the existence of information on the Internet does not equate with the accuracy or legitimacy of the information. It is important to remember that any individual can publish information on the Internet with a minimum of training and cost. While this availability has established an egalitarian atmosphere around the Internet, it has also invited numerous questionable sources to profit from individuals with less technical knowledge. A perfect setting for these questionable activities is the professional education domain. Since engineering professionals are facing increased continuing education requirements while having less time to attend formal courses, the incentives exist for individuals to fill this gap with electronic education modules. With increasing momentum to develop additional modules, the inevitable questions of quality, worth, and relevance will not be far behind.

■ The Common Issue—Dissemination

Although both informal and formal education methods provide viable organization alternatives, none of these methods will provide strategic advantage unless the organization aggressively addresses the need to disseminate this knowledge. Educating an individual member is a valuable component of organization development, but it does not serve the purpose of expanding the organization knowledge base beyond individual islands of knowledge. As discussed previously, the strength of an organization is greatly reduced when knowledge is isolated into individual islands. The customer may be unaware that additional knowledge exists within the organization, since exposure is limited to a single island of knowledge. With the development of new technologies and new project delivery mechanisms, organizations can no longer risk individual members not having access to the latest components of the organization knowledge base. As such, the dissemination of knowledge is the issue that binds each of the education alternatives together in a strategic management perspective.

From a strategic perspective, the investment in education must serve two constituencies, the organization and the individual member. For the organization, education initiatives must be designed to enhance the quality of service that the organization provides. This enhancement can only be guaranteed if the organization knowledge base is available to all members. Given this foundational requirement, the organization must approach the need for education with a constant focus on the distribution of knowledge from every individual involved in education programs. For these individuals, the perspective is slightly different. Where education was once considered a bonus required by organizations, economic prosperity requires organizations to provide greater benefits to attract the best and the brightest. Advanced education is becoming one of these standard benefits. However, in becoming a standard benefit, the emphasis on education as a component of organization enhancement has slowly shifted to an emphasis on education as a personal improvement bonus with little emphasis on knowledge transfer. Individuals are becoming accustomed to advanced education as a basic organization perquisite. The challenge to organization leaders is to develop knowledge dissemination mechanisms that retain a focus on individual enhancement while enhancing distribution to the overall organization.

The development of such knowledge distribution plans brings strong opinions from both engineering and education experts (Stata 1989; Senge 1990). The challenge to individual initiative versus organization requirements is a strong polarizer that is difficult to resolve. However, similar to education alternatives, dissemination options extend to both structured and unstructured possibilities. In terms of structured options (Table 6-6), the most common action selected by organizations is the lunchtime seminar series. Instituted by organizations as a required activity with an informal atmosphere, lunchtime seminars provide an opportunity for organization members to transfer knowledge by making short presentations of recently studied topics. The advantage of this structured approach is the individual enhancement that it provides to members making presentations. First, public speaking experience is gained in a controlled, non-client atmosphere. Second, individuals gain a greater sense of worth since they are considered the in-house expert on a new topic. Finally, leadership skills are enhanced as the individual gains experience moderating question and answer sessions. However, as with all options, the advantages associated with lunchtime seminars can quickly be lost when overzealous organization leaders attempt to change the informal atmosphere into an overstructured presentation. Placing formal presentation requirements, extensive preparation requirements, or extensive followup procedures can change the dissemination process into a resented corporate activity.

TABLE 6-6
Examples of Structured Knowledge and Information Dissemination Options and the Requirements to Institute These Options.

Structured options	Dissemination requirements
Internal course	Development of single or multi-session course given during business hours to multiple organization members.
Lunchtime seminar	One-time seminar to organization members during a preannounced lunchtime hour as a standalone presentation or part of a regular series.
Division presentation	One-time seminar to division members during a formal meeting.
Course report	A written report on the course objectives, lessons, and results. Organization member includes critique of the course as well as benefits and competitive advantages.

In contrast to these organized meetings, unstructured knowledge dissemination emphasizes the interaction between organization members through unstructured activities. Both interactive and asynchronous communications can be classified within the unstructured processes (Table 6-7). Of particular interest in this category is the rapidly expanding use of electronic knowledge bases to disseminate knowledge. Utilizing the strength of Intranet technology, electronic knowledge bases emphasize the categorization of knowledge into an electronic format that can be accessed by any member from personal computers. Implemented by engineering, consulting, and manufacturing organizations, these knowledge bases become effective when the organization places an emphasis on the continuous expansion of the knowledge. By giving each member access to the knowledge base, individuals can gain up-to-date information on topics that impact their individual disciplines. Gathered from individual experience, classroom activities, professional journals, and conference presentations, the knowledge bases can rapidly grow to represent the collective knowledge of the organization. While this technology is still in its relative infancy, the potential appears to be limitless in terms of knowledge transfer. Using multimedia technologies, the repositories can include video of project sites, interviews with project participants, electronic CAD models, and other materials that build a context for the captured information.

In summary, the selection of an appropriate dissemination methodology should not be viewed as a right or wrong proposition. A single dissemination method will not be appropriate for every knowledge circumstance. At times, structured methods may be required to ensure every individual is presented with essential knowledge, while unstructured methods may be appropriate to convey incremental changes in the knowledge base. However, the common theme that pervades the entire dissemination

TABLE 6-7

Examples of Unstructured Knowledge and Information Dissemination Options and the Finimal Formal Requirements to Institute These Options.

Unstructured options	Dissemination requirements
Knowledge group exchange	An informal, interactive presentation by a member who attended the course and other organization members who could benefit from the materials.
"Walk around" exchange	One-on-one interaction encouraged by management to disseminate course knowledge between organization members.
Course memos	Summary document sent to interested members that outlines course syllabus, benefits, and major topics.
Knowledge base entry	An on-line entry of course topics that can provide answers for other organization members facing issues addressed by course materials.

process is the need to extend knowledge beyond the individuals who originally possess the information. An organization cannot progress forward unless it moves forward collectively. Individual progress is admirable for individual initiative, but it ultimately creates imbalance in the overall knowledge base. Rebuilding this balance and retaining it throughout knowledge development is a fundamental component of creating a learning organization.

Organization Education Examples

The decision to implement an education plan should not make organization leaders feel like they are the first individuals to undertake such an endeavor. Thousands of organizations of all sizes have implemented formal and informal education policies. However, a few of these organizations have undertaken education initiatives that stand above the rest. Why? The answers are diverse and incorporate several factors. However, at the foundation of each of these efforts is a focus on the long-term education of the individual and a commitment by the organization to provide the best learning environment possible for every individual that joins the team. The following examples spotlight two very different approaches to professional development from two very different organizations and industries.

Arthur D. Little School of Management

One of the most unique corporate universities in the world is the Arthur D. Little School of Management. Founded in 1886, Arthur D. Little (ADL) is the oldest consulting firm in the world (Moore 1997). As a leader in all facets of business consulting, ADL has built a global reputation of providing quality services to many diverse industries. With offices in over 50 countries, ADL has the global presence to address both emerging and classic organizational issues. However, retaining this reputation and presence is not a trivial matter. The need to remain current with emerging technological, political, business, and societal changes requires existing employees to remain focused on lifelong learning, while new employees must be introduced to the ADL consulting perspective. To address these needs, ADL initiated one of the first corporate university efforts. In 1964, the Management Education Institute (MEI) was initiated to provide both employees and outside professionals with graduate-level business education.

By 1971, the MEI program achieved a level of consistency, quality, and demand to support ADL taking the unusual step to petition the Commonwealth of Massachusetts to become the first corporation to grant a Master of Science in Management degree. Given this opportunity, ADL embarked on a corporate education effort to provide all of their employees with a continuous inventory of courses on current consulting topics. By 1976, the ADL School of Management was accredited by the New England Association of Schools and Colleges to become the first accredited corporate university. Focusing on providing both employees and outside professionals with the highest-quality education, the School of Management quickly gained an international following.

Through 1997, more than 3,500 professionals from over 115 countries have participated in the School's programs. Each year, 60 participants from as many as 23 countries enroll in the Master of Science in Management program ("Arthur D. Little"). Focusing on an intensive 1-year program that is similar to the traditional 2 year-MBA curriculum, ADL emphasizes the latest business topics as classroom subjects. The ADL faculty brings a strong applied emphasis to the program by combining internal employees with outside consultants and university professors. Responding to the needs of a corporate clientele, the applied focus provides consulting professionals with a commitment to education while understanding the practical requirements of an operating business. Finally, by making the decision to undertake their own education initiatives, ADL retains complete control of the curriculum, educators, and students associated with the program. Although expensive, ADL is committed to the internal university concept as a strategic component for long-term competitive advantage.

U.S. Army Corps of Engineers

Diverging from the corporate consulting objectives of ADL is the U.S. Army Corps of Engineers. Formed with both military and civilian engineering objectives, the Corps of Engineers incorporates a broad range of responsibilities including water and wetland management, military facility construction, and operational support. Given this broad objective, the need for diversity in education has become apparent to the Corps over many generations. Responding to this need has become a central component of Corps operations. Focusing on lifelong learning as a direct complement to career advancement, the Corps combines internal and external education into a comprehensive package.

In terms of internal education, the Corps has initiated extensive education programs in all areas of professional activity. From contracts to advanced regulatory compliance, the Corps has established short-term courses that are offered throughout the country to personnel. Focusing on a response to immediate needs, the courses provide an intense introduction to a specific topic that is required to successfully complete assigned tasks. Recent course offerings including contract administration, procurement, and environmental analysis provide examples of this focus ("Corps of Engineers"). In addition to these short courses, the Corps also has a longer-term education option for officers progressing along the career path. To ensure that these individuals are provided with the greatest preparation prior to being assigned new command responsibilities, the individuals are required to attend 2–20-week command courses at the U.S. Army Engineer School at Ft. Leonard Wood, Missouri. With topics such as Engineering Skills and Leadership Training, these courses are designed to augment an attendee's current education with knowledge that is relevant to their new position. Through this process, the Corps ensures that each member of the organization receives standardized education opportunities as they enter a new position of responsibility.

With this investment in internal education programs, many organizations would prefer to limit their employees to these educational opportunities. However, the Corps of Engineers breaks from this mold by encouraging its officers to apply for outside education opportunities. Typically granted in their sixth to eighth year of service, the Corps provides officers with a 12–24-month time period to attend graduate school in a field related to their specialty. With heavy competition for these opportunities, the Corps balances its internal education emphasis with an acknowledgment that external graduate education is an equally important component of professional development. Sending students to the highest-ranked universities in the United States, the Corps makes a statement both internally and externally that expense is not a barrier to educating the best engineers in the field. Additionally, the Corps creates an atmosphere that promotes lifelong learning as a natural extension to the career advancement process. Combined, these commitments represent an investment in education that is not an afterthought to organization procedures, but rather acts as a central element of strategic management.

Lessons Learned

These examples represent two cases of very large organizations with large education budgets taking a close look at their education requirements and addressing them with a focused plan. Each organization has made a commitment to education as a central element in their strategic development. However, each organization has approached the issue from slightly different angles. ADL has retained all education activities internally, while the Corps of Engineers has divided the responsibility between internal and external efforts. Both have retained immediate needs within the control of in-house instructors, but the Corps has allowed individuals to pursue longer-term knowledge at external universities. In each case, the education effort has been analyzed for both its immediate and long-term impact. The lesson here for civil engineering organizations is that a large education budget, such as those used in the previous cases, is not a prerequisite for education activities. However, a commitment to the program and a focus on its goals and objectives is essential to overall success. Allowing a few individuals to obtain professional education without a strategic objective in place is equivalent to no education plan. Education is expensive for any organization, but for a smaller organization with limited resources, wasted education is an economic hardship that is not easy to absorb.

■ Justifying Education Costs

In conducting research for this book, a common refrain was given by civil engineering organizations: "How do we economically justify the education expense, and where is the return on investment?" Although organizations would prefer a clear response to this question, one does not exist as it does for equipment or other tangible investments. In the cases of these tangible investments, a clear investment return can be calculated based on factors such as productivity improvement, depreciation, and net profits. Unfortunately, education investments do not fall into such a clear return-on-investment category. Education is an intangible investment. Rather than a purchase of a physical object, an organization is investing in knowledge when it develops education opportunities for its members. Unlike a computer or soil-boring equipment, knowledge cannot be directly attached to a financial analysis. Justifying an investment in education requires organizations to take a nontraditional approach to return on investment. Specifically, an organization can adopt either a strategic objectives approach or an opportunity costs approach in order to justify the cost.

Strategic Objectives Approach

In the strategic objectives approach, retaining a focus on strategic objectives is the primary requirement when justifying education expenses. By arguing that an organization cannot meet long-term objectives without the knowledge of changing and emerging business issues, education plans are justified as the foundation for achieving strategic objectives. With this focus on the long-term benefit, justification based on strategic objectives emphasizes the need to have a strong organization knowledge base as a foundation to retain competitive advantage. In addition to the traditional requirements to remain current on core competency topics, the strategic objectives approach expands this focus to include knowledge of emerging issues, minimization of knowledge breaks, and preparation for personnel replacement.

Knowledge of Emerging Issues

At a strategic level, the single greatest threat to an organization is being left behind when a new business issue changes the competitive environment. Examples of these issues include computerization, alternative delivery methods, and emerging markets. In each case, the organizations that identify the issue early in its development have the greatest amount of time to prepare a response and implement it. Organizations that elect to postpone action, or miss the issue completely due to nonexposure, are forced to respond reactively once the issue takes center stage in the business environment. Working from a reactive stance, these organizations may never regain the ground lost to organizations that took a proactive approach to addressing the new issue. To illustrate this case, consider two hypothetical competitors in the transportation consulting field, Midwest Consulting and Northeast Consulting. Midwest Consulting is a 20-year-old firm with 50 employees who specializes in the design of highways. A successful firm, Midwest personnel enjoy a strong education benefit (paying 100% of tuition costs up front for each course) that allows them to pursue part-time graduate work as well as participation in professional education seminars on a regular basis. Many of the personnel working for the firm focus on these benefits as a factor in their decision to remain at Midwest. Recently, many of Midwest's public clients have become increasingly interested in alternative procurement strategies such as design–build and design–build–operate as methods for working with Midwest. A similar interest exists with Northeast's clients. Slightly larger than Midwest with 75 employees, Northeast is only 10 years old, but was founded by two partners with extensive highway design experience. However, Northeast has a different approach to education. While still an advocate of professional education, Northeast believes that employees should make a commitment to the organization in return for education opportunities. As such, Northeast employees receive a 50% reimbursement up front for part-time graduate courses, but only 10% a year for the next 5 years for the remaining cost of the course. In this manner, employees must remain at Northeast for 5 years if they desire full reimbursement for education costs. In addition, Northeast prefers to have employees give seminars for most topics rather than sending employees to outside seminars. Northeast executives believe that this is a more cost-effective method for continuing education.

The different approaches to education between Midwest and Northeast are representative of many civil engineering organizations. Midwest represents the more aggressive approach, while Northeast represents the conservative approach favored by many traditionalists. The difference between the two approaches may be difficult to see during daily operations. Although Midwest has more employees taking advantage of the educational opportunities, everyday operations are very similar in the two organizations. However, the difference does appear in the response to the emerging issue of alternative procurement. In the case of Midwest, senior organization members have been aware of the new approach, as they have attended seminars addressing the emergence of the issue. When approached by clients on the possibility of adopting such an approach, Midwest is able to respond with an analysis of the benefits and drawbacks of changing from a traditional contracting arrangement. Furthermore, the organization is able to present a cost analysis of the alternatives in the context of an informed client presentation. In contrast, Northeast returns to the office from such a request with a heightened state of urgency. Since fewer Northeast employees take advantage of the education benefits due to the 5-year reimbursement policy, minimal formal exposure has been obtained to outside speakers. As such, Northeast employees are aware of the issue, but are ill prepared to make an informed client presentation. The organization is left to learn quickly in a reactive mode, with little time to draw conclusions on benefits and drawbacks. Where once Midwest and Northeast were on equal terms, Midwest now obtains a competitive advantage that will be difficult for Northeast to overcome.

This example is a representative case of proactive versus reactive response to emerging issues. Whereas Midwest was able to leverage its investment in education to develop a proactive response to the emerging procurement issue, Northeast was left to develop a reactive response since it did not possess the internal knowledge to address the issue. This difference is not directly reflected in a single line on the balance sheet. However, it does appear as a component of overall profits. The organization that invests in education and obtains awareness of emerging issues as a result of that investment will witness greater overall performance. Specifically, the organization that makes this investment will obtain a competitive advantage by placing itself in a position to gain clients through greater access to information, increase productivity through access to new operating methods, and increase profitability by entering new markets prior to oversaturation by competitors. Each of these items has a revenue amount associated with it. For example, a 5% productivity improvement in design operations could reduce project costs by $50,000 on a $1,000,000 project. This $50,000 is directly reflected in the profitability of the firm, and thus a direct connection to education investment.

Minimization of Knowledge Gaps

Continuing a focus on customer interaction, the second strategic justification for education investment is the minimization of knowledge gaps. Similar to the previous category, the minimization of knowledge gaps emphasizes using education to reduce deficiencies in the organization knowledge base. As business development professionals advocate, winning a client is often based on the ability of the project presenter to generate a feeling of trust and relationship (Shapiro 1988). This trust is often a result of the client believing that the presenter possesses knowledge in their field. This basis is tenuous at best. The slightest reason to doubt this standing can rapidly destroy the trust that has been building between the service provider and the client (Jones and Sasser 1995). Understanding this tenuous position, organizations must prevent the dissolution of client trust and retain clients that have been selected as having strategic importance. A basic element in achieving this goal is ensuring that every member of the team that has contact with the client reflects the overall knowledge of the organization.

To justify this argument, it is important to remember that clients have a one-to-many relationship with an organization (Fig. 6-4). In this relationship, a client may speak to a senior person, a middle manager, a project manager, or a field or design engineer at different stages of the project. To the client, each of these individuals is a representative of the organization. Although each is expected to have knowledge at different levels, each is expected to have a command of the knowledge within their domain. Therefore, whether it is the person at the top, or the person at the bottom, strategic relationships can be permanently damaged if any of these individuals fails to demonstrate command of an expected knowledge area during interaction with clients. For example, a client working with Northeast Consulting to design alternatives for decaying urban infrastructure may be concerned about new wetlands regulations that reduce the amount of allowable wetlands area that can be reduced without a complete environmental impact statement. While this regulation is a new development in environmental policy, the client expects Northeast to be aware of the change. Therefore, a scenario may exist where the client is discussing the design with the project manager; the client mentions the regulatory change to the project manager with a concern for the impact on the new design. The project manager hesitates with a response and informs the client that he will check with their environmental group to get an answer. Although this may be a logical response, the trust previously de-

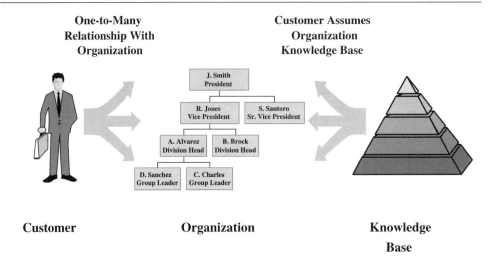

Figure 6-4 Customers have a one-to-many relationship with an organization. As such, each member of the organization is a reflection of the overall knowledge base that resides as a foundation for organization business initiatives.

veloped with the client takes an immediate blow. This is due to the fact that the client perceives the project manager as lacking in knowledge. Since the client was aware of the provision, it is reasonable for the client to assume that the organization he is hiring to perform work has a greater knowledge base in the contracted area.

In fact, the environmental specialists within Northeast may have the information requested by the client. However, since the project manager was not aware of the issue, the client perceives a gap in the organization knowledge base. Whether perceived or real, the gap becomes real to the client when it is identified. Preventing this circumstance is a primary goal of education initiatives. Every member of the organization must be educated on issues that are known by the organization to be of concern to clients. Whether it is regulatory, construction, or design, every employee should be provided opportunities to obtain knowledge if it is relevant to their interactions with clients. The lack of this knowledge will be reflected in the financial performance of the organization. Once a client begins to lose trust with an organization, the chances for return contracts are greatly reduced.

Preparation for Employee Replacement

Succession—the word often initiates bouts of anxiety, restlessness, and concern for employees, investors, and senior organization members. While generally discussed in terms of organization leaders, the topic of succession applies to individuals at all organization levels. Whether it is a design engineer or CEO, planning for the replacement of that individual is a priority item. To paraphrase one comment from a senior member of a large civil engineering organization, the organization is very stable as long as everyone stays where they are. Unfortunately, people do not stay where they are forever. This is not only unrealistic; it also leads to very unfulfilled employees. Therefore, an organization must plan for individuals leaving current positions either through promotion, lateral moves, or leaving the organization completely. A central part of this planning process is preparing the individuals who are in position to make this replacement.

Preparing individuals to advance encompasses several issues, including leadership training, organizational knowledge, and increased responsibilities, but at the core of these issues is education. An individual is not adequately prepared fully to execute the responsibilities of a new position unless they have the background for entering that position. For example, returning to the Midwest Consulting organization, a senior project manager has recently been promoted to Vice President of Engineering. Although this promotion is a significant increase in responsibilities, Midwest is confident about the individual's ability to perform the functions based on the extensive effort the individual has made toward professional development. Participating in professional development courses, graduate programs, and seminars, the new Vice President has built the knowledge base required for the new responsibilities. Correspondingly, the individuals below the new vice president are equally qualified for selection as a new Senior Project Manager. By making education a priority as well as an understood requirement for promotion, Midwest has created an environment that develops individuals for career advancement.

Civil engineering organizations need to focus on establishing similar environments. After establishing a strong technical foundation, the organization should focus members on obtaining advanced education in organization and strategic management functions. Recognizing the need to enhance an individual's ability to address long-term issues in addition to daily operational issues, education programs should be structured to address both organization and individual needs. As part of this education, the organization can focus on preparing individuals at all levels to progress through their careers. The lack of these educated successors will have repercussions throughout an organization. The time that it takes to educate successors is time that is lost due to reduced productivity. This reduced productivity is in turn reflected by reduced profits. In this manner, the learn-by-doing approach becomes increasingly destructive as individuals at senior levels are expected to make decisions impacting the entire organization.

Opportunity Costs

The second approach to justifying education costs diverges from the objective-based arguments to a focus on monetary ramifications. However, rather than a direct justification, the monetary justification presented here is based on the concept of opportunity costs. In contrast to a focus on returning benefits from a tangible purchase, opportunity costs are those revenue streams that are lost due to an organization placing assets into alternative projects. A concept introduced by economists (Mansfield 1985), opportunity costs must become a component of civil engineering vocabulary if organizations are going to achieve their stated objectives successfully. This relates to education in that every time an organization chooses to delay implementation of an education initiative in favor of alternative expenditures, the organization is opening itself to opportunity cost ramifications.

To understand this, we return to the Northeast Consulting example. As stated earlier, Northeast takes a conservative approach to education, offering education opportunities, but placing enough restrictions as to create a negative atmosphere for employees pursuing such opportunities. As a result of this atmosphere, fewer and fewer Northeast employees are pursuing new education opportunities. One such opportunity that has been overlooked is a series of recent seminars on market expansion. In these seminars, business development experts emphasized that over 200 billion dollars was spent in 1997 for architecture and engineering services. As part of this expense, an increasingly large amount is being spent on program development services such as site selection, value engineering, and permit compliance. The instructors spotlight the fact that large infrastructure projects can absorb costs exceeding 100 million dollars in these early stage costs. Furthermore, the instructors emphasize that this segment of the project lifecycle is wide open from a competitive business perspective. No apparent leaders currently exist, while owners are increasingly looking for assistance. Of greater importance is the increased profit margin that organizations are receiving for performing these services due to the perceived importance of the task by owners. Having chosen not to attend these seminars, Northeast is aware of the trend from professional journals, but has not focused on it as a strategic opportunity. Subsequently, Northeast fails to present a proposal to Metro Communications to assist in the development of a new $60 million cellular communications center.

The decision to pass on the Metro project is not directly seen on the year-end balance sheets. The loss of a $3 million contract to perform engineering program consulting services is not evident, since Northeast did not absorb any costs in developing a proposal. However, this same project takes on a completely different perspective from an opportunity costs perspective. In this perspective, the organization must examine what is lost compared to the costs that it placed in other projects rather than pursuing the proposal initiative. In this case, the cost of pursuing the project contains two parts, the education cost and the proposal preparation cost. The former of these costs can be roughly calculated by taking the cost of the seminar, $800 per person for two days. Most organizations attending the seminar sent two attendees; so a total registration cost of $1,600 would be required. In addition to this direct cost, the salaries of the two attendees must be absorbed for an additional $1,200 for two senior organization members. Finally, since the seminar was located at conference center 50 miles from the office, Northeast absorbs another $100 in expenses. The total of these direct costs is $2,900. Added to these direct costs would be the proposal preparation cost. Estimated at 0.25% of the total contract value, the proposal preparation cost for Northeast would be $7,500. When combined with the education costs, Northeast would have absorbed $10,400 to complete a proposal for Metro.

Where many civil engineering organizations make a mistake is looking at these costs and stopping due to the tangible costs that will be reflected on the balance sheet. What is overlooked is the opportunity calculation that should accompany the costs. In this case, Northeast has the potential to win a $3 million contract. Although some risk analysis must be performed to determine the potential of success, the fact that the course instructors emphasized the openness of the field cannot be overlooked. Neither can the opportunity associated with being one of the first organizations to enter a rapidly emerging market. Placed in a return on investment perspective, the $10,400 investment represents only 0.35% of the potential revenue from the project. However, Metro does not have an opportunity to obtain this revenue, since it made the decision to invest the human and monetary resources in alternative initiatives.

The difficulty in generalizing this example from a single instance to a standard methodology is not an inability create a structured procedure. Rather, the difficulty is focused on the fact that opportunities may never be quantified by the organization since the opportunities cannot be quantified unless they are known. For example, in the previous case we were able to develop a quantitative result based on the knowledge that the Metro project was a viable opportunity. However, an organization may not have this advantage, since they may not be aware of the opportunity in the first place. This creates a dilemma for organization members who desire to use a quantitative approach to justify education. Without knowledge of potential revenue streams, it is impossible to create a grounded financial argument. Therefore, the alternative for organization members is to analogize from previously completed projects to potential scenarios. For example, an argument to justify an education initiative focusing on new communication technology advances could be based on examples of other organizations leveraging technology to expand revenue streams. Based on this analogy, an argument can be made that equivalent revenue gains could be achieved if the organization pursued an equivalent education path. Equally persuasive is the argument that the organization will never be in the position to make this evaluation without the knowledge required to evaluate opportunities. While traditionalists may balk at this due to up-front education costs, the long-term strength of the organization and the ability of the organization to compete in changing business environments are dependent on establishing the knowledge base required to evaluate emerging opportunities.

Summary

The development of a strategic education initiative requires organizations to plan, justify, and implement in the same manner as a traditional project. Investments in time and money are both required to provide the greatest opportunity for success. However, rather than focusing on the resources required, organization leaders must focus on the potential negative results that may emerge without the investment. Whether it is a delay in achieving a strategic objectives or a penalty in opportunity costs the failure to integrate education into long-term strategy will be reflected on end-of-year financial statements.

Next Steps

The education plan is in place, the internal strategic objectives are set, and the vision is established. Now the organization is prepared to move from internal analysis to external competitive positioning. Preparing to move from this internal perspective requires an additional foundational element. Specifically, an organization needs to establish an economic foundation for pursuing strategic objectives. Building upon the vision, competency, resource, and education components, the next step is the development of a sound economic foundation to support new competitive ventures. Chapter 7 introduces the process required to build this foundation.

DISCUSSION QUESTIONS

1. Justifying education costs is a central issue for many civil engineering firms. As a result of this concern, many organizations elect to follow an education path that ensures competency in discipline fields. Given that additional management knowledge is required in today's professional environment, what mix of traditional and advanced education should be advocated by an organization?

2. Formal and informal education providers are in a constant state of competition in the professional education market. Each of these options has advantages and disadvantages for the civil engineering organization. Is there a single issue, or group of issues, that should guide an organization in selecting a formal or informal education method?

3. The move to corporate universities has received significant attention from management writers. However, much of the attention in this area has been on efforts of large organizations such as Motorola. What can smaller civil engineering organizations learn from these larger organizations that can be translated into a civil engineering corporate university?

4. One education difficulty in civil engineering organizations is the dissemination of knowledge to employees who spend considerable time in the field. What type of dissemination program could be developed to overcome this difficulty?

5. Two perspectives are presented in this chapter to justify education costs. In the minimization of knowledge gaps perspective, the relationship and contact with customers is emphasized as a driver for developing an education plan. The danger of an organization member at this level lacking knowledge is apparent, but what advantages can be developed through this same scenario?

Case Study	**Integrating Education into the Organization Environment**

The development of informal and formal education methods for civil engineering organizations is receiving extensive attention from all sectors of the industry. From associations to private companies and academic institutions, the reevaluation of education needs for civil engineers is a primary topic. Of particular concern in these discussions is the need to integrate education plans into existing organization environments. Specifically, the need to introduce new education opportunities to address emerging business environments is a significant challenge. Gilbert Southern Corporation provides an example of the issues associated with this challenge. Gilbert

Southern Corp. is a subsidiary of Kiewit Construction Group, Inc., consistently ranked as one of the 10 largest construction companies in the United States. Focusing on heavy civil construction work, such as roads, bridges, utilities, and airport runways, Kiewit had revenues of $2.8 billion in 1997. A 110-year-old company, Kiewit employs over 3,000 salaried employees and over 14,000 craft workers. Founded in Omaha, Nebraska, Kiewit now has offices throughout the United States and Canada as well as an international presence.

Operating as a decentralized organization, each district office is responsible for the operating results for a particular geographic area. Gilbert Southern is one such district office. Operating in seven states out of offices in Atlanta and Miami, Gilbert Southern has revenue in excess of $100 million and approximately 200 salaried personnel. Working in one of the fastest-growing regions in the United States, Gilbert Southern is leveraging Kiewit's extensive experience and personnel to compete for the rapidly increasing number of infrastructure projects being undertaken by local and state entities throughout the Southeast.

The Players

A central component of Kiewit's business approach is a decentralized management style. Each district office operates under the direction of a District Manager who is responsible for the operating results of that region and is equivalent to a President position in an independent office. The District Manager for Gilbert Southern is Steve Carlyle. Operating under the District Manager are several individuals who are responsible for specific areas of the operation. One such individual is Bob Berry, the District Engineer, or Vice President in a traditional business parlance. Berry's background is solidly based in civil engineering and heavy construction. Moving to Gilbert Southern 6 years ago, Berry has taken an active role in several areas of the organization, including organization education. Focusing on the Kiewit strength of employee development and mentoring, Berry has emphasized the continued development of recruiting and education programs within the Gilbert Southern region. Given the growth and economic patterns, Berry and his professional staff believe that recruiting, training, and mentoring individuals is the key to increasing market share within the region.

The Issue

As a leader in the construction industry, Peter Kiewit has also taken a lead in establishing education programs for its employees. Emphasizing internal advancement, Kiewit aggressively searches for individuals who demonstrate initiative in exploring new ideas and seeking better solutions. As part of this effort, Kiewit has instituted a broad training program that includes classroom and field instruction, training materials, and correspondence courses. Components of the training program include:

- Supervisory conference—A 5-day conference covering supervisory skills, safety, equipment, purchasing, project planning, and project controls. It provides newer employees an opportunity to learn more about Kiewit policies and procedures at an early stage in their careers.
- Superintendents' school—A 9-day program that focuses on the information required to manage a construction job. Employees are presented with subjects including leadership, maintenance, labor relations, and quality control.
- Management seminar—An intensive 3-week seminar, the management seminar focuses on management methods and operations. Kiewit policies and operating procedures, management practices, bidding procedures, and contract administration are presented to managers preparing to undertake increased responsibilities within the organization.
- District training—In addition to the centralized training efforts, individual district offices such as Gilbert Southern offer employees specialized training courses covering topics such as safety, affirmative action, estimating, equipment, and business manager training.

With a belief that these programs are an integral component of the organization's success, Kiewit includes development of staff as a component of District Manager evaluation. However, this education emphasis has developed into an area of concern for the Gilbert Southern office in general and Bob Berry specifically. Berry's concern focuses on the fact that the organization-training program needs to become a formal part of each employee's career development. Although individuals are strongly encouraged to attend training courses, and staff mentors encourage training, no requirement currently exists for individuals to attend training courses. As a result, several consequences result. First, the pressure of daily operations can supersede the need for individuals to attend training classes. Second, although District Managers and senior staff are evaluated on their development of employees, no formal performance measures are in place at the organization level to measure the progress of employees attending courses. Third, with the focus on informal training evaluation, a set budget for training at the district level is not in place. Therefore, although the organization courses such as those described above get significant attention, the individual needs of the districts are not given as high a priority. Finally, while Kiewit strongly espouses a decentralized management approach, the strong emphasis on centralized training development has resulted in district personnel, such as Bob Berry, believing that education needs are not being addressed in a timely manner.

The Proposal

Senior district personnel have brought the training issue to the attention of corporate Kiewit personnel. However, as often occurs in an organization as large and as geographically dispersed as Kiewit, bringing an issue beyond this stage can be difficult. As a result, Bob Berry is taking the initiative to develop a proposal that will transition the already strong voluntary Kiewit training program into a mandatory program that will serve as a centerpiece of organization development activities. The core of Berry's proposal focuses on the formalization of the existing training program. Berry proposes that the new training program address three critical issues:

1. Performance measurement—If training is going to form the core of the Kiewit employee development program, then employees should have formal training requirements. Whether it is based on time spent in the organization or position of responsibility, every employee should be required to attend courses based on a formal training plan.

2. Time allocation—The implementation of the formal training program requires organization members at all levels to acknowledge that time must be allocated for individuals to attend courses. As such, it should be policy that education courses are a priority for employees and attendance at the courses should override everyday operational responsibilities unless circumstances dictate otherwise.

3. Budget allocation—Training requires funding. To facilitate the formal Kiewit program, training should become an overhead line item in each district's business plan. The specified budget should cover expenses for employees to attend in-house programs, outside seminars, and the development of custom programs for region-specific issues.

With these three issues as a foundation, Bob believes that Kiewit can elevate its already strong training program to a level that would be a model for the entire industry.

The Issue

The Berry proposal represents the initiative and aggressiveness that many organizations look for in employees. The development of an education plan is an initiative that could impact the entire organization. However, similar to many organizations, the difficulty to propose a major initiative from the bottom up is significant. Although the plan appears solid, Berry is faced with the difficulty of selling corporate headquarters with a plan that was developed from a regional office. While Kiewit does emphasize a decentralized management structure, issues such as training are primarily controlled from the central office. As such, decisions in these areas become difficult to influence from outside the central office structure. This difficulty has brought Gilbert Southern to a point of decision. Bob Berry's responsibilities encompass a broader area than just education. As such, the organization cannot afford to have Berry focus all his efforts on the education issue. However, abandoning this initiative leaves Gilbert Southern with the same training issues that it had prior to Berry's efforts. Whether the impact is positive or negative depends heavily on the decisions made at both the regional and corporate level.

QUESTIONS

1. Placed in Bob Berry's position, how would you proceed with the training proposal?

2. Placed in Steve Carlyle's position, how would you use your position to enhance, encourage, or dissuade Bob Berry's efforts?

3. Looking at the current training program, would you incorporate additional levels of training into the program for areas such as senior management responsibilities? If so, how does this complicate the existing proposal?

4. Given the opportunity to present this proposal to senior Kiewit managers, what would be your justification for the organization investing in the proposal?

PART III

EXTERNAL STRATEGIC MANAGEMENT ISSUES

FISCAL MANAGEMENT 7

Putting A Focus On
Long-Term Bottom Lines

What financial risk analysis is in place to forecast and protect the organization from economic swings?

As the dawn of the twenty-first century appears, a new fiscal reality is appearing for the civil engineering industry. The evolution of the global economy, and the ripple effects that occur from changes within the economy, are drawing individual projects into a larger web of economic interaction. Consequently, the ability of an organization to isolate a project from changing economic conditions is diminishing as global economic changes impact resource price and availability. Responding to this global economy requires organizations to broaden their economic horizons and adopt economic practices introduced from outside industries. The global export outlook of manufacturing, the global resource outlook of professional service industries, and the global communication requirements of information technology organizations provide benchmarks for study. Each of these innovations brings a component of the new economic reality to the civil engineering industry. Beyond accounting, civil engineering organizations require exposure and understanding of finance, economic forecasting, and corporate financial analysis. Through this exposure, financial analysis techniques, including the economic evaluation of clients, economic forecasting to survive fiscal cycles, and the economic analysis of potential partnering opportunities can be incorporated into operational procedures. Given this opportunity, the strategic manager can address stock market fluctuations, currency exchanges, and global interest rate changes as they affect the organization.

In this chapter, the development of an expanded fiscal management strategy is examined in the context of economic stability. This chapter emphasizes the need to develop expanded economic outlooks as building blocks for establishing an economically secure organization. The roles that strategic financial analysis can perform within strategic management functions are explored as the basis for expanding the financial outlook of civil engineering organizations. Combining existing fiscal practices with expanded economic analysis, this chapter emphasizes the multiple opportunities available to incorporate economic analysis as a basis for expanding strategic positioning and competitive positions. However, establishing the impetus to implement this long-term economic focus sets the challenge for this chapter—*to examine the strategic economic opportunities within the organization and set the economic objectives required to create long-term financial stability.*

■ The Three Perspectives of Financial Numbers

The development of a broader financial perspective within civil engineering organizations is based on a single overriding factor, the need for organization leaders to examine financial data beyond the boundaries of current projects. While project data continue to remain a critical component of the overall organization health, project-based finances should not be exclusively relied upon as indicators of long-term direction. Specifically, the success or failure of an individual project may have little relation to the strategic objectives laid out as long-term milestones. For example, an organization that elects to focus on health care facilities as a central component of its market sector may lose money on its initial foray into the field. However, this loss should not be equated with the long-term potential for succeeding in this sector. Rather, just as in every activity, a learning curve exists, and sometimes a loss is required to learn the nuances of specific market sectors. Observed from a strategic perspective, this loss could be entirely justifiable if it leads to a foothold in the market sector and enhanced opportunities for long-term asset gain for the organization.

The difficulty with adopting this broader financial focus once again returns to the pressures of project operations. With a focus on man-hours and client billing, it often becomes difficult for organization personnel to observe the broader ramifications of an individual project. The budget for that individual project becomes the ultimate evaluation mark for success. If the project meets the budget, then the project is considered a success. If the project exceeds the budget, then questions are raised regarding who is at fault and how to prevent the problem from recurring. While this perspective is essential to preventing the organization from losing money on a consistent basis, it is also too narrow in its emphasis. Transferring the focus of the project to a broader strategic perspective, the individual budget is only one aspect of the project's value. On a broader scale, an organization moving to a strategic management focus needs to examine financial concerns from three distinct perspectives: the project, the organization, and the overall economy. These three perspectives represent three distinct spheres of concern, a project concern, an inward organization concern, and finally, an outward economic concern (Fig. 7-1). The following sections review the fundamentals of these financial spheres as a basis for developing an expanded approach to incorporating finances into strategic management and strategic decision making.

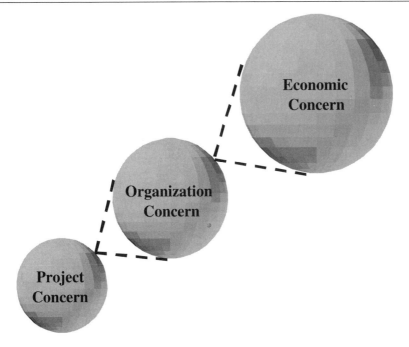

Figure 7-1 Civil engineering organizations must view financial numbers from a project, organization, and overall economic perspective.

■ Project-Based Finances—A Project Perspective

The first sphere of financial interest for the organization is the project perspective. This traditional finance model emphasizes the planning and control of finances on individual projects as the cornerstone of organization health. Civil engineers have traditionally been taught that the organization with the greatest capacity to plan and control project finances will ultimately emerge as the strongest player within a given market sector. As a result of this emphasis, project planning and control receives extensive academic and professional attention by civil engineers. In particular, the estimating and budget components of project planning receive extensive attention. With this extensive publication base available to civil engineers, this topic will not receive extensive coverage in this forum. Rather, the reader is referred to one of many excellent publications on project planning including Halpin and Woodhead (1998), Peurifoy and Oberlender (1989), and Kerzner (1998) to review in-depth planning procedures.

In contrast to the extensive focus on planning and controls, the value of project funding is often undervalued in the civil engineering profession. A common perspective often heard in the classroom is that project funding is the purview of the project developer and outside the responsibility of the civil engineer. The perspective is reinforced by the fragmentation witnessed within the industry that often relegates the civil engineer to the role of engineering specialist and constructor rather than an integrated component of the project development team. However, the strategic emphasis on service expansion is inexorably changing this role to one that brings civil engineers into the early stages of the project development process. Consequently, the project financing component will increasingly become important as civil engineering organizations adopt nontraditional project roles, including project developers, project operators, and members of own, operate, and transfer consortiums. In these roles, civil engineers require broader knowledge of the financial data required to obtain project financing from individual, institutional, and public investors.

Private Funding

In the private domain, financial data supporting the efficacy of a project are a fundamental component in obtaining individual and institutional loans. Although investors are interested in development projects due to the potential for profit, they are also aware of the risk involved with speculative building projects. Whether it is a commercial complex, residential complex, or private energy generation project, investors are taking a calculated risk when investing in the project. The responsibility of the civil engineer as project developer is to develop a financial picture that validates the project and minimizes the risk to the investors. As illustrated in Figure 7-2, this validation process requires the developer to present data on potential revenue, expected costs, and forecasted demand.[1] Is the commercial project illustrated in Figure 7-2 an acceptable risk for an institutional lender? The projected income from tenant leases anticipates a positive cash flow of 32% on an annual basis. However, the assumptions that led to this development are heavily dependent on the continued demand for the commercial space. Fluctuations in the economy and the accompanying demand for commercial space could significantly change these expected numbers.

Assuming that the demand numbers remain stable, the project appears viable. However, a lending institution typically limits a commercial development loan to 75% of the total economic value of the project. Given the $2,500,000 value of the proposed project, this translates to a maximum loan of $1,875,000. At an interest rate of 7.75%, this commits the project developers to a monthly payment of $22,257. The developers must demonstrate an ability to make this payment over an extended period of time. Specifically, the lending institution will be concerned about the ability of the project developers to meet the overall project requirements beyond just loan payments. If the developer is unable to meet maintenance, util-

---

[1]Although not illustrated in Figure 7-2, other financial considerations impacting an investment decision include components such as the capitalization rate and tax liabilities. The reader is advised to review an introduction to finance for further reading on these matters, including Halpin and Woodhead (1998).

Estimated Revenue Detail	Amount
Market rent for proposed property:	
20 Standard Office Suites @$750/mo.	180,000
30 Deluxe Office Suites @$1,100/mo.	396,000
Estimated Annual Income	576,000
Less vacancy factor of 8%	(46,080)
Adjusted Gross Income	529,920
Less Estimated Expenses @26%	(137,779)
Net income before debt service	392,141
Economic Value	2,500,000
Loan/value ratio	75%
Requested Loan	1,875,000
Annual Loan Payments @7.75% Interest	267,079
Return Margin	32%

Figure 7-2 Private financing requires developers to demonstrate financial viability to investors, including the capacity of the project to return a margin that is sufficient to cover all expenses and retain a monetary buffer for the developer.

ity, and tax requirements, then the ability to retain tenants is questionable. Therefore, the developer must demonstrate a capacity to meet the complete range of project commitments.

A second component of this financing issue is the ability of the project developers to finance the construction of the project. Although a long-term loan as described above is provided by the lending institution to the owner based on the economic viability of the project, the project has little value to the lender until it is completed. Therefore, a project developer will not receive the cash from the mortgage lender at the beginning of the project. Rather, the mortgage lender will issue a commitment to provide the mortgage funds once the project is completed. The developer is thus placed in the position of obtaining funds to construct the project prior to receiving the cash from the mortgage commitment. The manner in which these construction funds are obtained is through a short-term construction loan. In a similar manner to obtaining a mortgage commitment, a developer requiring a construction loan must demonstrate to the commercial lender that the project is viable and the developer has the resources to repay the loan. For the developer, this demonstration is made somewhat easier since the mortgage commitment exhibits to the construction loan provider that a cash source is available to repay the loan once the project is completed. However, this commitment does not relieve the developer of demonstrating a stable financial position to qualify for the construction loan. Additionally, the developer faces the issue that a construction loan will generally be issued for only 75–80% of the projected long-term mortgage. Therefore, the developer is faced with obtaining private financing for a component of the project. In terms of the example, Figure 7-3 illustrates the gap created during the loan process. For a project to be ultimately viable and successful, the developer must account for how this gap is going to be filled and how the individual investors are going to receive profits on their investment.

Estimated Revenue Detail	Amount
Market rent for proposed property:	
20 Standard Office Suites @$750/mo.	180,000
30 Deluxe Office Suites @$1,100/mo.	396,000
Estimated Annual Income	576,000
Less vacancy factor of 8%	(46,080)
Adjusted Gross Income	529,920
Less Estimated Expenses @26%	(137,779)
Net income before debt service	392,141
Economic Value	2,500,000
Loan/value ratio	75%
Requested Loan	1,875,000
Project Cost	2,100,000
Construction Loan @75% of Mortgage	1,406,250
Construction Loan Interest Addition (2%)	28,125
Total Construction Loan	1,434,375
Project Financing Gap	665,625

Figure 7-3 The financial commitment for a private project often requires a developer to obtain additional financing beyond that which an institution will provide. As illustrated, a financing gap exists because lending institutions prefer to retain a 75% cap on development and construction loans.

Public Funding

Similar to the civil engineering organization attempting to become a member of a project development team, the organization focusing on expanding its role in the public sector must be familiar with the nuances of public financing. In contrast to private projects that are only limited by the amount of time it takes to develop a viable financial plan for private and institutional lenders, public projects are dependent on legally mandated methods for obtaining financing. In most instances, these methods are limited to either government appropriations or public bond offerings. In terms of the appropriations process, construction projects face the same regulatory process as that established for every other public interest project. A construction project must first find an advocate within the regulatory body responsible for budgeting public revenues. Once this advocate is identified, a funding proposal must be developed as a component of an annual budget. Given the numerous competing interests for public financing, a clear and compelling reason must be identified for the project to be initially included in a budget proposal. Fortunately for civil engineers, infrastructure projects have inherent advantages, including the need to ensure water quality, the need to keep roads and bridges safe, and the need to keep schools in satisfactory repair. Each of these projects is visible to the voting public. However, budget revenues are a finite resource, and civil engineers have historically failed to develop political influence equal to groups such as medical and legal professionals. As such, a segment of infrastructure projects are often cut from the budget as it moves through budget committee hearings. Additional cuts are often made during the final budget development debates among the full regulatory body. In the end, a finite amount of financing, less than that originally proposed, is available to complete construction projects.

In many situations, public revenues cannot cover the projected costs of public projects. Major building efforts for schools, correctional facilities, parks, and roads are often outside the ability of the local population to finance. In these cases, the regulatory entity is forced to seek additional financing. For such entities, the avenues to pursue this funding are often limited to either a tax increase or a public bond sale. Since a tax increase is often unpopular to the voting public, the required finances are often pursued through bond sales. In this situation, the governing entity asks the public to approve through a vote the sale of bonds to cover the costs of building specific projects (Zipf 1995). These bond referenda often invoke heated discussions as voters debate both the merits of the proposal and the ability of the governing entity to pay off the bonds without risking other revenue areas or increasing taxes.

As illustrated in Figure 7-4, the concern of voters may be well founded if the interest rate offered on the bonds becomes significant due to pressures from other investment options. Specifically, if a 10-year bond is offered at an 8% annual interest rate, then the governing entity is committing to repay the bond purchaser an 8% return every year for 10 years. At the end of the 10 years, the bond purchaser can redeem the bond for the purchase price. As such, the bond is a formal loan taken by the governing body from the general public at a set interest rate with a fixed due date. If the governing body cannot afford to repay the bond, and is not in the position to refinance the bond, then it risks defaulting. Traditionally, this was considered a minimal risk due to the revenue generating possibilities of local, state, and Federal jurisdictions. However, the bankruptcy of Orange County, California, in 1994 changed this equation ("Outlook" 1997). After this incident, voters have become increasingly wary of bond offerings. This hesitancy may have a direct impact on civil engineering organizations. Without alternative funding capabilities, public-sector projects may be limited, thus reducing the overall revenue available to civil engineering organizations.

Bond Category	Amount
Total Bond	250,000,000
Interest Rate	8%
Maturity Term	10 Years
Annual Interest	20,000,000
Total Interest	200,000,000
Total Bond Commitment	450,000,000

Figure 7-4 Public financing through bond sales commits the issuer to significantly higher monetary returns than the original selling price.

■ Organization Finances—An Internal Perspective

The project perspective provides civil engineering organizations with a long-standing comfort area from which to make financial decisions. However, the project perspective represents a limited foundation from which to make strategic management decisions. Since project finances focus solely on the ability of an organization to turn a profit on an individual project, they fail to recognize the second sphere of financial focus, the financial status of the organization. This focus is often overlooked by project-focused individuals, since the erroneous equivalency is made that project success equals organization success. In reality, an individual project can be very successful while the overall organization is facing financial difficulties. How is this dichotomy possible? The direct answer is that a project represents a snapshot of an organization's overall activities. While that project may be doing well in relation to its projected budget, it does not represent the success of other projects, the success of the organization in entering new markets, or the success of the organization in developing a hedge against economic cycles. Only broader, organization-level financial indicators can address these factors.

Balance Sheets

Organizational finance indicators have become a popular topic among the general public as the Western economy has experienced an extended growth pattern over the past decade. Investors looking for sound investment opportunities have moved financial interests from the business press to mainstream media. A central focus of these financial analysis reports is the organization balance sheet. The balance sheet is the organization equivalent of the project summary sheet. Detailing the financial health of the organization, the balance sheet provides the basis for several economic decisions, as detailed later in this chapter. Once again, extensive publications exist in the field of financial management, and this is not the forum to repeat these presentations. However, to support the strategic management decisions presented later in this chapter, a brief review of the components within the balance sheet is presented.

Figure 7-5 illustrates the end-of-year balance sheet for a representative civil engineering firm, Superior Engineering. Superior is a large civil design firm with approximately 2,500 employees performing design and project management work in the industrial sector. Superior, based in Chicago, has been in operation for over 35 years, consistently growing in annual revenue until it now ranks among the top firms in the region with current revenue over $2 billion. The firm has been a public corporation for the past 5 years, listed on the New York Stock Exchange under the symbol SENG. The stock price has fluctuated between $25 and $37 per share, with a projected target of $42 per share.

Balance Sheet—Superior Engineering	(in 1,000s)
Assets	
Current Assets	
Cash	83,567
Accounts Receivable	425,100
Contracts in Progress	203,153
Prepaid Income Taxes	23,085
Total Current Assets	734,905
Land, Buildings, and Equipment	573,673
Less Accumulated Depreciation	126,325
Net Book Value	447,348
Investments	63,235
Intangible Assets	149,123
Deferred Income Taxes	10,514
Total Assets	1,405,125
Liabilities and Stockholders' Equity	
Current Liabilities	
Accounts Payable	190,080
Bank Loans	26,745
Income Taxes	10,756
Advance Payments by Customers	49,456
Estimated Costs to Complete Contracts	301,225
Total Current Liabilities	578,262
Long-Term Debt	474,664
Deferred Income Taxes	17,074
Employee Benefits	84,606
Total Liabilities	1,154,606
Stockholders' Equity	
Common Stock	
No par value; authorized 2,000,000	105,534
shares, 800,000 shares issued	
Retained Earnings	144,985
Total Stockholders' Equity	250,519
Total Liabilities and Stockholders' Equity	1,405,125

Figure 7-5 A representative balance sheet for a civil engineering organization including asset and liability components.

The Components

The balance sheet illustrated in Figure 7-5 is representative of any such financial statement issued by an organization. The balance sheet consists of a list of assets, liabilities, and owner equity as of a specific date. As the name implies, the balance sheet is one component of the double-entry accounting system used to record income, expenses, assets, liabilities, and owner's equity. In accordance with the double-entry system, all debit entries must have equal and offsetting credit entries. This basic system is summarized in the balance sheet through the equation:

$$\text{assets} = \text{liabilities} + \text{owner's equity}$$

Assets The assets of an organization are its resources, liquid and invested, valued at fair market or cost depending on the particular asset. On the balance sheet, these assets are reflected as either current assets or other assets. Current assets are items such as cash, short-term investments, and other assets that are expected to be converted to cash, sold, or used up within a year or less in the normal operations of business (Nickerson 1986). As illustrated in Superior's current assets category, the organization has four types of current assets: cash, accounts receivable, contracts in progress, and prepaid income taxes.

Cash The cash and cash equivalents category reflects the unrestricted cash accounts on a given date. Items within the category include cash in the bank, foreign currency, petty cash, and investments with maturation of 3 months or less. The category reflects the liquid position of the organization.

Accounts receivable The accounts receivable category reflects work that has been billed to customers, but has yet to be collected. This category is included within current assets based on the assumption that invoices will be paid within a standard 30–90-day period.

Contracts in progress A category not found in many balance sheets outside the engineering industry, but prevalent within the industry, the contracts in progress category represents the revenue recognized from work completed but not yet billed to the client. In the typical percentage-of-work-completed system used in the industry, this category reflects a percentage completion at the date of the balance sheet over and above that currently billed. For example, in Superior's balance sheet, they are claiming that the revenue recognized from current projects is $203,153 million above and beyond that currently billed to their customers.

Prepaid income taxes The final category typically found in a civil engineering balance sheet is that of prepaid income taxes. Since organizations oftentimes pay income taxes on a quarterly basis, prior to the end-of-the-year due date, the cash paid to the government is considered a prepaid, current asset.

The second component of the assets category is assets that may not be available within the upcoming year. As reflected on the Superior balance sheet, these items may include real estate and equipment holdings, long-term investments, intangible assets, and deferred taxes.

Real estate and equipment holdings Land, buildings, and equipment owned by the organization are considered long-term assets. These assets can include everything from the office building owned by the organization to a backhoe used on projects. These assets are important to investors, as they represent tangible items that have flexibility in their financial use. They can be used as collateral for loans and bonding, they can be mortgaged for additional funds, or they can be sold to increase cash holdings.

Long-term investments Investments for a civil engineering organization can include a number of elements, including traditional securities investments, investment in subsidiaries, or investments in other businesses. The difference between this category and the current asset category is the outlook that the investment will not mature within the next 12 months.

Intangible assets Intangible assets are long-lived assets that are useful in the operations of a business, are not held for sale, and are without physical qualities (Warren, Fess, and Reeve 1996). These assets include items such as patents and trademarks that are anticipated to provide the organization with monetary return over their lifetime, but are not included in the previous categories. For example, Superior Engineering has a patent on a new GPS-based locator for construction equipment that enables project managers in the home office to monitor the operations of all vehicles from a single computer. Although Superior is not manufacturing the item, the patent is available for others to manufacture the product with a royalty given to Superior.

Deferred taxes Generally the net income for tax purposes and the net income for financial statement purposes are different because of timing differences between when income or expenses will be recognized for tax purposes versus financial purposes. If more taxes are actually paid than the financial statement indicates should be paid, then the deferred taxes are prepaid and recorded as an asset. If fewer taxes are actually paid than the financial statement indicates should be paid, then the deferred taxes are recorded as a liability.

Liabilities and Stockholders' Equity The second primary component of the balance sheet is the liabilities and stockholders' (or owners') equity category. Reflecting the balance equation (assets = liabilities + owners' equity), this second half of the balance sheet must equal the assets calculated above. The division into the liabilities and stockholders categories is due to the fact that these financial components can either be owed to an outside creditor or represent ownership positions. In the case of liabilities, outside creditors can include suppliers who are owed payments or banks that hold loans. In the case of owner or stockholder equity, this can include investors who hold ownership rights through stock or owners who have invested money directly in the organization.

Focusing on liabilities, the division of liabilities follows the pattern set by the asset categories. Similar to assets, the organization liabilities are divided into current liabilities and other liabilities. Current liabilities are those debts or encumbrances that will fall due within the next 12 months. As illustrated in the Superior balance sheet, civil engineering organizations typically have a minimum of five current liability categories: accounts payable, bank loans, income taxes, costs to complete contracts, and advance payments by customers.

Accounts payable Known to the layman as bills, accounts payable are invoices that have been received by the organization and are due within the next 30–90 days. Since the balance sheet reflects the overall financial position, this category includes both project expenses and organization expenses that are currently owed.

Short-term notes payable The short-term notes payable category includes all principal payments that are due to the lending institutions within the next 12 months. This includes two components: short-term loans and the short-term component of long-term debt. The former includes any loans such as bridge loans, which were taken from a lender to finance short-term purchases or cash flow shortfalls. The latter component represents the principal payments that are due during the next 12 months of long-term loans such as mortgages or equipment loans.

Taxes payable   The taxes payable category represents the amount of taxes (income, payroll, sales, etc.) that the organization owes but has not as yet paid.

Costs to complete contracts   Every organization makes a significant effort in estimating the costs of a project. However, even the best estimate may have errors due to unforeseen circumstances. The costs to complete contracts category reflects the anticipated losses that the organization must absorb to complete a project that is going to run over budget. The category reflects the total loss as a current liability to reflect that the loss is the best estimate at the current time.

Advance payments by customers   Also called deferred or unearned revenue, this category represents payments received for work not yet completed.

In addition to current liabilities, organizations have debts that extend over longer periods of time, including loans and benefit commitments. The remaining liability categories illustrated in the Superior example represent such long-term commitments.

Long-term notes payable   The most common long-term liability category is long-term notes payable. Focusing on items such as mortgages or large equipment loans, this category reflects the long-term debt that the organization has undertaken as part of its initial establishment or continued growth.

Deferred income taxes   See definition under assets.

Employee benefits   Some organizations elect to provide members with benefits such as retirement pensions that require the organization to make periodic payments into a central fund over an extended period of time. The employee benefits entry reflects the liability owed by the organization to the fund.

With the calculation of the organization liabilities complete, the balance sheet now reflects the total organization financial resources and commitments. However, to achieve the required balance in the financial document, a final category is required, owner or shareholder's equity. This category represents the sources of capital that are invested in the organization and the net profits that are retained from the organization earnings.

Stockholder equity   Assuming the organization is a corporation, the first of the equity categories is the stockholder's equity. This category reflects the capital raised by the organization as a result of issuing and selling stock. If the company is not a corporation, then the account is instead referred to as the capital account.

Retained earnings   In layman terms, retained earnings represent the profits (or loss) that the organization has accumulated since its inception. This amount does not reflect available funds, but rather indicates the owner's profits and losses resulting from organization operations from its inception.

With the calculation of owner's equity, the balance sheet is complete. As a basis for the financial health of the organization, the balance sheet represents a snapshot in time of the organization finances. As seen later in the chapter, this statement also provides a basis for making strategic decisions on customers, partners, and markets.

Income Statement

Although the balance sheet provides a snapshot in time of the organization's financial health, the balance sheet is generally not provided as the only financial document available to investors. Rather, the income statement typically accompanies the balance sheet. The value of this second statement lies in its succinct presentation of the revenue and expenses experienced by the organization during the current fiscal year and the net profit or loss that resulted from this income and expenses. While civil engineering organization members may not be familiar with the components of the income statement, the final number indicating a profit or loss is one to which any individual can relate. Additionally, the industry practice of ranking organizations based on their annual revenue places the income statement at the forefront of industry interest on an annual basis (ENR 1998). The combination of these influences places the income statement in a position of relevance that must be acknowledged by organization leaders in developing a strategic perspective. Potential partners and customers will negatively receive the organization that fails to develop a strategy to enhance the income statement on a regular basis.

Components

Figure 7-6 illustrates the income statement from Superior Engineering. The income statement is divided into three primary components: revenues, costs of revenues, and other income and expenses. The combination of these categories provides the organization with its earnings before income taxes. The subtraction of the income tax expense provides the net earnings number that is often anticipated and analyzed by investors. Finally, for a public organization, these earnings are divided by the number of currently issued shares to generate an earnings per share number. Tracked over a number of years, these earnings numbers provide investors with a pattern of success, or lack of, to indicate the potential stability and profitability of the organization.

Revenues   The revenue category incorporates all income generated from operating units within the organization. In the case of Superior Engineering, this encompasses the engineering design and project management divisions. All income derived from contracts originating within these divisions is reflected in the income statement. The numbers do not reflect the expenses required to complete the project. Rather, the revenue number is a gross income number and not a profit number.

Cost of revenues   This category is often taught as cost of goods sold in accounting and management courses due to its origins in the manufacturing world as a reflection of the costs required to manufacture a product. In the engineering domain, the costs of revenues category reflects the costs incurred by the organization to complete projects. For Superior Engineering, this category includes all the project costs associated with project revenues reflected in the revenue category.

Income Statement—Superior Engineering	(in 1,000s)
Revenues	
Engineering	1,509,536
Project Management	678,125
Total Revenues	2,187,661
Cost of Revenues	
Engineering	1,376,873
Project Management	443,988
Total Cost of Revenues	1,820,861
Other (Income) and Expenses	
Corporate Administrative and General Expenses	232,080
Interest Expense	16,510
Interest Income	(1,835)
Total Other Income and Expense	246,755
Earnings Before Taxes	120,045
Income Tax Expense	13,436
Net Earnings	106,609
Earnings Per Share	$0.13

Figure 7-6 The income statement for Superior Engineering illustrates the focus on revenues, cost of revenues, and overall profit for the organization.

Other income and expenses   The final component in the income statement is the other income and expenses category. Rather than focusing on project-level revenue and expenses, this category reflects expenses absorbed at the organization level together with income generated by non-business-center investments. In the case of expenses, this category represents the administrative and operating expenses required to run the organization. For example, expenses in this category may include the senior executives, marketing, and office rental. In each case, the expense is not directly related to a particular project, but is associated with organization operations. Interest expense is separated from these operational expenses to reflect the organization's financing commitments.

The combination of these categories results in the earnings for a fiscal year. In contrast to the common practice of evaluating organizations based on revenue, the income statement illustrates that revenue is only one component of the overall financial picture. An organization can be very successful in generating revenue, but if the cost of the revenue is greater than the revenue, then the organization is still going to fail. In the end, it is the net earnings that determine the profitability of the organization. Thus a 100-person organization making $10 million in net earnings on $200 million in revenue is actually doing worse on a percentage basis than a 100-person organization making $20 million in net earnings on $200 million in revenue. Although each organization is recorded as equal in the publications, the latter organization has achieved superior performance. Therefore, the first step in focusing on strategic numbers for a civil engineering organization is to stop focusing on revenue and start putting greater emphasis on net earnings.

Budgets

The final financial tool that serves as a foundational component of organization analysis in strategic management is the budget. Whereas the balance sheet and income statement are primarily reflective tools on the surface, a budget is a predictive statement that portrays anticipated revenue and expenses. Although not usually published by organizations as part of the annual financial statement, the budget is a powerful tool in developing market entry decisions. Specifically, a sound budget provides the organization with a prediction of the costs and revenue associated with conducting business in a particular market segment. The budget can assist in predicting whether a profit or loss should be expected prior to commencing operations. Given a solid negative or positive analysis, the organization can make a decision to either pursue or disregard business opportunities.

Figure 7-7 illustrates a summary budget for Superior Engineering. As illustrated, the budget incorporates three primary categories: revenues, fixed cost of services, and operating expenses. The first category incorporates projected revenues that will be generated from each operating division. The second category incorporates the known costs that exist for project operations. For example, projects have salaried employees conducting the actual work. Revenues from project contracts must cover the salaries of these employees. Similarly, equipment, such as computer-aided-design workstations, is designated as project-specific components. To pay for these workstations, projects must be obtained that can absorb equipment charges. The final category represents the costs of operating Superior Engineering. As seen in the table, these costs range from the salaries of nonproject employees to travel and marketing costs. Each of these costs exists in addition to the costs associated with specific projects. If the profit from individual projects does not cover these organization costs, then the organization will absorb losses even if an individual project is profitable.

As with the income statement, the budget statement highlights the need for the organization to generate revenue that leads to a profit. The fact that an organization is generating revenue is worth very little unless it is leading to revenues that exceed expenses. The budget statement is a prediction of the revenues required to meet this goal.

■ Industry and Regional Economics—
An External Perspective

The final perspective for strategic finance turns the focus outward from the organization to the economic conditions of the industry and the geographic region in which

Budget—Superior Engineering	(in 1,000s)
Revenues	
Design Services	1,509,536
Project Management Services	678,125
Total Revenues	2,187,661
Fixed Cost of Services	
Permanent Salaries	202,500
Equipment	15,465
Insurance	2,085
Resources	5,675
Total Fixed Cost of Services	225,725
Operating Expenses	
Salaries	1,835,554
Advertising	545
Facilities	301
Equipment	2,376
Travel	1,743
Insurance	245
Consultants	2,467
Marketing	565
Total Operating Expenses	1,843,796
Total Expenses	2,069,521
Projected (Loss) or Gain	118,140

Figure 7-7 The budget statement for Superior Engineering illustrates the focus on estimating costs and revenues to determine the projected loss or gain in the upcoming budget period.

the organization operates. In contrast to the project and organization perspectives that emphasize an analysis of organization-controlled factors, the external economics perspective emphasizes the analysis of financial matters outside organization control. Whereas project items such as budgets and schedules, and organization items such as debt and investments, are under the guidance of the organization, the economic world is outside the influence of the organization. Rather, economics provide organizations with indicators of the future business environment in which the organization will operate. Through these indicators, civil engineering organizations can accomplish two specific objectives: (1) take proactive steps to establish advantageous market positions, and (2) anticipate economic downturns. In terms of the former, the organization that is able to identify the next investment area will have the opportunity to establish a marketing, technical, and managerial infrastructure prior to the competition. In terms of the latter, the organization that is able to anticipate economic downturns will have the opportunity to establish defensive positions. In both cases, the information required to make these decisions will not be found in project or organization numbers. Rather, the information required to make strategic market decisions originates from the outside economy.

Economic Forecasting

The fundamentals of micro- and macroeconomics should be familiar to civil engineers based on undergraduate requirements. The concepts of supply and demand, inflationary pressures, and perfect competition are all components of basic economic theory. As such, numerous textbooks are in print that provide excellent introductions to this material, including publications by Mansfield (1985), Dornbusch and Fischer (1987), and Skaggs and Carlson (1996). While these concepts are critical to an understanding of the business environment and its underlying operating theory, strategic management requires methods to predict long-term trends and cycles. In response to these needs, the field of economic forecasting focuses on the establishment of mathematical models to predict long-term economic relationships and trends (Levenbach and Cleary 1984). Since the early twentieth century, economists have focused on forecasting business and economic cycles (Kacapyr 1996). In these efforts, economists have attempted to provide businesses with indicators of future market and sector fluctuations, the fundamental components of strategic business positioning. Given this relevance to strategic management functions, it is becoming increasingly important for civil engineers to understand how economic forecasts have a direct impact on the ability of organizations to make informed market decisions.

Time-Series Charts—A First Model Approach

A fundamental component of traditional economic forecasting is the study of economic data over an extended period of time. These time-series models provide economists with data curves that reflect the positive and negative business environments over time. As illustrated in Figure 7-8, the focus of economic curves is on three primary elements: the duration of a cycle, the depth of a cycle, and the dispersion of the

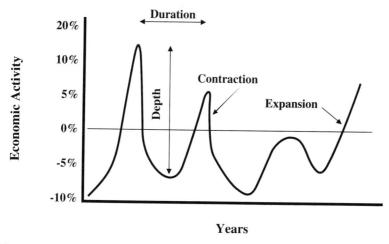

Figure 7-8 A time-series chart provides a long-term perspective on economic expansion and contraction and the relative strength of individual economic periods.

cycles. The first of these elements, duration, refers to the length of time for an individual peak or trough period. Given a number of individual peaks or troughs, averages can be determined for particular economic cycles. For example, looking at the overall economic conditions of the United States since World War II, the data indicate that growth periods have lasted for an average of 50 months, while contraction periods have lasted an average of only 11 months (Kacapyr 1996). The second cycle element, depth, refers to the magnitude of the individual cycle. Although the economy moves through regular cycles, the magnitude of the growth or contraction cycles will vary. This means that every growth period will not be a boom cycle, nor will every contraction period result in a recession. Unfortunately, it is not possible to predict this magnitude a priori, thus requiring organizations to hedge their positions to reduce the possibility of getting caught in an extreme cycle. Finally, the dispersion of the cycles refers to the breadth of impact that an individual cycle has on the business environment. Although the economy is often reflected in a single number such as the Gross Domestic Product (GDP), the economy is actually composed of many different sectors. As such, an individual cycle may impact one or more sectors during an expansion or contraction period. For example, in the late 1980s the contraction cycle focused heavily on the management and service sector, while the contraction period in the late 1970s focused greater on the manufacturing sector.

The value of time-series graphs becomes apparent when the cycles emerge from collected data. Given a set of data over an extended period of time, predictions can be made concerning the regularity of the cycles and the next potential time period for the cycle to repeat. Taken a step further, individual sectors or economic factors can be plotted to develop theories on the cyclical nature of specific components of the economy. Examples of theories that have emerged from this type of analysis include the Kitchin cycle, which focuses on inventories as economic indicators, the Juglar cycle, which relates durable equipment sales to the economy, and Kondratieff cycles, which focus on prices as economic indicators (Kacapyr 1996). However, of particular interest to the civil engineering industry is a time-series cycle referred to as the Kuznet cycle. Kuznet cycles focus on the investment in structures as indicators of economic activity. Based on the concept that investments in structures reflects issues such as interest rates, employment figures, and overall economic strength, the Kuznet cycle compiles structure spending data since 1960 (Fig. 7-9). What emerges from these data is that structure cycles occur on the average of every 6–8 years, with growth for about 4 years and contraction for 2–4 years. Using this as a basic forecasting tool, it could be postulated that spending for structures should see growth patterns in 2001 and 2007, while experiencing contraction periods in 2006 and 2011.

Although time-series models provide a basic economic forecasting tool, organization leaders should be aware of their limitations. Specifically, the cyclical graphs reflect data, but do not necessarily explain the interdependencies between the factors that led to the cycles. However, this should not dissuade the civil engineering organization from using time-series data as one component of an overall economic analysis. Although it may not provide the final answer, a time-series graph can provide the basis for further analysis of a potential trend at a future point in time.

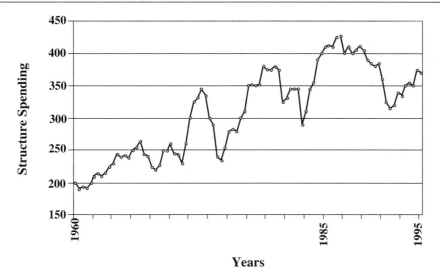

Figure 7-9 A specific time-series chart based on spending within the building sector is the Kuznet cycle. This cycle profiles long-term economic emphasis on investment in structures.

Leading Economic Indicators—Compiled Statistics

Although an individual time series may lack the explicit relationships that engineers have been trained to analyze, a collection of time-series charts provides a broader perspective on economic conditions. Specifically, time-series charts can be divided into three types of indicators: leading indicators, coincident indicators, and lagging indicators (Sobek 1973). Leading indicators are those that predict upcoming economic trends. Coincident indicators provide a measure of current economic conditions, such as the GDP and employment figures. Finally, lagging indicators, such as long-term unemployment, are those that confirm an economic change has occurred. Of these three categories, the organization interested in future economic developments would focus on the leading economic indicators, since these data provide the basis desired for making long-term strategic decisions. To assist organizations in obtaining these data, the United States Commerce Department's Bureau of Economic Analysis publishes a series of leading economic indicators (LEIs) for the U.S. economy.

Currently, the LEIs are composed of 10 indicators as listed in Table 7-1 that increase or decrease 6–12 months prior to overall economic movement. The "Composite Index of Leading Economic Indicators" is revised and published monthly and is calculated by using a weighted average of the LEIs. The composite index is based on a relative score to a baseline number of 100 that was established in 1987. For example, a composite index of 110 today indicates that the index is 10% higher than it was in 1987. The changes in the index are used to map future economic activity. An increase of 0.5% or higher in a given month is an indicator of an improving economic condition. Similarly, three consecutive declines may indicate a recession within the next 12 months. Figure 7-10 illustrates the composite index from 1970 to 1998.

TABLE 7-1
The Leading Economic Indicators (LEIs) Are Compiled by the U.S. Government to Provide a Broad Foundational Basis for Developing Economic Prediction.

Indicator	Description	Standardization Factor
1	Average weekly hours, manufacturing	0.222
2	Average weekly initial claims for unemployment insurance	0.025
3	Manufacturers' new orders, consumer goods, and materials	0.047
4	Vendor performance, slower deliveries diffusion index	0.026
5	Manufacturers' new orders, nondefense capital goods	0.012
6	Building permits, new private housing units	0.017
7	Stock prices, 500 common stocks	0.031
8	Money supply, M2	0.293
9	Interest rate spread, 10-year Treasury bonds less Federal funds	0.310
10	Index of consumer expectations	0.017

Although economists have developed forecasting models based on the LEIs, the reliability of these models may be difficult to determine for a specific business sector such as civil engineering. However, the volume of data that serves as a basis for the models provides a solid basis for economic projection (Stiltner and Barton 1990). Given this statistical basis, civil engineering organizations can use the composite index to determine appropriate time frames to either increase organization resources or consolidate resources as part of a defensive positioning against economic contraction. Specifically, if the composite index indicates continued growth for the next 12 months, then focus and resources should be placed on gathering information on emerging market opportunities. However, if the composite index indicates an upcoming contraction, then a retrenchment into established market sectors may be in order.

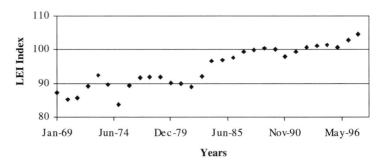

LEI Index 1969-1998
(Averaged to LEI of 100 in 1992)

Figure 7-10 The LEI index over the past 30 years has shown both expansion and contraction periods. However, an overall positive trend has been in evidence since 1983.

Econometric Models

While time-series data provide a historical study approach to economic analysis, civil engineers may feel uncomfortable with the questionable reliability associated with this form of economic forecasting. In response, a final forecasting methodology is introduced, econometrics. Literally defined, econometrics is economic measurement. However, Kacapyr (1996) gives a more robust definition by stating, "econometrics applies statistical techniques to the relevant data in order to discern the relationships among economic variables." In terms of the previous discussion on time series, econometrics moves beyond the analysis of cycles to the study of the underlying data to determine the mathematical relationships that drive the creation of the data points. In full-scale econometric models, hundreds of equations may be used to model a complex economic system. Each of the equations in the model serves to establish one relationship in the overall system. In this manner, the system is similar to the function of a structural frame. In the case of the frame, each connection and load has an individual mathematical relationship that defines the behavior of that connection under specific load conditions. When consolidated into an overall design, the individual relationships define the behavior of the frame under potential loading conditions. Econometric models work the same way. Each economic relationship is defined by a single statistical formula. In combination, the individual formulas establish the system-level operation of the economic environment. Given these explicit mathematical relationships, data can be analyzed to develop economic forecasts based on specific changes in economic variables. For example, relationships may be developed that link the changes in steel prices to the overall increase in building costs, or the impact of interest rate hikes to commercial development permits.

The development of these econometric models is based on statistical regression models. To review regression techniques briefly, regression models determine the relationships between two or more variables. For example, Table 7-2 illustrates a table of numbers reflecting interest rate changes and housing starts over a 20-year period. Although the numbers in the table reflect a relationship between the two variables, it is difficult to determine the exact relationship by just "eyeballing" the data. The graph in Figure 7-11 gives greater clarity to the relationship, but still does not provide the mathematical relationship required to forecast future numbers. To achieve this relationship, a line must be drawn that fits between the observations plotted on the graph. Regression analysis provides the statistical basis for fitting this line on the graph. Specifically, regression analysis provides the basis for developing a relationship between the variables that will be reflected in a slope intercept equation of the form:

$$Y = a_1 + a_2(X)$$

where a_1 is the value of the vertical intercept and a_2 is the value of the slope of the line. Similarly, Y is the dependent variable in the equation since its value depends on the value of X. X is called the independent variable since its value changes independently of any other values in the equation.

TABLE 7-2
A Listing of Housing Starts and Interest Rates Provides Little Direct Correlation at a Surface level.

Date	Interest Rate	Housing Starts
Jun-78	9.71	216,000
Jun-79	11.04	191,800
Jun-80	12.71	116,400
Jun-81	16.70	105,800
Jun-82	16.70	91,100
Jun-83	12.87	173,200
Jun-84	14.42	184,000
Jun-85	12.22	163,200
Jun-86	10.68	183,600
Jun-87	10.54	162,900
Jun-88	10.46	150,200
Jun-89	10.20	143,200
Jun-90	10.16	117,800
Jun-91	9.62	103,400
Jun-92	8.51	117,800
Jun-93	7.42	128,500
Jun-94	8.40	136,400
Jun-95	7.57	123,400
Jun-96	8.32	138,000
Jun-97	7.69	140,400
Jun-98	7.00	160,000

The focus of regression analysis is to find the values of a_1 and a_2. By deriving these values, predictions can be made on the impact of housing starts (Y) by interest rates (X). As illustrated in Figure 7-12, a least-squares method is used to derive the values of a_1 and a_2. The least-squares method measures the distance between the data points to a line that is the closest fit between the data points. The results of the measurement are values for the constants in the slope–intercept equation. In this example, a_1 is equal to ($-1,816$) and a_2 is equal to 164,376. Given these values, predictions can be made on future housing starts based on changes in interest rates. If the interest rate decreases 0.75%, it can be anticipated that there would be an associated increase of 165,738 housing starts based on the equation:

$$165,738 = -1,816(-0.75) + 164,376$$

Although this analysis provides a numeric predication capability for related data, the quality of the prediction is limited by the data provided. Specifically, regression analysis creates a best-fit line for the data included in the least-squares method. This line may not explain every data point in the set. One measurement used to determine the percentage of data points addressed by the regression line is the r^2 measurement,

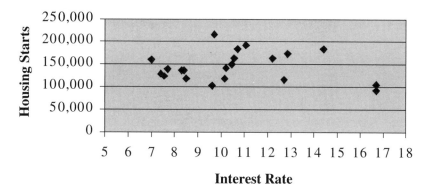

Figure 7-11 Graphing the housing starts and interest rates provides little additional information to develop a specific correlation.

referred to as the coefficient of determination. Calculated by first dividing the covariance by the product of the standard deviations of x and y, and then taking the square of the result, the r^2 measurement returns a percentage representing the number of data points described by the regression line. For example, if the coefficient of determination returns 0.93, then 93% of the values in the data set can be explained by the regression formula. Using this measurement, regression formulas can be evaluated for their accuracy. Specifically, the higher the coefficient percentage, the greater the fit of the developed equation.

Econometric models build upon this basic theory to create complex equation systems that reflect the many variables and relationships that exist within an economic system. In advanced systems such as those developed by Georgia State University and the University of Michigan, the models include hundreds of equations and variables that interact to model economic factors over extended periods of time (Christ 1993). However, to demonstrate this concept in a simple model, assume that investment in infrastructure projects is based on housing starts and labor prices, as reflected in the following equations:

$$A = a_1(C) + a_2 \tag{1}$$

$$B = a_3(A) - a_4 \tag{2}$$

$$D = A + B \tag{3}$$

where A is the previously illustrated housing start relationship; B is change in work force numbers based on minimum wage and stock market averages; C is interest rate change; and D is the projected change in infrastructure spending in thousands.

Equation (3) is not a regression equation. It is known as an "identity" since it does not include any constants to calculate. Equations (1) and (2) are regression equations since they model how housing starts and labor prices affect infrastructure investment. These equations are referred to as "structural equations" since they embody the structure of the model. Once the model is developed, the constants in Eqs. (1) and

Date	Interest Rate (x)	Housing Starts (y)	x^2	y^2	xy
Jun-78	9.71	216,000	94.2841	46,656,000,000	2,097,360
Jun-79	11.04	191,800	121.8816	36,787,240,000	2,117,472
Jun-80	12.71	116,400	161.5441	13,548,960,000	1,479,444
Jun-81	16.7	105,800	278.89	11,193,640,000	1,766,860
Jun-82	16.7	91,100	278.89	8,299,210,000	1,521,370
Jun-83	12.87	173,200	165.6369	29,998,240,000	2,229,084
Jun-84	14.42	184,000	207.9364	33,856,000,000	2,653,280
Jun-85	12.22	163,200	149.3284	26,634,240,000	1,994,304
Jun-86	10.68	183,600	114.0624	33,708,960,000	1,960,848
Jun-87	10.54	162,900	111.0916	26,536,410,000	1,716,966
Jun-88	10.46	150,200	109.4116	22,560,040,000	1,571,092
Jun-89	10.2	143,200	104.04	20,506,240,000	1,460,640
Jun-90	10.16	117,800	103.2256	13,876,840,000	1,196,848
Jun-91	9.62	103,400	92.5444	10,691,560,000	994,708
Jun-92	8.51	117,800	72.4201	13,876,840,000	1,002,478
Jun-93	7.42	128,500	55.0564	16,512,250,000	953,470
Jun-94	8.4	136,400	70.56	18,604,960,000	1,145,760
Jun-95	7.57	123,400	57.3049	15,227,560,000	934,138
Jun-96	8.32	138,000	69.2224	19,044,000,000	1,148,160
Jun-97	7.69	140,400	59.1361	19,712,160,000	1,079,676
Jun-98	7	160,000	49	25,600,000,000	1,120,000
	222.94	3,047,100	2,525.467	463,431,350,000	32,143,958

Mean of x	$\bar{x} = \dfrac{222.94}{21}$	$x = 10.62$
Mean of y	$\bar{y} = \dfrac{3,047,100}{21}$	$y = 145,100$
Standard deviation of x	$s_x = \sqrt{\dfrac{1}{n}\left(\Sigma x^2 - n\bar{x}^2\right)}$	$s_x = 2.75$
Covariance	$s_{xy} = \dfrac{1}{n}(\Sigma xy = n\bar{x}\bar{y})$	$s_{xy} = -204,636$

$b = \dfrac{s_{xy}}{s_x^2} \qquad b = -1816 \qquad a = y - bx \qquad a = 164,376$

$y = -18163x + 164,376$

Figure 7-12 The least-squares method of statistical analysis provides a structured method for determining a regression line equation between two variables. As illustrated, the least-squares method determines that a change in interest rate will result in a change in housing starts equal to 164,376 + (the interest rate * − 1816).

(2) must be calculated. These calculations are completed using the least-squares method, as illustrated previously in Figure 7-14. Once these constants are calculated, an economic forecast can be developed based on projected values for B, C, and D.

To illustrate, suppose the values of the constants were calculated as; $a_1 =$ (−1816), $a_2 = 164,376$, $a_3 = 5.75$, $a_4 = 1,450$. Also, let us suppose that the values of A and C are 165,738 and (−0.75), respectively, as per Figure 7-14. Given this data point, the values for B and D can be calculated using simple algebra. Substituting Eq. (1) into Eq. (2) gives:

$$B = a_3(A) - a_4$$

$$B = 5.75(165,738) - 1,450$$

$$B = 951,544$$

Solving for this substitution gives:

$$D = 165,738 + 951,544$$

$$D = 1,117,282 \text{ (in thousands)}$$

Thus it is forecast that if interest rates are decreased by 0.75%, then 951,544 new jobs will be created and infrastructure investment will be increased by $1,117,282,000.

In this manner, econometric models have the ability to forecast large-scale systems of equations for individual values. How does this relate back to the discussion on strategic management in civil engineering? The answer to this question is knowledge. An increasing number of private companies and universities are conducting research into econometric models (Christ 1993). Having the knowledge to interpret the results of these models provides a potential strategic advantage for civil engineering organizations. Understanding the relationships between economic variables provides an organization with the ability to project future economic developments, and consequently, the ramifications on the civil engineering industry. For example, if an econometric model indicates that industrial investment is tightly aligned with consumer confidence levels, then a strategic opportunity for a civil engineering organization lies in the observation of consumer confidence numbers reported by the government each month. A discernable upward trend in the numbers should lead an organization to anticipate increased industrial investment in facilities. Placing the organization in the position to respond to this investment prior to its occurrence gives the organization an advantage over competitors. Similarly, if the confidence levels indicate a downturn, then the organization can transfer its resources to a different market sector prior to the business being faced with a sudden loss of clients and no fallback position. In this manner, the organization is proactively directing business efforts rather than allowing the marketplace to direct operational efforts.

■ The Strategic Application of Finance

The three perspectives of financial data provide a foundation for creating a strategic perspective on financial application. In this perspective, the emphasis changes from analyzing financial and economic data for the purpose of gauging internal health to a perspective on using these same data as indictors for strategic movement. Viewed as an additional tool within the cadre available to organization leaders, financial numbers can provide the numeric basis to justify strategic decisions. From developing internal objectives to selecting potential partners, the financial numbers emerging from the above perspectives can quantify the rationale used to make these decisions. However, to achieve this potential, civil engineering organizations must change their perspective on the role financial personnel play within the organization. Rather than viewing accounting personnel as checkpoints on projected budget schedules, these same individuals need to be working with the broader team to develop investment-based strategies. Specifically, the organization needs to adopt some of the investment analysis techniques used by financial analysts to assist in making long-term business decisions. For example, rather than analyzing the standard qualifications document of a potential partner that focuses on project experience, organizations should additionally analyze the past 5 years of financial statements to develop income–expense ratios as an indicator of efficiency and fiscal management. This information will complement experience-based qualifications by providing a broader picture of the manner in which the organization conducts business.

The use of financial data to assist in strategic decision making can be extended throughout the civil engineering organization. The following sections focus on three specific areas: customers, partners, and economic stability. However, these three should not be viewed as the definitive domain of financial input possibilities. Rather, these discussions should serve as a starting point for adopting financial investment strategies within the overall decision making process.

Finance as a Basis for Selecting Customers

The first area of focus for incorporating financial analysis is the customer. Since the customer ultimately serves as the difference between success and failure of an organization, identifying, winning, and satisfying customers is the underlying objective of every organization. Unfortunately, in the daily drive to win customers, organizations may overlook the difference between a financially robust customer and one that is struggling to stay afloat. Understandably, many organizations may disagree with this statement. This disagreement is not based on the use of financial data in making customer decisions, but rather on the need to analyze customers. Given the competition within the civil engineering industry, many organizations are faced with the daily prospect of simply remaining in existence. For these organizations, any customer is a needed customer, and the thought of rejecting that customer because it does not meet a particular financial ratio is not a feasible action.

In truth, replying to such perspectives is difficult. The civil engineering industry follows many of the components of a perfect competition model (Skaggs and Carlson 1996). First, many small firms in relation to the size of the market characterize the industry. Second, the product produced by industry organizations is relatively similar. Although variations exist between competitors, a customer could argue that the final product is almost identical. And finally, very few barriers exist for organizations to enter or leave the industry. Although an established customer base is a strong asset, there is nothing inherent in the civil engineering industry from preventing a new organization from entering the business. Given these characteristics, market demand is the only fluctuating component in the civil engineering industry. Therefore, declining a project is placing the organization at risk, since finite demand exists and declining work gives a competitor an advantage by increasing their share of the market sector.

Although this argument is compelling, in fact, the civil engineering industry is not based on perfect competition. First, although many small players characterize the industry, the percentage distribution of business between the top organizations and the small organizations is significant. Therefore, resources between the organizations are not evenly divided. Second, although a final product may be similar, the development of that product within an organization may be very different. Cost efficiencies, management effectiveness, and experience all play a role in the ability of the organization to produce a final product. And finally, the diversity of customers who require civil engineering services makes each project a semiunique endeavor. Emerging from these differences is an industry that requires its members to create an individual identity to attract customers. While this is perhaps obvious and follows the requirement of almost every other industry, the distinction for organizations that desire to prosper, rather than just sustain, goes a step further. Specifically, the fragmentation and competition within the civil engineering industry requires organizations to establish advantages over the competition. While it is true that the members of the organization provide an initial advantage, even the greatest work force cannot turn a customer with limited resources into a long-term cash cow. Rather, customers that can provide long-term organization benefits are an integral component of long-term success. As such, given two customers and limited resources, the selection of which project to pursue should not be limited to the immediate value of the project, but rather, the analysis should also include consideration of the strategic benefits provided by the selected customer.

Selecting Financially Sound Companies

Stating that an organization should analyze the financial stability of a potential customer is often easier said than done in the civil engineering industry. Organizations that are not accustomed to finding or analyzing financial data may need to enhance existing resources to accomplish the task. Additionally, organizations that focus primarily on public works may need to modify traditional balance sheet analysis to incorporate public institution analysis techniques. However, these conflicts are resolvable. As long as some component of financial data is available, the opportunity exists to leverage those data into strategic decisions. With this as a basis, the use of balance and income sheet data to analyze customers is introduced.

Balance and income sheets provide an indication of the financial health of an organization at any given time. While this information is not definitive, it does provide a snapshot that subsequently can provide the input required to make an inference as to the strategic desirability of a customer. The key component of this decision-making process is the adoption of an investment perspective toward the customer. To illustrate this perspective and provide a basis for the discussion, the Superior Engineering example is revisited. As stated previously, Superior focuses primarily on design and project management as related to industrial projects. Based on Superior's experience and reputation, they are prequalified to make proposals to most of the well-known chemical producing corporations. As such, Superior has had the luxury in recent years to select the projects they wish to pursue. Unfortunately, this selection process has not always achieved the anticipated results. Focusing primarily on man-hour commitments, division-level managers have argued persuasively for the need to win projects that provide the greatest opportunity to keep the Superior staff billable to client projects. Although this tactic has successfully kept Superior profitable, the organization has begun to notice that overhead costs are beginning to increase in proportion to revenue. In particular, the cost of getting projects is steadily increasing. On further examination, Superior's managers have found that the organization is spending too many resources on one-time customers. These customers, whether from large or small companies, are contracting with Superior to build projects, but are not providing Superior with the repeat business that leverages initial business development expenses.

This example is similar to the situation encountered by many in the civil engineering industry. Focusing on individual projects, appropriate time is not always allocated for focusing on long-term potential. However, this focus does not necessarily require extensive financial analysis. A few basic techniques can be employed to perform an initial analysis of the financial health of an individual organization or the comparative attractiveness of several customers. Using the data available in the financial statements, the first analysis that can be developed is the current ratio analysis (Bernstein 1993). Focusing on the ability of an organization to meet financial commitments, the current ratio analysis emphasizes the liquidity of an organization. Specifically, the ratio is based on the ratio of current assets to current liabilities. Returning to the Superior Engineering example, Figure 7-13 illustrates two balance sheets from potential chemical producers. Extracting the current asset and current liability information from the two balance sheets, the following current ratio calculations can be developed:

$$\text{customer A} = 41,874/14,070 = 2.98$$

$$\text{customer B} = 9,044/6,044 = 1.5$$

Given these calculations, the surface analysis might place customer B in a stronger financial position than customer A. However, applying the current ratio test reveals a different picture. As seen by the ratio, customer A has a 3:1 ratio of assets to liabilities, while customer B is set at 1.5:1. This ratio has significance for the civil engineering organization. First, although customer B has more current assets, the liabilities that offset these assets place in question the fiscal responsibility of customer B. Second, the excess liabilities must also be questioned for the ability of customer B to meet its cash flow demands. Third, the margin of safety that customer B enjoys against unforeseen cost overruns is questionable. In combination, the current ratio analysis opens the door to question the ability of customer B to retain liquidity over

Balance Sheet—Customer A	1,000s
Assets	
Current Assets	
Cash and Equivalents	1,146
Accounts Receivable	35,740
Merchandise Inventories	4,070
Deferred Income Taxes	521
Prepaid Expenses	397
Total Current Assets	41,874
Land, Buildings, and Equipment	54,284
Less Accumulated Depreciation	30,701
Net Book Value	23,583
Investments in Affiliates	3,477
Other Assets	4,008
Total Assets	72,942
Liabilities and Stockholders' Equity	
Current Liabilities	
Accounts Payable	3,007
Bank Loans and Lease	6,154
Obligations	
Income Taxes	593
Other Accrued Liabilities	4,316
Total Current Liabilities	14,070
Long-Term Debt	
Notes and Lease Obligations	15,929
Other Liabilities	8,919
Deferred Income Taxes	2,084
Total Liabilities	41,002
Stockholders' Equity	
Common Stock	
No par value; authorized 1,800,000,000 shares, 1,150,000,000 shares issued	23,574
Stock held in trust for employee benefits	1,693
Retained Earnings	9,389
Total Stockholders' Equity	31,270
Total Liabilities and Stockholders' Equity	72,942

Balance Sheet—Customer B	in 1,000
Assets	
Current Assets	
Cash and Equivalents	166
Short-Term Investments	979
Merchandise Inventories	1,174
Accounts Receivable	5,585
Prepaid Expenses	1,140
Total Current Assets	9,044
Land, Buildings, and Equipment	49,357
Less Accumulated Depreciation	28,814
Net Book Value	20,543
Investments	2,099
Long-Term Receivables	803
Total Assets	32,489
Liabilities and Stockholders' Equity	
Current Liabilities	
Accounts Payable	3,026
Bank Loans	218
Short-term obligations	751
Income Taxes Payable	1,264
Other Accrued Liabilities	785
Total Current Liabilities	6,044
Long-Term Debt	
Notes and Debentures	4,639
Capitalized Leases	80
Income Taxes	2,868
Other Liabilities	2,408
Total Liabilities	5,276
Stockholders' Equity	
Common Stock	
No par value; authorized 800,000,000 shares, 483,000,000 shares issued	2,568
Foreign currenty translation	(149)
Retained Earnings	13,900
Total Stockholders' Equity	16,319
Total Liabilities and Stockholders' Equity	32,489

Figure 7-13 Two representative balance sheets from prospective customers provide an economic perspective on the long-term potential relationship between the customer and the civil engineering organization.

an extended period of time. Given the focus on retaining strategic customers for strategic alliances, customer A appears to be a better choice based on this initial analysis.

Although the Current Ratio analysis provides an indicator of liquidity for civil engineering organizations, the ratio may be skewed by different factors. One factor of interest to civil engineering organizations is the impact of inventories. Since organizations such as the industrial customers that Superior serves may have significant assets locked in inventories, the current ratio can project a skewed vision of liquidity. In response to this potential problem, a second ratio known as the acid-test ratio may be employed. This ratio is calculated in the same manner as the current ratio except inventories are excluded from the calculation. Using the same data from Figure 7-13, the following acid-test ratio is developed:

$$\text{customer A} = 34,804/14,070 = 2.47$$
$$\text{customer B} = 8,570/6,044 = 1.42$$

As seen in this calculation, the difference between customers A and B becomes slightly smaller. The emphasis on inventory by customer A reduces the ratio from 3:1 to 2.5:1. However, this is still a stronger position than the 1.42:1 of customer B.

Building on this focus to sustain long-term benefits for the civil engineering organization, a second area of interest is the debt accumulated by the potential customer. Given that the reason the customer is focusing on a capital asset project is the need to either expand or replace infrastructure, the need for the project is established. The possibility that the customer has made a faulty judgement in requiring the capital investment is ignored for the moment. With this need in place, the customer will either need to increase debt or reduce cash reserves to pay for the project. In either case, the ratio of liabilities to assets will increase since one side or the other will have a modification. However, rather than automatically assuming that the organization with the ability to fund capital projects out of cash reserves is in a superior position, it is important to analyze the organization's commitment to short- and long-term debt repayment. In terms of short-term debt, organizations that undertake extensive short-term debt are in a questionable position from a strategic perspective. Since the organization is required to repay the debt within the coming year, the organization is going to be pressured into making decisions that respond to this need. This pressure may result in product or management decisions that have short-term benefits, but long-term consequences. An example of this is the reduction of payroll expenses to meet debt payments. Although the elimination of jobs provides additional cash flow for the repayment of short-term debt, the demand for good personnel in the industrial sector precludes an organization from assuming that equally talented personnel will be available in the future. Thus the organization has placed long-term effectiveness behind short-term financial commitments as a priority.

In contrast to these forced decisions required to meet short-term debt payments, long-term debt may be accepted for leveraging purposes. Since long-term debt is spread over an extended period of time, the payments associated with the debt may be inconsequential in comparison to the revenues generated by the investment. This financial leveraging is reflected in an index calculated as follows:

$$I = \frac{E}{A}$$

where I is the financial leverage index; E the return on common equity; and A, the return on total assets.

In this analysis, two customers with the same assets will have a different financial leverage index based on their ability to use long-term debt to finance operations. As long as the index is greater than 1, the organization is leveraging debt. However, when the index falls below 1, then the debt is costing the organization assets and thus may be subject to further scrutiny. From the civil engineering perspective, the ability for a customer to leverage debt demonstrates financial foresight. The customer is anticipating the investment will return revenues, so is less concerned about raising cash through traditional stock offerings. This emphasis illustrates a strategic outlook that would be comparable to the one being advocated here for the civil engineering organization.

A final analysis tool presented here is intended as a caveat to the short-term versus long-term debt discussion. Although long-term debt as an investment tool may be a strong positioning tool for an organization, there are limits to this effectiveness. Specifically, if the total debt incurred by an organization creates a burden, then the efficacy of the long-term debt justification may be questioned. Similar to the previous analysis numbers, the total debt calculation is stated as follows:

$$T = \frac{D}{C}$$

where T is equal to the debt versus capital ratio, D is equal to the total debt, and C is equal to the total capital plus the debt. Given the balance sheets illustrated previously in Figure 7-13, the following ratio can be calculated:

customer A = 41,002/(31,270 + 41,002) = 0.57 or 57% of capital

customer B = 5,276/(16,319 + 5,276) = 0.24 or 24% of capital

As seen in these calculations, customer A has a percentage of 57%, while customer B has a percentage of 24%. The question for Superior Engineering to address in this situation is whether or not the organization believes that the 57% debt ratio of customer A is a burden on their ability to sustain a profitable position. As a rule of thumb, anything over a 50% debt ratio may be considered problematic by investors (Bernstein 1993).

In summary, the selection of customers as strategic partners goes beyond looking at the size or revenues of an organization. Even the largest organization may be unable to invest in capital projects if the liability or debt burden exceeds their ability to invest in such revenue intensive projects such as new plants, distribution facilities, or office complexes. The civil engineering organization that is analyzing organizations for long-term, repeat business potential must look beyond the glossy revenue numbers to the underlying financial health of the organization. Simply put, the civil engineering organization must adopt the financial investor's perspective and objectively rate the customer on its ability to meet strategic objectives.

Finance as a Basis for Partnerships and Alliances

The use of financial numbers for strategic customer analysis can be translated to a second concern for civil engineering organizations, establishing sound partnerships. While the selection of customers for strategic purposes can be characterized as a preference, selecting a proper partnership is essential. In contrast to a manufacturing operation that produces thousands or millions of copies of a product, a civil engineering organization creates a limited number of final products. As such, the civil engineering organization cannot afford to have a bad product. Even one bad project can result in significant loss of business due to a damaged reputation. This danger extends to the partnerships established by the organization. The selection of a partner that produces poor-quality work is a direct reflection on all project constituents. Extending this line of thought, a partner that is unable to meet their financial obligations reflects poorly on the entire team. Whether it is payments to a subcontractor, payments to suppliers, or payments to financiers, the recipients of this poor financial pattern may be unwilling to discern which organization was responsible for the problem. In this case, the entire team may absorb a negative result.

Similar to the financial analysis conducted for customers, financial statement data can be used to assist in reducing the potential for this negative result. Following the analysis introduced for selecting strategic customers, potential partners can initially be analyzed based on their ability to meet liquidity requirements. As illustrated in Figure 7-14, potential partners will have different balance sheet profiles. Using the same current ratio analysis introduced previously, the following calculation can be applied to the balance sheets.

$$\text{partner A} = 2,225,886/1,990,684 = 1.12$$
$$\text{partner B} = 2,596,823/1,145,568 = 2.27$$

Similar to the customer analysis, the partner analysis focuses on the ability of the partner to fulfill financial obligations. As seen by the calculations, partners A and B have very different liquidity positions. Given these data, debt considerations should also be taken into account to provide an additional level of financial analysis. However, prior to performing this analysis, civil engineering organizations should consider the strategic compatibility of the financial statements. Compatibility can occur in several areas. Potential partners can have asset compatibility. In this scenario, a potential partner could have a strong liquidity position and a weak investment position, but that may be compatible to an opposite financial position. In this scenario, each organization brings a financial advantage to the table for potential investor consideration. Similarly, a partner with a strong receivables account and weak liquidity may complement an organization with strong liquidity and minimal contracts in progress. The underlying component in each of these cases is the focus on balance. While much is written about compatible cultures (Sillars 1998), partnering still requires a benefit for each participant. Financially, this complement is a financial position that the organization requires to both qualify for a current project and a strong basis for long-term partnerships.

Balance Sheet—Partner A	in 1,000s		Balance Sheet—Partner B	in 1,000s
Assets			**Assets**	
Current Assets			**Current Assets**	
Cash and Equivalents	299,324		Cash and Equivalents	846,964
Accounts Receivable	930,104		Accounts Receivable	561,490
Merchandise Inventories	175,448		Merchandise Inventories	73,927
Deferred Income Taxes	58,039		Deferred Income Taxes	50,157
Contract Work in Progress	691,395		Contract Work in Progress	561,490
Other Current Assets	61,487		Other Current Assets	52,360
Total Current Assets	2,225,886		**Total Current Assets**	2,596,823
Land, Buildings, and Equipment	2,173,106		Land, Buildings, and Equipment	1,418,326
Construction in Progress	766,999		Construction in Progress	33,020
Less Accumulated Depreciation	1,001,315		Less Accumulated Depreciation	30,403
Net Book Value	1,938,790		Net Book Value	1,677,662
Investments	96,549		Investments	108,107
Goodwill	158,399		Goodwill	84,772
Other Assets	278,216		Other Assets	284,362
Total Assets	4,697,840		**Total Assets**	3,951,726
Liabilities and Stockholders'			**Liabilities and Stockholders'**	
Equity			**Equity**	
Current Liabilities			**Current Liabilities**	
Accounts Payable	878,187		Accounts Payable	204,186
Bank Loans	61,886		Bank Loans	29,916
Advance Billings on	525,518		Advance Billings on Contracts	445,807
Contracts			Accrued Salaries and Benefits	290,426
Accrued Salaries and Benefits	303,490		Other Accrued Liabilities	175,233
Other Accrued Liabilities	221,603		**Total Current Liabilities**	1,145,568
Total Current Liabilities	1,990,684		Long-Term Debt	
Long-Term Debt			Other Liabilities	90,833
Other Liabilities	598,859		Deferred Income Taxes	42,632
Deferred Income Taxes	66,739		**Total Liabilities**	1,279,033
Total Liabilities	2,656,282			
Stockholders' Equity			**Stockholders' Equity**	
Common Stock			Common Stock	
No par value; authorized	621,699		No part value; authorized	2,125,406
150,000,000 shares, 83,700,000			100,000,000 shares, 75,000,000	
shares issued			shares issued	
Translation Adjustment	(40,645)		Translation Adjustment	(33,239)
Retained Earnings	1,159,996		Retained Earnings	577,559
Total Stockholders' Equity	1,741,050		Total Stockholders' Equity	2,669,726
Total Liabilities and	4,697,840		Total Liabilities and	3,951,726
Stockholders' Equity			Stockholders' Equity	

Figure 7-14 The strength of a balance sheet can provide tangible evidence for making a decision on strategic partnerships beyond that provided by cultural and management correspondence.

Continuing with the financial analysis, the debt ratio analysis can also be applied to potential partners. Based on the data in Figure 7-14, the following calculations can be completed for partners A and B:

$$\text{partner A} = 2,656,282/(1,741,050 + 2,656,282) = 0.60$$
$$\text{partner B} = 1,279,033/(2,669,726 + 1,279,033) = 0.32$$

As seen in these calculations, the debt ratio of partner B is significantly less than that for partner A. This debt ratio is a particular concern to civil engineering organizations. The existence of an excessive debt ratio could be detrimental to receiving a contract since financiers will have a negative response to a significant debt holding. Placed in the perspective of developing a strategic partnership, the existence of a heavy debt ratio can forecast potential financial problems during a long-term partnership. Although the organization may be solvent at the current time, if the debt payments begin to create a financial drain on profits, then the partner will need to retrench financially. This could result in the partner not being available to participate in a strategically important project due to lack of financial resources.

Although a high debt ratio may indicate potential financial difficulties, an opposing viewpoint may exist. The existence of a high debt ratio may also indicate an investment in new technology, or an investment in new office facilities. For a construction firm, the debt ratio may also be the consequence of an equipment purchase. In either case, the debt represents an investment in resources that may not be available in the current organization. This dichotomy is the essence of financial analysis. The financial statements are a summary of current operations. The ratios only indicate issues that may require further analysis. However, without performing the ratio calculations, the organization will never be in a position critically to evaluate the strategic financial position of a possible partner.

Management Team

It is important to note that many external factors can influence a financial statement. Many of these factors, such as the economy, regional competition, and inflationary pressures, are beyond the explicit control of an organization. However, one financial influence within the control of an organization is the management team. The management team is a vital component of financial health. Ultimately, it is the management team that sets the financial direction for any organization. As such, stability and knowledge within this team are prerequisites for achieving financial stability. As an organization developing strategic relationships, analyzing the management team of a potential partner is as important as studying its technical abilities.

As stated, the two elements of interest in this study are the stability and financial knowledge of the management team. The former of these elements, stability, has a direct bearing on organization finance; stability creates a greater assurance that an organization will adopt a consistent approach to financial issues. Based on the fact that external perspectives of financial health are often limited to financial statements, consistency in these statements is essential for reassuring constituents that the organization is following a fiscally responsible path. Although investors and financiers are sometimes willing to overlook a single set of inconsistent data as an aberration, persistently changing financial numbers creates an atmosphere of uncertainty around the organization. This uncertainty brings with it a hesitancy to either invest in, or finance, operations. As documented by leading management analysts, a primary cause of this uncertainty revolves around a stable management team (Katzenbach 1997). This connection is based on the evidence that a constantly changing management team is unable to develop coherence in the decision-making process. Therefore, decisions are often delayed or made without thorough discussion of the ramifications that may result. In particular, decisions with significant financial impact are often delayed due to internal management struggles. This delay is subsequently reflected on financial statements that lack consistency from one year to the next.

The second element of management team study is the financial knowledge. In this context, financial knowledge is the ability to develop financial policy and objectives for long-term fiscal growth. In contrast to the short-term focus of projects, this financial emphasis is on the development of financial objectives that are reflected in positive changes on the balance sheet as well as in overall organization strength. For too long, civil engineers have relied on the language of engineering to communicate with the outside world. While the engineering community understands this language, it is not the language of the business environment. Rather, the business environment revolves around the language of finance. Civil engineers must embrace finance as a language of their own, or risk miscommunication with external constituents. In the context of partnerships, developing a strategic partnership with an organization that is incapable of speaking this language is equivalent to developing a partnership with an organization that has an insufficient engineering foundation. Making a commitment to understand and communicate in a new language is a time-intensive task. Making this commitment and then partnering with an organization that fails to do the same places the partnership on an uneven footing and places the long-term balance in doubt.

A specific example of management attention to finance and its impact on civil engineering operations is in the allocation of resources. While project managers focus on allocating resources to an individual project, senior management must allocate resources including finances across all projects undertaken within the organization. This strategic resource management must focus on the rate and timing of value generation and with rates of resource consumption (Birkett 1995). Major questions that must be addressed in this allocation include:

- How are financial resources to be used best in supporting strategies that will in turn generate positive financial growth?
- How are financial resources used in, or recovered from, change processes that support organizational strategies?
- How are financial resources to be deployed, leveraged, and redeployed over time?

The link here between strategy and finance is the interrelated concepts of financial return, competitive advantage, and resource allocation. Specifically, organizations need to examine the length of time that an investment in resources will take to both generate a positive financial return as well as contribute to the retention of a competitive advantage position. In a financial context, the competitive advantage period is the number of years a company is expected to exceed returns on investments (Mauboussin and Johnson 1997). Most organizations conduct strategic planning forecasts for a period of 3–5 years. In contrast, the competitive advantage period for most company's return on invested capital is between 10 and 15 years. This contrast can be a challenge for organizations desiring to establish a strategic alliance. If the management team in the proposed alliance organization does not understand the long-term financial benefits of investing in resources, then the organization may not wish to collaborate on investments that exceed the traditional payback period. In this manner, the partnership may fail to realize the benefits that could be accrued from long-term resource investment and leveraging.

Although financial knowledge and management stability may not be an initial consideration in developing strategic partnerships within the civil engineering industry, it is important for organizations to understand the role that financial health plays in long-term relationships. While the ability of the partnership to perform the required engineering tasks is fundamental to success, the component that may differentiate one team from another is the financial stability of the partners. Projecting a strong financial status is not only reassuring to potential customers, but also provides each member of the partnership with a reassurance that fiscal responsibility will not be an issue during the completion of a given project.

■ Economic Forecasting: Surviving Economic Roller Coasters

The final area of opportunity for strategic finance decisions is based on the previously introduced concept of economic forecasting. Although civil engineering organizations do not need to become experts in topics such as econometrics and economic modeling, the civil engineering sector is constantly buffeted by economic changes. Everything from material prices and labor availability to project starts and financing are impacted by economic swings. However, the focus on these external influences is often neglected due to their lack of immediacy. Often it is not until economic trends affect a request for proposal (RFP) that an organization pays attention to economic changes. Even at that point, the immediacy of the RFP may force organizations to focus on immediate cost impacts and neglect the forces that are causing the changes. The danger of this neglect lies in the arrival of unexpected regional, national, or global economic changes and their impacts on projected projects.

To illustrate the effects of these unanticipated impacts, the 1996 Olympic Games in Atlanta can be used. With 5 years of planning available, the organizing committee had a tight, but feasible, schedule to plan, design, and construct facilities. With a projected budget of over $500 million for Olympic facilities in the city, the engineering–construction effort was one of the largest and most concentrated at that time in the United States. The local organizing committee was confident that the schedule could be met and appropriately brought in some of the most experienced project managers in the country to oversee operations. The combination of effective management, a definitive end date, and a detailed statement of needs by the respective athletic associations provided the basis for design and construction planning. This basis was translated into requirements for primary competition venues, training facilities, athlete housing areas, and numerous media support facilities. Through the implementation of several unique project management components, the organizing committee produced plans for each of the required facilities. The unanticipated economic impact occurred when these plans were analyzed for final budget numbers. Unforeseen by the designers and builders was the inflationary pressures placed on the microlocation of Atlanta. With the extensive construction tasks being concentrated in a single area, prices for labor and materials quickly climbed. Labor, concrete, and masonry prices became especially variable, rising to almost double pre-Olympic levels in some cases. Placed in microeconomic terms, the demand for labor and materials quickly exceeded the available supply. This excess demand resulted in the inflationary pressures outlined in all basic economics textbooks. With the resulting increase in prices, the organizing committee was forced to revisit some designs and make cost-cutting decisions that significantly changed some venues. For example, the Aquatics Center that was originally planned to be an enclosed facility after the Olympics was never enclosed due to increased prices resulting from Games-related construction. Similar changes occurred in many of the completed projects.

Although the Olympics are an extreme example, in actuality, they are not necessarily outside the parameters seen in many other projects. For example, notable projects, including the EuroTunnel, the I-90 highway expansion in Colorado, and the Third Harbor Tunnel in Boston have all seen the effects of microarea inflationary pressures. However, these pressures are generally not reported, since the projects are either modified to meet budgets, or budgets are increased to meet the new prices. In either case, it is generally treated as an estimating miscalculation rather than an economic result.

Anticipating Trends: Breaking the Reactive Management Tradition

The response to the reactive mode issue lies in many different areas, as discussed throughout this book. However, in terms of the current discussion, civil engineering organizations need to adopt a greater focus on economic indicators as a predictor of upcoming economic developments. Specifically, organization leaders need to adopt a greater focus on the trends that are predicted by items such as the Leading Economic Indicators and the econometric model indices. By focusing on these elements, organizations have an opportunity to reduce their vulnerability to changing conditions as well as predict localized inflationary pressures. For example, a comparison between the LEIs and construction spending finds that construction spending follows closely behind the LEI changes. Given this simple comparison, civil engineering organizations can already make some predictions of upcoming economic changes.

Building upon this emphasis on economic indicators and historical data, a civil engineering organization can develop an individual analysis for its own financial data. The same time-series chart that can be developed for housing starts can be developed for individual organization items such as revenue and costs. Time-series graphs can be constructed based on the historical data found in the balance sheets. These time-series graphs can indicate cycles for the organization business. However, when combined with a time-series graph of an economic indicator such as the LEI index, the organization can begin to make relationships that apply specifically to the organization's market focus. Given these relationships, the organization can begin to develop predictive relationships. For example, as illustrated in Figure 7-15, the relationship between the LEI index and Superior's revenue appears to have a lead time of approximately 12 months. Translated from a historic perspective to a predictive perspective, these data can be used to predict upcoming changes in Superior's revenue. Of greater importance, the predictive function can be used to develop hedges against economic change prior to the realization of an economic downturn.

This example of economic analysis is characteristic of a proactive approach to economic cycles. As discussed in the beginning of the chapter, techniques such as

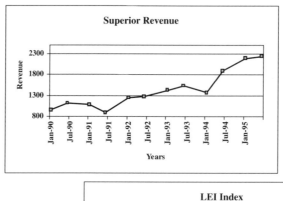

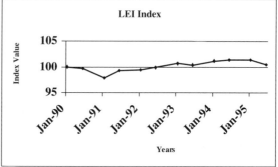

Figure 7-15 Basic economic forecasting can be accomplished at an individual organization level by methods such as comparing accumulated financial data with economic indicators.

time-series graphs and econometric modeling are providing organizations with numerous predictive data points. However, these data points are useless if an organization elects not to apply them to individual operating scenarios. In the case of the civil engineering industry, these operating scenarios include both resource supply and customer demand. Responding to these various market inputs requires organizations to develop economic plans that include multiple market segments to hedge against a focused recession, anticipating increases in prices to retain profit margins, and the development of strategic customers that have the economic stability to withstand recessionary pressures.

■ Return on Management: An Emerging Financial Concept for Civil Engineering

As a final component in this chapter, an emerging management and finance concept is introduced that is relevant to the operationally focused civil engineering industry, return on management (ROM). As introduced by Simons and Davila (1998), managerial energy is an organization's most scarce resource. As such, if managerial energy is misdirected or diffused over too many opportunities, even the best strategies stand little chance of being implemented and translated into financial value. In this concept, the amount of time that a manager spends away from core organization concerns is time that is not being invested into future financial growth. This concept is reflected in the ROM ratio of:

$$\text{ROM} = \frac{P}{T}$$

where P is equal to productive organization energy released and T is equal to management time and attention invested. P and T may be quantified as the amount of time focused on core issues versus overall time expended on all activities, respectively. P and T may also be stated qualitatively based on the intuition of the individual manager. The actual numbers are less important than the overall message developed by the ratio. If an organization accepts the pretense that management time is a valuable resource, then the ideal focus would be on achieving a ratio as close to one as possible. In this preferred scenario, all management activities would be focused on core or financially beneficial issues. In the opposite case, as the ratio returns a number close to zero, it indicates that excessive time is expended on noncore issues.

To illustrate this concept, consider two managers in a large engineer–procure–construct (EPC) organization, manager A and manager B. The organization is built upon a hierarchy that permits the two managers to run their individual divisions in relatively autonomous states. Manager A retains a tightly structured schedule that emphasizes the completion of management responsibilities. Manager B prefers a less structured approach to the management process, preferring to have the flexibility to respond to opportunities as they arise. Examining the ROM ratio of each manager, it is found that manager A spends about 70% of his time in productive management tasks. In contrast, manager B is spending only half of that time, 35%, on core management issues. As a result, manager B is working additional hours to accomplish the same tasks as those performed by manager A. Of greater importance is the difference in focus on external business prospects that the two managers display. Given the focus of manager A, he is able to translate a structured approach into aggressive market analysis sessions. In contrast, manager B spends extensive time discussing multiple opportunities, but fails to arrive at a focus on any single market segment. In this manner, manager B is expending extensive time on marketing issues without expending productive energy. Placed in a financial context, manager B is expending a high rate of financial resources for the resulting revenue return.

The previous example provides an introductory context to the ROM concept. However, to assist organizations in retaining a maximum ROM focus, five components can be analyzed to determine the emphasis an organization places on beneficial management tasks. First, the organization must be focused on its intended opportunities to pursue. Companies with high ROM ratios reduce the number of initiatives and clients that will be pursued outside strategic market sectors. Second, organizations must focus on potential success and not fear of failure. In situations that promote a fear of failure, managers will focus on preventing failure rather than on creating success. Third, has the organization specified management diagnostic measures that each manager understands? Low-ROM organizations may be concerned with too many diagnostic measures. High-ROM organizations make managers responsible for a limited number of diagnostic measures to reduce confusion about where energy should be focused. Fourth, high-ROM organizations make a focused effort on reducing paperwork and processes. Repeated meetings and progress reports eliminate the time a manager can spend on strategic issues. Finally, organizations need enough resources to ensure that managers are not spread too thin. In the scenarios where managers are spread too thin, the possibility significantly increases that managers may spend excessive time on inconsequential items since the manager does not have the opportunity to differentiate primary and secondary issues.

Although the ROM concept does not produce a dollar or financial figure such as those developed for financial statements and economics, the ROM ratio is integrally tied to the financial world for civil engineering organizations. In an industry that focuses heavily on project requirements, the amount of time that management is provided to focus on strategic issues becomes increasingly reduced. However, from a strategic management perspective, the latter effort is the one that ultimately results in positive financial growth and health. Placing management personnel in the position of continually addressing nonstrategic issues reduces the likelihood that the manager is focusing on financially beneficial tasks. Therefore, the greater the ROM ratio for an organization, the greater the likelihood that a strong financial position will be established. In this manner, the ROM ratio is an indicator of future financial directions and a potential warning sign for possible financial difficulties.

◼ Summary

Summarizing the financial aspects of a civil engineering organization is a multifaceted task. From project finances to organization finances and economic projections, the financial status of an organization is intertwined with the economic environment in which it operates. However, rather than adding an additional level of complexity, this environment can be used to strategically enhance the long-term viability of the operation. By using financial numbers to assist in identifying customers, partners, and market opportunities, an organization can anticipate both short-term and long-term trends within the industry. However, to achieve this potential, the organization must alter its focus from an exclusive emphasis on revenue to a greater emphasis on profits. Given this broader focus, the organization will be better equipped to make long-term marketing decisions.

◼ Next Steps

The establishment of a solid financial future is the basis for strategic success in any organization. The lack of positive financial indicators creates potential negative scenarios with customers, partners, and financiers. In contrast, the existence of a strong financial position places the organization in the position aggressively to pursue strategic objectives. A central component of these objectives is the exploration of new market opportunities to expand current income streams. Chapter 8 builds upon this premise by exploring these new market opportunities and the risks associated with new ventures.

DISCUSSION QUESTIONS

1. With a strong emphasis on project finances as a basis for civil engineering organizations, the move to an emphasis on organization and economic concerns as a basis for strategic management may be difficult to accomplish in some organizations. What are the primary barriers to accomplishing this change in financial perspective? Are these barriers different for large and small organizations?

2. Given the balance sheet illustrated for Superior Engineering in this chapter, would Superior Engineering make a good investment decision for investors? Why are most civil engineering firms private rather than public organizations?

3. Economic forecasting is one tool used by manufacturing industries such as aerospace and automotive to predict future orders. In contrast, civil engineering organization leaders often dispute the value of forecasting due to the industry's reliance on current market requirements. What current or past economic forecasting models could be used to dispute this claim?

4. Selecting financially sound organizations is presented as a basis for long-term strategic success. What other factors beyond financial resources should be considered when selecting customers?

5. Partnerships are touted as the AEC project arrangement of the future. Financial analysis of potential partners is emphasized in this chapter as a hedge against poor partnership decisions. How does an organization leader inform a potential partner that the organization wants to conduct a financial analysis without risking a professional insult?

6. If an organization conducts an economic forecast and determines that its current market segment may be susceptible to an economic downturn in the near future, to what degree should the organization change its current market emphasis?

<table>
<tr><td>

**Case
Study**

</td></tr>
</table>

Short-Term versus Long-Term Financials

The balance between demonstrating short-term financial returns for investors or owners and developing long-term strategic plans is an issue that transcends industry boundaries. Whether the organization is in the manufacturing domain, the hospitality domain, or the engineering service domain, short-term financial pressures can often overtake the need to focus on strategic objectives. Specifically, the need to demonstrate market share or revenue increases on a monthly or quarterly basis can place organization leaders in a position that prevents individuals from taking risks that may adversely affect these numbers. For individuals placed in positions of leadership in a distributed organization, this pressure develops an additional component. Since financial results are combined to form an overall organization financial statement, the pressure on regional offices to meet individual goals can be extreme.

Placed in an executive position within such a regional office, an individual that desires to change this financial model faces a significant challenge. To highlight this challenge, the case of ICF Kaiser Engineers and Constructors (ICF), a division of ICF Kaiser International, is examined. Based in Fairfax, Virginia, ICF is one of the largest engineering, construction, and consulting companies in the United States, with 5,000 employees and 70 offices worldwide. With company roots dating back to 1914, the organization has established itself as a worldwide competitor. With revenue of $1.1 billion, ICF ranks as the twelfth-largest design firm and the fifteenth-largest construction management firm in the United States. Founded on the principles of leadership, performance, partnership, breadth, and reliability, ICF now provides engineering and construction services in the infrastructure, process, and microelectronics sectors. Divided into geographic regions, both in the United States and internationally, ICF provides individual offices with the autonomy to pursue a diverse range of projects within the ICF business sectors. For example, current projects include a refinery in Western Australia, process plants in the Midwest United States, and steel mills in the Czech Republic and China. Given this sector and geographic diversity, ICF believes it has positioned itself to hedge against future economic cycles.

The Focus

At the center of the ICF Kaiser financial case is Tom Bartlett, head of a district office. With over 30 years in the construction business, Bartlett has experienced all facets of the design and construction industry. From field supervision to project management and senior management, he has experienced both project and organization financial perspectives. With this background, Bartlett understands the need to meet project expectations and the need to meet organization objectives. With 6 years at ICF, Tom believes that changes should be implemented in the strategic approach to finance. Specifically, the emphasis on short-term earnings is overshadowing the focus on long-term strategic objectives. With a constant pressure to produce results in quarterly increments, the focus on expanding market focus and investing in strategic resources is often placed behind the need to demonstrate earning potential to Wall Street investors. As a career construction professional and a senior manager, Bartlett believes that this emphasis is contrary to the evolving needs of the engineering and construction industries. In particular, the extended time periods required to design and construct projects, the extended development phases of projects, and the cyclical nature of the industry create an environment that should promote extended analysis of strategic positioning. However, with an emphasis on quarterly earnings, the decision to pursue projects to demonstrate earning potential often outweighs the need to determine the long-term benefits of the project.

Given that ICF is a publicly traded company, Bartlett understands that investors ultimately determine the financial strength of the organization. As such the need to release quarterly figures such as those in Note A are an integral part of the business process. In addition, the need to demonstrate stock value over an extended period of time are critical to attracting future investors (Note B). These financial demonstrations are basic requirements for public companies. Whether the organization is in the manufacturing, banking, or service industries, the demonstration of quarterly returns is a basic tenet of the financial system. However, in Bartlett's perspective, this emphasis becomes clouded within the engineering–construction industry. Specifically, the corporate structure of the organization creates an environment that precludes strategic thinking. In this corporate structure, the parent company is responsible for reporting earnings for Wall Street investors. However, since projects are predominantly awarded at a regional level, the parent organization is essentially the administrative component of the organization. As such, the parent organization relies on the individual divisions, regions, and districts to generate the earnings that, when combined, demonstrate the strength of the overall organization. In the case of ICF, this downward focus results in ICF International setting goals for ICF Engineers and Constructors, which in turn sets goals for its individual regional offices. Each region in turn sets goals for specific districts. It is at this level that Tom Bartlett receives his annual and quarterly objectives.

Although this hierarchical process is similar to many industries, Bartlett believes that the engineering–construction industry is ill served by this model. Since Bartlett has minimal input into the final objectives, when they are outlined for him, he is placed in the position of focusing heavily on obtaining projects to meet the financial

NOTE A

(in thousands, except per share amounts)(Unaudited)	Three Months Ended March 31	
	1998	1997
Gross Revenue	305,256	265,957
Subcontract and direct material costs	(199,020)	(162,266)
Equity in income of joint ventures and affiliated companies	1,027	287
Service Revenue	107,263	103,978
Operating Expenses		
Direct labor and fringe benefits	69,435	70,000
Group overhead	22,089	19,827
Corporate general and administrative	4,944	5,184
Depreciation and amortization	2,335	2,369
Operating Income	8,460	6,598
Other Income (Expense)		
Interest income	466	339
Interest expense	(4,821)	(4,353)
Income Before Taxes and Minority Interests	4,105	2,584
Income tax provision	1,149	568
Income Before Minority Interests	2,956	2,016
Minority interests in net income of subsidiaries	2,511	1,969
Net Income	445	47
Foreign currency translation adjustments	554	(142)
Comprehensive Income (Loss)	999	(95)
Basic Earnings Per Share	$0.02	$0.00

NOTE A ICF Kaiser International consolidated statements of income.

objectives. This gives Bartlett little opportunity to explore the strategic options open to him in his specific market. For example, his market is currently experiencing a strong move into alternative procurement strategies. Bartlett believes that ICF's strengths could place the organization into a competitive early position for this market change. However, the continued emphasis on short-term returns is limiting his ability to investigate this strategic direction. With directives coming down the hierarchy, the strategic planning that initiated the original market goals at the corporate level is changed into specified targets for individuals at Bartlett's level.

The Issue

Given the traditional financial scenario at ICF, Tom is faced with a serious issue. If Bartlett continues to pursue projects to demonstrate short-term earnings, he will

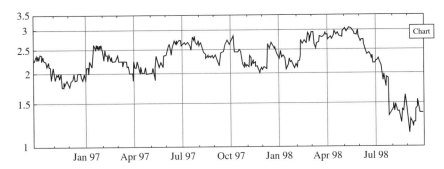

NOTE B Two-year price per share of ICF international stock.

successfully meet his corporate objectives. Meeting these objectives will reward him financially as well as place his district in a favorable position at the corporate level. In other words, Bartlett will have successfully achieved his corporate objectives. However, he questions whether this is the appropriate goal for the office. After spending time investigating the region, he believes that additional competition is creating an environment that will demand companies to differentiate themselves in terms of services to the client. Failing to develop new capabilities will place traditionally strong organizations in a defensive position against new competitors. Although Bartlett believes that ICF is a strong organization, he has witnessed the opening of district offices for most of ICF's national competitors within his district in the past 5 years. Furthermore, these national organizations have targeted his district as a primary growth opportunity. With reputations, experience, and work forces equal to or greater than ICF's, these organizations are placing themselves in direct competition for Tom's potential revenue sources.

Recognizing that his district has a finite capacity to support engineering–construction projects, Bartlett believes that a focused effort on developing long-term financial growth opportunities is essential. He would like to investigate the strategic opportunities in the district beyond the traditional ICF markets as well as invest in resources that would further differentiate ICF from its competitors. However, faced with short-term revenue goals that are increasing on an annual basis, Bartlett does not believe that he can pursue these strategic objectives without taking away from the revenue focus. Placed in this position, he sees that he has few choices. He can elect to stay on his current path and focus on meeting his short-term revenue goals. He can attempt to convince his regional vice president, Stan Lindsey, that his district needs to invest in strategic analysis and thus his profit margin should be lowered for the quarter or year while he invests in personnel. Finally, Bartlett can elect to discuss the situation with the corporate office in Virginia directly, with the intent to persuade ICF senior management that a greater emphasis on strategic finance is important to long-term organization health. Bartlett is not particularly satisfied with any of these options, but he believes that a decision has to be made sooner rather than later.

QUESTIONS

1. Has Bartlett overlooked any of his options to pursue strategic objectives while retaining his emphasis on revenue expectations?

2. Placed in Bartlett's position, what approach would you take in this situation?

3. Placed in Stan Lindsey's position, if Bartlett approached you with his idea, what would be your response and why?

4. Placed in the position of senior ICF manager, how would you address stockholders at an annual meeting if the pursuit of strategic opportunities results in a noticeable drop in profits or revenues?

EVALUATING MARKET OPPORTUNITIES

8

What market analysis procedures are in place proactively to identify new and expanded organization opportunities?

The examination of new market opportunities is not a new concept for civil engineering organizations. The expansion by Bechtel, Black & Veatch and M. W. Kellogg into the global petroleum market is not an uncommon example in the history of the industry. However, these examples have predominantly followed the reactive model of market expansion. By building upon existing project procedures and reacting to economic developments, the organizations have placed themselves in the position to succeed in new markets. While this continues to be a viable approach for organizations that dominate narrow markets, the vast majority of organizations need to identify emerging markets strategically in a proactive model of market development. In this proactive model, organizations need to combine economic, financial, and risk analyses into a comprehensive approach that identifies market opportunities within public, private, and nontraditional domains. Through the combination of quantitative analysis and market identification, organizations possess a potential for growth that is limited only by internal organization barriers. Concurrently, the opportunity to break from market niche constraints is limited only by the expanse of the vision established as the organization roadmap.

In this chapter, the development of a market development strategy is examined in the context of market analysis and innovation. This chapter emphasizes the need to develop creative and analytic approaches to market development as entry points to expanded market opportunities. The contrast between reactive and proactive approaches to market development is examined, together with their benefits and limitations from a strategic perspective. Additionally, the contrast between public and private markets is introduced, together with the risks associated with entering previously unexplored market niches. Finally, an introduction to risk management is presented as a tool to reduce the barriers associated with entering unexplored markets. The combination of these topics sets the challenge for this chapter—*to examine evolving mar-*

ket opportunities and establish an expanded organization vision that encourages exploration and innovation.

■ Why Market Analysis and Expansion

Whether an organization is a multi-billion-dollar international engineer–constructor or a specialty engineering firm with less than a million dollars in annual revenue, the thought of market expansion can exact similar concerns. Although each organization is different, these concerns can be summarized in two common areas, comfort zone and control. The first of these reasons, comfort zone, emphasizes the daily routine that an organization adopts over an extended period of time. While organizations scramble to react to daily emergencies, real or imagined, these daily escapades are a part of a comfortable routine. The players in the operations understand the extent of their authority, the extent of their project responsibility, the competition, and the owner expectations. Similarly, the organization members responsible for business development understand the clientele, the benefits of hiring the organization, the market sector of interest, and the unwritten rules of the game concerning bidding, suppliers, subcontractors, and consultants. Expanding into a new market segment threatens each of these comfort zone elements. Members must acknowledge new operational rules and expectations, while leaders familiarize themselves with project responsibilities that were never addressed in lower-level responsibilities. Additionally, business development personnel must learn the intricacies of new client relationships and competition. In summary, every long-held belief concerning operational procedures becomes threatened when the concept of expanding the current market focus is introduced.

Similar to the comfort zone challenge is the challenge to operational control. In today's business environment where civil engineering firms expand and merge into national and international organizations, the concept of control may seem outdated. Specifically, the concept that an individual, or small group of individuals, can directly control all aspects of operations on a daily basis appears to be losing ground as quickly as organizations are expanding. However, in contrast to this surface appearance, the authors have found through personal interviews that civil engineering firms demonstrate a strong affinity for closely held organizational control. Management structures within the organizations tend to reflect closely the management styles of senior management. Similarly, the market focus reflects the experience of the senior personnel. In this manner, senior personnel retain the ability closely to guide the organization direction and procedures. Introducing a new market segment threatens this control. The organization may require outside experts to address the new market while at the same time requiring changes to procedures that will allow personnel more flexibility in addressing client requirements. Additionally, the perceived position of the organization in the existing market segment cannot be automatically transferred to a new market segment. Where once the organization was a leader directed by strong control procedures, the organization may now find itself playing catch-up to previously entrenched players and facing the need for procedures to adapt to the new market position. This change is a direct challenge to organization traditions and leadership.

Given these threats to existing procedures and comfort zones, the question of why an organization should expand its current market focus must have a convincing answer. In strong market and economic conditions this convincing answer is difficult to answer in the civil engineering industry. While fragmented, strong economic conditions provide individual organizations with the profit margins and volume that sustain and grow operations. However, the twenty-first-century civil engineering business environment is changing these parameters. Relying on the existence of a strong economy by itself will no longer be sufficient to retain market position. Put simply, an organization that elects to retain its position based on past performance will quickly find itself bypassed by aggressive competitors with greater vision and strategic concepts. Therefore, to answer the question of why an organization should pursue new markets, three justifications are presented: a numeric-based approach, a strategic-based approach, and a client-based approach.

Expansion Needs by the Numbers

Strategic management is a combination of people skills and numeric analysis. As demonstrated by the spectrum of topics introduced up to this point, developing a strategic management focus requires organizations to address elements as diverse as human collaboration and financial analysis. However, the civil engineering tradition remains steeped in a numeric basis. Civil engineering is the application of physical concepts in the built environment. With this as a basis, civil engineers will often choose a numeric-based explanation over a "soft" management-based decision. As such, a numeric-based justification is initially presented to explore market expansion opportunities. Specifically, the numeric-based justification focuses on revenue and competition.

As introduced in Chapter 2, the economic indicators present the civil engineering industry with a positive outlook for future operating conditions. Unfortunately, these statistics do not tell the complete industry story. In reality, while the overall industry is healthy, individual organizations face daunting revenue challenges. As illustrated in Figure 8-1, the general and heavy construction component of the civil engineering industry had an aggregate total of $311 billion in revenue in 1995 (the last year of available data). On the surface, this number appears healthy, as it represents a continuation of the 4–9% increases seen in recent years (U.S. Census Bureau 1997). However, the $311 billion number is deceiving. To illustrate this contradiction, a closer examination of the construction sector is required. As of 1992 there were approximately 206,000 general and heavy contractors in the United States (U.S. Census Bureau 1997). If these 206,000 contractors divided the market evenly, each contractor would have revenues approaching $1.5 million. This number is hardly enough to keep many of the firms afloat, and it certainly is not sufficient to weather recessionary times. But, of course, in a free-market economy every organization does not divide the market share equally. This is clearly evident in the construction industry. When analyzed further, the economic prospects for many construction firms must be seriously questioned. As illustrated in Figure 8-2, of the $311 billion spent on con-

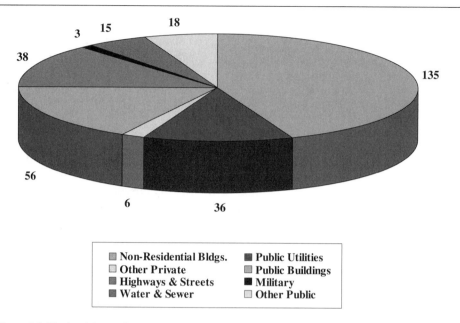

Figure 8-1 The breakdown of construction expenditures within the civil engineering industry shows broad strength in numerous sectors.

struction, the top 400 contracting firms accounted for $122 billion of the revenues (ENR 1996a). This represents a 10.8% increase from the previous year. Placed in the context of the overall industry, these numbers translate to 0.2% of the contractors receiving 39% of the annual revenue. Placed in terms of average revenue, the top contractors average $305 million annually compared to an average of $920,000 annually for the remainder of the organizations, or a 300:1 ratio. Similar to the construction numbers, the engineering design side of the industry follows a similar pattern. While the entire design component of the industry had revenues of approximately $77 billion (U.S. Census Bureau 1996), the top 500 design firms accounted for $29.4 billion (ENR 1996c). This translates to almost 38% of the revenue being accumulated by less than 1% of the industry.

These revenue numbers paint two distinct pictures for the market expansion argument. For the firms in the top echelon of the industry, the comfort zone exists because the competition is well known and the playing field is well defined. For every project that becomes available, a limited number of players have the ability to pursue the project. Knowing this dynamic, each organization in this segment can predict competitive forces. This understanding breeds familiarity and comfort. It can also breed complacency. While understanding the competition is fundamental to strategic success, competing with the same organizations for the same revenue in a limited market segment places the organization at risk. Ultimately, the organizations become dominant in a limited market that leads to specialization and dependency on the health

Industry Revenue Disbursement

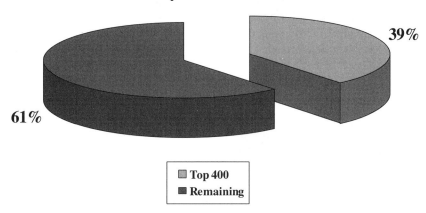

39%

61%

Top 400
Remaining

Figure 8-2 The broad strength in the civil engineering industry is heavily weighted towards the top 400 contractors in the industry.

of the market segment. To illustrate the danger associated with this dependency, consider the late 1980s and the decline of the nuclear market. During the peak of the nuclear market in the 1970s, a few firms such as Stone & Webster became dominant in this market. During this time, the firms witnessed revenue increases and growth previously unknown in the industry. The combination of cost-plus contracts and speculative building acted as bait for the organizations. Responding to the bait, the organizations focused on the nuclear market and established expertise in the area. Believing the market to have unlimited potential, the organizations tied their revenue projections to the market segment growth. In the early 1980s, the inflated market came to a crashing halt amid the ruins of Three Mile Island and 300% cost overruns. However, since the public was paying the development costs, the utility companies did not suffer a significant setback. In contrast, the contractors that were tied to the limited market segment absorbed significant damage. Organizations such as Stone & Webster absorbed 50% cutbacks and more. They were simply not in a position to compensate for the revenue loss, since their resources were focused on a narrow market segment. They had become comfortable, complacent, and vulnerable.

For the smaller organizations, the revenue issue paints a similar, but distinct, picture. Rather than establishing a dominant position in a small market and then becoming complacent as the organization becomes comfortable with a revenue stream, the small firm is faced with the position of competing for a significantly smaller revenue stream from significantly smaller clients. Given the fact that 205,000 contractors and 68,000 design firms compete for an average revenue stream of $920,000 and $1,125,000, respectively, little room is left for focusing on long-term plans, since extensive effort must be placed in retaining the limited revenue. For these organizations, it does not take the collapse of an entire market sector to inflict serious damage. Rather, with limited revenue and significant competition, these firms are vulnerable to numerous localized fluctuations. By retaining a focus on a limited market, an organization becomes dependent on the continued ability of that segment to support current operations. The failure of the segment to retain that capability results in an immediate drop in revenue for dependent organizations, and as a result, organization instability. Breaking from this dependence requires an organization to expand its focus to a broader market segment.

Strategic Differentiation

Why should a customer choose one civil engineering organization over another? The answers frequently cited by organization leaders can be recited by many in the industry. Issues such as quality, cost, customer focus, and experience are all cited as factors that separate one organization from another. Are these issues enough to differentiate one organization from another? Or are these issues actually surface differences that mask the underlying similarities between the majority of firms in the industry? This is a difficult question for organization leaders to respond to honestly. In fact, an honest answer may result in questioning the fundamental premise on which an organization pursues clients. Merely being less expensive or having a stronger quality control program is not strategic differentiation. These issues are important, but represent surface qualities that any organization can match if the decision is made by organization leaders to pursue such a direction. With this as a tenuous hold on market positioning, organization leaders must address a follow-up question, "Can the organization survive in a market leadership position over an extended period of time based on surface differentiation?" The organization making an honest assessment will find that the answer is no.

Establishing strategic differentiation requires an organization to have a market presence that is fundamentally different from the competition. For example, given two civil design firms, one that focuses exclusively on site design, and one that offers both traditional civil design services and environmental impact analysis for projects in environmentally sensitive areas, a strategic difference exists in the services offered. The latter firm presents the customer with multiple services extending over multiple project phases. Whereas the former firm focuses on a traditional approach to civil design, producing a product that is time tested and familiar to customers, the latter firm extends the industry boundary. In this case, the customer observes strategic differentiation. For the organization, the multiple services offered by the second firm place it in a position to pursue projects from multiple perspectives. It can be a one-stop service provider for environmentally sensitive site permitting, it can compete on a traditional civil design basis, or it can emphasize its environmental services for a third distinct market. In contrast, the first firm is limited to emphasizing surface differentiation such as experience, quality, and cost. While these differences are important, they are a temporary differentiation. Not only can the second firm claim equal capabilities in these areas; it also has the expanded market focus that the former firm has no ability to defend. The latter firm obtains a strategic advantage that is difficult for the former firm to threaten immediately.

As seen in this example, strategic differentiation requires an organization to offer services that competitors are not currently in the position to offer. As in the example, the traditional firm had not cultivated innovative services and as a result was unable to compete with the more innovative competitor. Over the next decade, the strategic differentiation concept will become increasingly important as the era of competing solely on price or experience slowly ends. However, to establish this differentiation, organizations must explore market segments that fall outside their traditional purview. Rather than exclusively focusing on the incremental improvements that can be achieved in existing practices, organizations need to place some of this effort into exploring new market and service opportunities. The bottom line for these organizations is straightforward—remaining in a single market segment with a single service will ultimately result in being surpassed by organizations exploring expanded strategic opportunities.

The need for this differentiation does not imply that an organization should haphazardly adopt new services and market focal points. Expanded services should complement existing strengths. The new services and markets should be a logical expansion of the current services. As illustrated in the site example, the expansion into environmental permitting is a natural extension to the traditional site design services. Any customer that desires to develop property in an environmentally sensitive area will require environmental consulting services. Placing the organization in the position to provide both services demonstrates a seamless expansion of existing services. In this manner, the organization achieves strategic differentiation while concurrently establishing a market position in a related project component.

Client-Based Expectations

The final justification for expanding market focus diverges from the internal financial and service analysis previously presented to focus on emerging customer expectations. From sources as diverse as video games, television specials, newspaper stories, and internal experience, customer organizations are becoming increasingly familiar with the role that civil engineering organizations play in the development of projects. Whether it is a project development consultant, permitting and analysis consultant, or construction firm, customer organizations understand that knowledge of these entities is crucial to the financial success of a project. When compounded by the increased wariness of financial organizations to invest money into speculative projects due to past economic disasters, as well as stockholders who demand positive returns from capital investments, customer organizations are under increasing pressure to develop projects that are financially sound. This situation appears to be no different from the demands that have always existed. The existence of shareholders and cyclical financial situations is fundamentally the same as it has been for decades. Why is today's customer any different from a strategic and market viewpoint? The answer to this question returns to the increased knowledge of the customers. Whereas customers previously tended to hire an architect or contractor with little regard for nontraditional services and then removed themselves from the inner workings of the project, today's customer has a greater understanding of the building process and the parties associated with the process. As such, the customer is making greater demands of the civil engineering organizations associated with the project. The first of these demands is the reduction in the number of parties involved in the project.

Viewed from the customer perspective, the greatest waste of project funds is in the multiple overhead percentages that must be paid to specialty consultants. When faced with ten or more consultants on a project, the customer is faced with overhead and profit expenses from each participant. Given greater knowledge of the project process and the participants associated with the process, it is understandable that the customer would prefer to reduce these redundant costs by reducing the number of project participants. Real or perceived, this difference is tangible to the customer. As such, those organizations that respond to increasingly knowledgeable customers will have a greater opportunity to obtain a strategic market advantage.

Reinforcing this increased demand for service consolidation is the increasing number of customers who are pursuing design–build options. In 1997, design–build revenues reached almost $40 billion in revenue (ENR 1998). This increase is a direct result of two catalysts. First, from a regulatory standpoint, the Federal government is passing legislation such as the Two-Phase Design–Build Selection Procedure bill (Federal Register 1997), to encourage public owners to pursue design–build options. Second, private owners are recognizing the savings that are available through design–build procurement arrangements. In areas such as industrial production, single-source development options such as design–build are providing the delivery time improvements required for reduced design-to-market time frames. The civil engineering organizations that have aggressively pursued design–build options have established strategic positions that place competitors in a catchup position.

One should not forget the lesson of the U.S. automotive market in the 1970s. When domestic customers decided that economy cars were the preferred mode of transportation, the automotive industry questioned this decision and waited to react. This delay cost the industry significant market share and almost placed Chrysler and Ford in bankruptcy. It took over a decade for these organizations to recover from the mistake and remember that customer preferences will always drive organization success. Civil engineering organizations must take this lesson to heart. Remaining tied to a narrow market segment when customers are demanding a greater spectrum of services is a failed business waiting to materialize. It is up to the organization as to whether or not it wants to test the logical conclusion of the market forces or put in place the market analysis procedures required to change the outcome.

Summary

In summary, the previous three discussions provide three distinct justifications for exploring alternative market opportunities. However, making a decision to explore a new market represents only the first step in the process. The prepared organization builds upon this decision by understanding the opportunities, barriers, and risks associated with potential market opportunities. The following sections address these issues to establish a strategic foundation on which to pursue expanded market opportunities. First, to establish an understanding of the direction markets pursue and the appropriate responses to these market turns, the next section contrasts traditional market responses with a new focus on proactive market response. Second, to focus on opportunities, a discussion is presented on the emerging market focal points within the civil engineering industry. Finally, to assuage organization leader concerns on the risks associated with exploring new opportunities, a quantitative look at risk management is presented that can be applied to new market analysis.

■ Reactive versus Proactive Market Responses

The civil engineering industry has only recently begun to break from a conservative approach to market response (Arditi and Davis 1988; Mitchell 1995; Bady 1995). External inputs such as competitor moves have traditionally guided market direction to the point where reactive market response is standard operating procedure. Reinforcing these external inputs is the overwhelming perception that daily project operations must take priority over long-term market analysis. In combination, these factors create a classical business environment that uses reaction as the accepted planning process. In this business model, organizations elect to postpone market expansion or growth until external forces demand a market move. For example, a construction organization may recognize that the market is opening opportunities in the program management project phase. However, the pressures of project management may override the opportunity to pursue this market course. It may not be until competitors begin to pursue the new opportunity that organization leaders will recognize the need to pursue the identified course. This hesitation and reaction to external influences characterizes a reactive approach to market response.

The strategic management approach to market analysis requires civil engineering organizations to break from this reactive market response. Although civil engineering business practices and the conservative personalities of engineers encourage conservatism, the changing business environment is limiting the luxury of this approach. This responsive approach must be replaced by an aggressive market analysis approach. Specifically, civil engineering organizations must adopt a long-term vision for market positioning. At the core of this long-term vision is a proactive market approach that emphasizes the identification, aggressive pursuit, and reinforcement of new market services and sectors.

Traditional Market Outlook: Reacting to Current Industry Forces

Although the identification of the traditional reactive environment is a necessary element in breaking from the reactive mold, identification does not equate with understanding. For an organization to adopt an aggressive market development strategy, an understanding of the reactive inputs is necessary. For civil engineering organizations, these inputs include competitors, customers, new entrants, and substitutes (Fig. 8-3) (after Porter 1975).

Competitors

The first input that impacts market response is the industry competition itself. In contrast to industries such as aerospace, defense, and computer software, where a few organizations represent the dominant players in the field, the civil engineering industry is primarily a fragmented industry with numerous small organizations. The difference in this characterization is the impact on strategic advantage. In an industry dominated by a few players, competition between those players is intense, but the exclusivity of the situation permits the organizations to retain profitability margins as long as each player retains a sizeable market share. Given this market share, the organizations are not forced to expand the share by thoughts of bankruptcy or obsolescence. For example, in the auto industry, the big three United States auto manufacturers divide market share in double-digit increments. In contrast, the civil engineering industry is characterized by thousands of individual organizations competing for market share. This competition results in organizations needing to both protect limited market share and respond to competitors that threaten existing customers.

With this requirement to respond to competitors, organizations are forced to react continuously to competitor moves within the marketplace. An organization's opportunity to focus on differentiation is limited due to this constant focus on reacting to competitor moves. Compounding this difficulty is the pressure to keep controls on

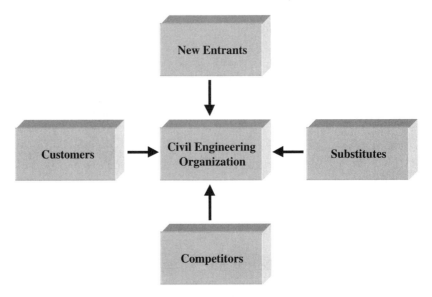

Figure 8-3 Market forces include both internal and external influences that each have the potential to threaten an organization's customer base.

costs to remain competitive. The focus to keep project costs closely controlled in accord with reduced profit margins reinforces the project-based emphasis at the core of today's civil engineering industry. In contrast to industries with large market shares distributed among a few players, the profit pressure in the civil engineering industry diminishes the opportunity to obtain discretionary funds that can be used for exploratory market expansion.

With these budgetary and financial pressures, it is understandable that civil engineering organizations predominantly focus on reacting to competitor moves. The strategic management issue that must be addressed is where to find the financial opportunity and the time to break from this pattern and change a defensive pattern of market response to an offensive positioning that preempts competitor moves.

Customers

The second impact on reactive market behavior is the focus on customers. Responding to the traditional business adage that the "customer is always right," the civil engineering industry carries this adage to the next level of "the customer determines the business." The civil engineering industry not only caters to the customer, but also tends to hesitate to offer new services until the customer demands them. Rather than actively anticipating needs and implementing services or expanded market focal points that will address the expected needs, the customer-reactive organization postpones implementing these decisions until a customer requests the service or expanded market focus. In this manner, the customer drives the market strategy rather than the organization setting strategic direction. Although this approach provides a safety net by ensuring that the customer is interested in the product, it has a negative effect in that every competitor has the same opportunity to provide the service. By reacting to the customer rather than anticipating need, the organization places itself on a level playing field with the competition rather than establishing an initially advantageous position.

In defense of the industry, this conservatism is grounded in concern for public safety. Convincing a customer that a new material or construction process should be incorporated into a project invites risk that is often unacceptable to the organization. In contrast to products such as automobiles, microwaves, and power tools, customers do not have the option of reading a consumer advocacy magazine to determine the safety factor of a civil engineering design. The customer is dependent on the qualifications of the civil engineering organization to produce a project solution that not only meets the stated requirements, but also meets the implicit requirement that the project establish a safe environment for the public. This expectation does not create significant room for experimentation and innovation.

New Entrants

The third component impacting reactive markets is the introduction of new entrants into the civil engineering marketplace. In contrast to manufacturing-based industries, service industries erect few barriers to organizations entering the marketplace. Initial capital investment is relatively small, a factor that is significantly different from heavy manufacturing. Marketing costs are relatively small in comparison to product manufacturers, and name recognition and loyalty is not a critical factor as it is for designers of items such as appliances and computers. The only barrier that realistically keeps a new entrant from entering the civil engineering marketplace is experience. At this point in time, any individual with experience in a civil engineering discipline can open a business venture with little resistance from existing industry entities.

The consequence of this open market is that existing organizations must continually focus resources on gaining knowledge of new entrants in the field. In the reactive market, existing organizations can choose to either ignore new entrants with the perspective that the new entrants have little chance to impact existing clients, or choose to react to new entrants in a defensive position. In this defensive maneuvering, existing organizations are forced to compare their tradition and experience with the new player, in essence, defending their right to continue to exist within the marketplace. In this mode, business development personnel must continuously establish differentiation between the existing organization and the new entrant. The concern is that, other than highly specialized types of work such as industrial design or dam construction, an aggressive new entrant can prepare proposals or presentations that are equal to or better than the products produced by existing organizations.

The lack of barriers to new entrants creates a decision point for existing civil engineering organizations. If an organization intends to focus on growth opportunities, then the organization must aggressively establish an individual identity that creates service opportunities that are beyond the capacity of new entrants. However, if this same organization decides to adopt a defensive market position, then the aggressiveness required for individuality will be lost in a constant series of moves countering the introduction of services by new entrants.

Substitutes

The final impact on reactive market forces is the introduction of possible substitutes into a market sector. In this scenario, an organization that provides an alternative service but has not traditionally competed in a market sector challenges existing organizations by introducing an alternative service to the sector customers. To illustrate, take the example of construction management services. Traditionally, construction management services including estimating, scheduling, and project controls were either provided by a contractor or construction consultant. The emphasis of these organizations was on the knowledge and experience required to develop a capital project successfully. With customers having limited knowledge of the construction industry, the construction-based organizations had the ability to present a convincing argument based on the experience theory. In fact, very few new entrants entered the competitive market due to the understanding that owners were satisfied with the existing arrangement. However, the increase in knowledgeable owners has begun to change this dynamic. Whereas previously owners were content to leave capital expenditures within the domain of facility managers, today's owners are increasingly viewing infrastructure expenditures as strategic components of long-term planning. Appropriately, these owners are demanding that projects coincide with strategic objectives and development. In response, nontraditional competitors, such as business consultants, are providing owners with a substitute for traditional construction management. Specifically, by providing construction management services that also emphasize components such as tax protection, depreciation, facility leasing, and bond disbursements, these substitutes are addressing the language of the owner. In this language, the substitutes emphasize the strategic picture with which the owners relate rather than the project picture with which the facility managers relate.

With this slow, but growing, influx of industry substitutes, civil engineering organizations must make a decision as to whether or not they view these players as threats to existing market segments. Currently, interviews with organization leaders have found the reaction to these substitutes to be primarily dismissive (Goodman 1998). With a perspective that these substitutes lack the tradition of mainstream civil engineering organizations, organization leaders are expending little energy to counter entering substitutes. Unfortunately, this lack of reaction will ultimately lead to a similar reactive response as that seen with new entrants. By giving the substitutes the opportunity to establish market recognition by a few customer organizations, existing organizations are implicitly taking a defensive market posture. Extending this posture out for an extended time period places the existing organizations in the same position as they face with new entrants. Rather than establishing individual identities that validate the use of civil engineering organizations to perform civil engineering tasks, the organizations are permitting customer-knowledgeable organizations to redefine the market playing field.

As the last element in the reactive market forces, substitutes may appear to be the least threatening of the four market factors. However, the threat from substitutes should not be taken lightly. Organizations such as business consultants have a stronger history of responding to customer needs than civil engineering organizations. The revenue associated with the civil engineering industry is far too great for these substitute organizations to ignore. Given the substitutes' experience in establishing new services, civil engineering organizations will not be able to afford to retain a reactive position and stay ahead of these substitutes.

Creating a Strong Foundation: Setting the Market Direction

The existence of a reactive market approach within the civil engineering industry does not necessarily lead to the conclusion that civil engineering organizations are forced to continue this approach. Rather, for organizations adopting a long-term, strategic management focus, the transition to proactive market development is an essential element. At the core of this transition is an acknowledgement that an organization must proactively develop new market services rather than postpone decisions until external forces require a response. This alternative perspective acknowledges that civil engineering organizations exist within a competitive environment and not in an exclusive club that requires extensive experience for membership. Given this competitive environment, the organizations that emerge as leaders will be the ones that take the steps to translate strategic objectives into specific market maneuvers. The following sections provide the first steps toward making this transition.

Competitors: Differentiation Rather than Competition

The core element of any change in market perspective is the manner in which an organization views its competition. If an organization elects to view competition as a mirror of itself and an equal provider of services, then the organization has little opportunity to separate itself from similar service providers. In this perspective, the organization will remain content to react to moves made by the competition rather than initiating market moves due to the belief that competition will remain in a well-defined sector definition. A proactive market approach challenges this passive market response. In a proactive approach, civil engineering organizations adopt a strategy of strategic differentiation rather than strategic comparison. In essence, the organizations can answer the question of why a customer should select one organization over another and what tangible differences exist between the competitors. In this alternative approach, the market rewards service innovators with a strategic advantage in terms of market segment and customer acknowledgment. For example, the general contractor that extends its market share to include urban loft development is establishing a new market service that is fundamentally different from its competitors. While any contractor can enter the loft building arena, the first organization to provide design, finance, and engineering services, together with the building services obtains a strategic advantage.

The transition to a proactive approach to market development centers on scenarios that are similar to the loft example. Developing an identity that is unique from the competition requires an organization to establish differentiation. This differentiation may take the form of services, project approach, integration, collaboration methodologies, or technology innovation. Whatever differentiation method is selected, the emphasis remains the same: The organization should be adopting an approach that is fundamentally, rather than incrementally, different from existing organizations. Incremental advancements such as providing budget analysis services in addition to estimating services may enhance revenue, but they do little to establish differentiation. Differentiation occurs when an organization elects to advance a service that is notably different from existing services, and preferably, different from existing competitor services.

Customers: Anticipated Rather than Reactive Needs

The transformation of customers from a reactive force to a proactive force is no less significant than the change in perspective associated with competitors. With the traditional perspective firmly entrenched to respond to changes in the market at the behest of the customer, changing the perspective to one that anticipates customer needs will appear foreign to many organizations. However, market experts almost unanimously argue that anticipation of customer needs is a key component of strategic advantage and success in almost any business sector (Levitt 1975). Brought into the civil engineering domain, this anticipation opens up numerous possibilities as the industry addresses the changing business climate of the twenty-first century. With the expansion of design–build as an accepted project delivery process, the opportunity to expand services throughout the project lifecycle is increasing at a steady rate. Central to this expansion opportunity is the ability for civil engineering organizations to establish market positions in areas where the customer has traditionally engaged experts from outside fields.

To illustrate the proactive expansion concept, a hypothetical structural design firm is introduced. The firm, Concrete Design, Inc. (CDI), has been in business for over 10 years focusing on the design of concrete structures and in-fill systems. Serving primarily as a consultant to architectural firms, CDI believes that it has the experience to change the architect–engineer relationship. Rather than serving as the engineering consultant, CDI wants to enter the multifamily housing market as an integrated AE firm. With an increase in interest in urban rehabilitation and housing development growing throughout the country, CDI anticipates that an increase in focus by customers on the rehabilitation of structures will be demonstrated in its geographic area. With this concept, CDI embarks on expanding its core service to include multifamily design services. By complementing its existing structural capabilities with design capabilities, CDI will have the combined strength to present comprehensive solutions to urban developers.

This example demonstrates two market-focused concepts adopted by CDI. First, the organization is demonstrating strategic differentiation from its competitors by adopting a combined design–engineer service package. Second, the organization is anticipating the increasing demand by customers for streamlined urban development options. Although there is no guarantee that the gamble will result in financial gains, CDI is taking the proactive risks required to place itself at the forefront of a competitive market.

New Entrants: Block Rather than React

Changing the market model for the third market force, new entrants, diverges from the previous two. In contrast to the focus on anticipating needs and establishing strategic differentiation, the focus on new entrants has two components: (1) the need proactively to establish barriers to block new entrants from threatening market share, and (2) establishing new market focal points that anticipate where new entrants are likely to penetrate. The first element, establishing barriers, is the easier of the two in terms of practical implementations, but the more difficult of the two in terms of management perspectives. Expanding upon this thought, ease of implementation is addressed first. Remembering the fact that few barriers currently exist in the civil engineering industry to dissuade new entrants from entering markets, the goal of existing organizations should be to make this entry more difficult. How is this accomplished? The simple answer to this question is the development of intangible assets that are difficult for a new entrant to emulate. In terms of market, these assets focus on market diversity and client expectation. Whereas it is comparatively easy for a new entrant to enter a market where a single service or product is being delivered, entry becomes exponentially more difficult when the new entrant must address multiple services or products (Porter 1975). In terms of the civil engineering industry, this translates to the proactive expansion of services to include multiple points of expertise such as design, regulatory permitting, and testing services. The comprehensive nature of these services implicitly creates barriers that are significantly more difficult for new entrants to overcome. However, while these barriers may be comparatively easy to erect, establishing the strategic management perspective to put the barriers in place represents the key stumbling point. If the management team is not committed to viewing the market arena as a competitive domain with constant threats from new entrants, then the momentum to erect barriers will be lost in the midst of daily operational pressures.

The second of the two new entrant components, new market focal points, differs from the barrier requirements in that it emphasizes outmaneuvering new entrants to anticipate where the new entrants may choose to enter the market. Given the spectrum of services that are encompassed within the civil engineering domain, the opportunity to select an area that is not currently saturated with competition is not as difficult as it may seem (see below). However, new entrants that are focusing on the task of finding these opportunities may detect these opportunities earlier than organizations that are currently engaged in retaining an established market sector. This difficulty is counted upon by new entrants to provide the time that is necessary to establish operating procedures and market share. By operating in an emerging area, the new entrant can establish a reputation for providing services that existing organizations are failing to address.

The goal for established civil engineering organizations is to attempt to identify potential market entry points prior to the emergence of new entrants. Of course, identifying all possible entry points is extremely difficult, if not impossible. However, the chances of identifying these entry points are greatly improved if the organization changes its perspective to one that analyzes the market in the same manner as a potential new entrant. Rather than examining the market as an established player, civil engineering organizations need to reexamine their market sectors and the overall market periodically to identify where the trends are leading in terms of new opportunities. By placing the organization in the role of a potential new entrant, the perspective is changed to one that is looking for proactive opportunities rather than reactive responses. In this role, the organization can expand its own, perhaps much greater, resources to take advantage of the market opportunity and arrive in the new sector prior to the emergence of the new entrant.

Substitutes: Value Added

The final market force that requires a change from reactive to proactive perspectives is the issue of substitutes. Substitutes represent the most difficult of the four market forces to anticipate proactively since they appear from nontraditional sources. Rather than anticipating existing competitors or new entrants, the civil engineering organization that is preparing to anticipate substitutes must determine where the market appears promising to individuals who have been trained in areas outside civil engineering. Returning to the construction management example, the idea that business consultants would attempt to enter the market is a difficult phenomenon to anticipate. The long-standing construction tradition, combined with the low profile of construction compared to manufacturing and banking, makes this field an unlikely focus for substitutes. However, looking beyond these surface-level traits makes the entry of substitutes less surprising. First, the revenue volume attached to the construction industry makes it difficult to ignore. Second, the increasing focus by customers on the strategic importance of capital projects has brought construction out of the facility management domain. Third, the relationships established between business consultants and customers tend to be more integrated than those between constructors and customers. And finally, the expertise required to enter the construction management market is fairly easy to obtain and integrate into existing consulting services. Given these factors, the entry of substitutes into the construction domain is a clear extension of existing consulting practices.

The response to this issue focuses on one element: value-added services. The expansion of established civil engineering organizations into market segments that historically fall outside their domain such as design–build–operate creates a problem for substitutes. Specifically, if established civil engineering organizations can provide customers with services that are recognized as value beyond the traditional service, then a new threshold is created for substitutes. In this scenario, the substitute has to match both the traditional service and the value-added service in order to establish an even standing in the market place. Without this even standing, the customer may be less receptive to examining the alternative services offered by the substitute organization.

Similar to the new-entrant scenario, the civil engineering organization is faced with the task of establishing a presence in a new market segment prior to the emergence of the substitute threat. The organization must once again anticipate substitute threats by analyzing the industry from a third-party perspective. The opportunities perceived by substitutes need to be identified by established organizations prior to the substitute organizations having the opportunity to put in place a comprehensive market plan. This proactive approach may be a challenge to long-held civil engineering market approaches. Identifying potential substitute threats requires market analysis that is uncommon for many civil engineering firms. However, as demonstrated below, developing procedures to undertake this analysis is far less daunting than organizations may perceive.

■ Determining Market Opportunities

The change to a proactive market perspective places a civil engineering organization in the position systematically to examine new market opportunities. In this expansion mode, the critical component becomes the identification of opportunities that provide the organization with an advantage over the competition. This perspective is important based on the fragmentation factor. If an organization attempts to expand by offering services that overlap in a market segment that is already saturated due to fragmentation and competition, then the effort expended in the expansion process will have little long-term value. Rather, an organization must focus on the opportunities for expansion into less explored market sectors. The following sections provide a basis for following this premise by focusing on the nuances that are unique to both the public and private markets.

Private Clients: Opportunities by Growth and Expansion

Civil engineering organizations that focus on private clients face distinct challenges in developing a proactive market analysis approach. First, the private sector is composed of numerous distinct entities that emerge from various social, economic, geographic, and industry locations. Private projects can include, among others, traditional commercial projects, specialized industrial plants, and privately financed infrastructure projects. This diversity creates the scenario that organizations must develop specialization to compete effectively in any of the market segments. Although large civil engineering organizations have the resources to expand into multiple project type specialization, the majority of firms must emphasize a significantly reduced area of specialization. However, the need to focus on a specific area does not equate to remaining focused on a single task within that market segment. As discussed previously, demands and threats by customers, competitors, and new entrants are creating opportunities for organizations to build upon existing strengths to expand into complementary services. Rather than developing a completely new area of expertise, the return on investment may be significantly higher for the organization to extend its services within the context of an existing area such as commercial or industrial structures. However, the issue in this expansion returns to the basic element that the first player into the new segment will have the greatest opportunity to establish a foundational customer base. Understanding the indicators that spotlight these initial opportunities is the emphasis of strategic market management.

Opportunities Related to the Balance Sheet

The first indicator that plays a critical role in strategic market expansion returns to the financial factors discussed in Chapter 7. Far too often, civil engineering organizations fail to remember that the global business economy is based on public investment. Although the majority of civil engineering organizations remain private entities, the same rule does not follow for the majority of clients that the industry serves. Manufacturing, commercial, industrial, and infrastructure companies require significant capital investments. The primary source for this investment is public stock offerings. This investment process provides civil engineering organizations with a significant advantage if they choose to pursue it. Specifically, publicly financed corporations are required to provide stockholders with annual financial statements. Many such organizations publish these documents on the World Wide Web as part of their company information pages. Additionally, many of these companies publish the quarterly income reports they provide to investors as midpoint updates. These documents are equivalent to a strategic roadmap through the private client forest.

The move to a proactive market stance requires organizations to identify market opportunities prior to the competition. However, customer intent cannot be fully calculated until a formal project announcement is delivered. The issue thus becomes how an organization can use published financial information to anticipate these project requirements. Although the answer to this issue will never have complete certainty due to the many forces that act upon a final decision, financial indicators can be used to establish strategic market positions. To illustrate this process, compare the two balance sheets in Figure 8-4. In balance sheet A, Southeast Markets, a grocery store chain operating five states, demonstrates a strong cash and asset position with little short-term debt. In balance sheet B, Southern Clothing, a public clothing chain holding company with name-brand outlets in seven states, demonstrates a weaker asset position than Southeast Markets, but still retains a financially sound position. However, the debt carried by Southern Clothing is substantially higher than that carried by Southeast Markets on a percentage basis. On further analysis of Southern's financial statements, it is found that Southern has recently invested heavily on a new distribution center to facilitate merchandise distribution to its new stores opening in a number of new shopping malls. Although this expansion is encouraging to stockholders, who have rewarded Southern with a 20% stock price increase in the past 4 months, the outlook reads very differently for a civil engineering organization.

Taking a proactive market position, a civil engineering organization that provides engineering design or construction services has a greater opportunity to anticipate moves with Southeast Markets than with Southern Clothing. The key to this analysis is the debt comparison between the two organizations and the recent investment by Southern in a new distribution center. An organization that has heavy debt loads will have a significantly harder time obtaining financing for capital investments since the lending institution must take into account the existing interest and debt payments that the organization is facing. This factor will impact any decisions to expand the physical assets of the organization until the balance sheet returns to a favorable standing. In contrast, an organization that is undergoing a strong growth curve while retaining a minimal debt load on its balance sheet is in the position to expand. Whether it is customer-oriented centers, administrative facilities, or production facilities, an organization with a strong balance sheet and upward-trending growth curve is facing a short-term decision on capital investments.

The civil engineering organization that is able to identify the customer organizations within a market sector that fall within their specialty domain is in a position proactively to pursue market opportunity. Once an economically viable organization is identified, the civil engineering organization can examine the potential needs of the customer as indicators for market expansion. For example, to sustain its current growth pattern, Southeast Markets will have to expand its distribution capabilities beyond its

Balance Sheet—Southeast Markets	in 1,000s
Assets	
Current Assets	
Cash and Equivalents	975.5
Accounts Receivable	190.4
Merchandise Inventories	1,020.8
Prepaid Expenses	165.4
Total Current Assets	2,352.1
Land, Buildings, and Equipment	5,275.5
Less Accumulated Depreciation	1,545.3
Net Book Value	3,730.2
Investments in Affiliates	85.1
Intangible Assets	545.6
Prepaid Pension	250.4
Total Assets	6,963.4
Liabilities and Stockholders' Equity	
Current Liabilities	
Accounts Payable	355.5
Bank Loans and Debentures	150.3
Accrued Salaries and Wages	98.5
Current Lease Obligations	15.6
Other Accrued Liabilities	350.6
Total Current Liabilities	970.5
Long-Term Debt	
Notes and Debentures	2,110.6
Obligations Under Leases	153.7
Deferred Income Taxes	180.6
Accrued Claims	450.6
Stockholders' Equity	
Common Stock	
$.01 par value; authorized 250,000,000 shares, 150,000,000 shares issued	1.5
Additional Paid in Capital	3,430.5
Retained Earnings	635.9
Total Stockholders' Equity	4,067.9
Total Liabilities and Stockholders' Equity	6,963.4

Balance Sheet—Southern Clothing	In 1,000s
Assets	
Current Assets	
Cash and Equivalents	480.5
Short-Term Investments	135.4
Merchandise Inventories	550.8
Prepaid Expenses	115.4
Total Current Assets	1,282.1
Land, Buildings, and Equipment	1,910.6
Less Accumulated Depreciation	750.3
Net Book Value	1,160.3
Investments	38.1
Lease Rights	125.8
Total Assets	2,606.3
Liabilities and Stockholders' Equity	
Current Liabilities	
Accounts Payable	114.7
Bank Loans and Debentures	200.7
Accrued Expenses	252.6
Income Taxes Payable	90.5
Other Accrued Liabilities	7.6
Total Current Liabilities	666.1
Long-Term Debt	
Notes and Debentures	1,540.3
Deferred Lease Credits	120.7
Total Liabilities	1,661.0
Stockholders' Equity	
Common Stock	
$.05 par value; authorized 50,000,000 shares, 31,000,000 shares issued	1.6
Additional Paid in Capital	890.5
Retained Earnings	53.2
Total Stockholders' Equity	945.3
Total Liabilities and Stockholders' Equity	2,606.3

Figure 8-4 An economic analysis of potential customers can provide civil engineering organizations with insight into future capital investments that the customer organizations may be contemplating.

current geographic boundaries. In a proactive market approach, civil engineering organizations should be approaching Southeast with a comprehensive qualifications statement that includes preplanning services to address the expansion possibilities. By meeting with facility managers and company executives, the civil engineering organization can set the stage for expansion into new market services. In this scenario, opportunities exist in the development of infrastructure studies, economic studies, site selection studies, and environmental impact studies.

The underlying message here is that the organization that invests the time to pursue financial statement information will be rewarded with insights into the potential movements of clients. However, by waiting for traditional project announcements, the civil engineering organization relinquishes the strategic advantage and returns to the traditional competitive scenario.

Opportunities Related to Economics

The second area of interest for proactive market development is the use of economic numbers to develop anticipatory plans for private clients. Building on the financial basis introduced with financial statements, economic opportunities expand to focus on a broader, national economic scale. Specifically, civil engineering organizations can take a greater role in analyzing national economic data as a basis for anticipating customer requirements. Although this may sound like an easier statement to make than it is to implement, the reality is that economic information is far more readily available than many organizations realize. Statistics from both public and private organizations are published monthly, quarterly, and yearly to assist individuals and organizations in the development of strategic initiatives (Federal Reserve Board 1998). In each case, the statistics provide insights into the relative strengths of geographic regions, including population movements, investment changes, and investment patterns. Although a surface analysis of these data may reveal little direct connection to the project-focused civil engineering industry, a deeper analysis reveals multiple opportunities to expand operational focus, including both market and service expansion.

Table 8-1 and Figure 8-5 illustrate a compilation of these data on a national level. Within this abbreviated databank lie several key components for strategic expansion, including:

1. *Housing starts*—Housing data are a fundamental component of the economic indicators for civil engineers. Although civil engineers have traditionally separated their services from the housing market, this number affects every component of the civil engineering industry. Where housing starts are on the increase, investment by private organizations in support facilities will quickly follow. Developers focusing on low- and mid-rise commercial buildings, shopping centers, gas stations, and a multitude of other facilities will follow the housing trend in a closely bunched pack.

2. *Unemployment rates*—Similar to balance sheet data, unemployment numbers convey the economic strength of industry in a given area. The lower the unemployment numbers, the stronger the commercial base in the region. Given this equality,

TABLE 8-1
National Economic Statistics Provide Indicators of Potential Needs For Infrastructure, Commercial, and Residential Investment Throughout the Country in the Near Future.

Statistic	Example Change (May–June 1998)
Housing starts	+6.0%
Manufacturers' orders	+0.1%
Retail sales	+0.1%
Industrial production	−0.1%
Disposable personal income	+0.3%
Household income	+1.2% (annual basis)
Unemployment	+0.3%
Productivity	−0.2%
Interest rates	+0.7%
Business starts	+1.7% (annual basis)

regions with consistently low or reducing unemployment numbers are strong candidates for commercial expansion.

3. *Interest rates*—As interest rates fall, developers and owners invest in capital projects. While interest rates reflect a national trend, local variances reflect positive or negative conditions in each geographic area. The civil engineering organization that recognizes interest rate trends for the next 12–18 months will find itself at the beginning of a development curve rather than at the end of the development curve.

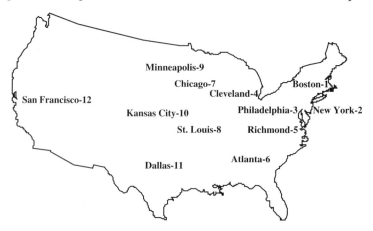

Economic Statistics from the Federal government includes regional data on consumer spending, manufacturing, construction & real estate, non-financial services, banking & finance, agriculture & natural resources, labor markets, and wages & prices.

Figure 8-5 Economic figures are divided into regional areas, providing civil engineering organizations with indicators for expansion and focus opportunities. Alaska and Hawaii are part of the San Francisco district.

4. *Business starts*—Similar to housing starts, the number of new businesses opening in a given geographic region is a reflection of the overall economic strength. Where entrepreneurs believe business has an opportunity to succeed, they will aggressively pursue capital investment. Civil engineering organizations that recognize the movement of business operations from one geographic area to another will have the opportunity to build on that movement by addressing the capital improvement needs of those new industries.

The common theme between each of these elements is the focus on anticipating needs based on regional economic trends. For example, the increase in new business development within a geographic region indicates a requirement for support facilities. In particular, new business ventures require office space, communications support, interior facility reconstruction, and meeting facilities. Although the timing for these needs may differ depending on the region, ultimately these facilities will be required. In response, civil engineering organizations can take a proactive approach to anticipate these needs ahead of the competition. An example of this proactive approach is the response to the recent increase in commercial construction in Region 1 (Boston). With a builtout urban core, suburban commercial construction is significantly increasing. Subsequently, a commensurate increase in support facilities can be anticipated. Of particular interest in this increase is the urban nature of Region 1. With this characteristic, the focus is expanding to the rehabilitation of aging structures. Capitalizing on these opportunities, civil engineering organizations can examine the fit between the urban focus and their current market emphasis. For example, structural engineering firms can expand into interior rehabilitation services, construction firms can expand into environmental remediation services, and general civil engineering firms can expand into planning and preconstruction services.

These responses to the economic indicators represent a divergence from long-held civil engineering industry traditions. Fragmentation and specialization have established an environment where organizations assume economic cycles are a given part of the business environment. However, analysis of economic indicators can break this cycle. The establishment of alternative service focal points can reduce the vulnerability to localized economic swings and provide alternative revenue streams.

Opportunities Related to Management Interests

The final area of interest to strategic market analysis in the private sector establishes a direct link to the stated interests of potential customers. The civil engineering customer is impacted by developments in multiple areas of the business community. From research conducted in university labs to legal rulings and stock market fluctuations, civil engineering customers are buffeted by forces that cover a broad range of topical areas. With each development in the customer marketplace, the customer organization is forced to make a responsive move to address changing developments. For example, a strong concern about food safety and the distribution of food products that contain harmful bacteria is a serious development for food producers. These producers face, at the minimum, a significant financial loss, and possibly bankruptcy, if they do not address the concerns of the general public. Embedded within these concerns is a need to redesign, expand, or develop existing facilities that encompass the ability to produce products that meet increasingly stringent guidelines. These concerns result in investments in capital projects and civil engineering services.

Unfortunately, the time for civil engineering organizations to focus on these product issues is often limited due to project pressures. Subsequently, the majority of market response offered by civil engineering organizations is often too late to capture the initial wave of opportunity. Relating this to the financial community, it is often said on Wall Street that by the time an investment comes to the attention of the general population, the opportunity to make significant gains on the investment has passed. The same can be said in the civil engineering industry. By the time a topic such as innovative processing facilities becomes a national issue that is documented by the leading industry publications, the opportunity to establish strategic leadership positions has already passed. Therefore, the organizations that are going to establish leadership positions must identify market opportunities and trends prior to the establishment of these trends as mainstream industry focal points.

The first step in identifying these indicators is identifying where these indicators exist. To begin this search, one needs to start with the understanding that the industry responds to client requirements. Strategically, the opportunity exists to establish initiatives that anticipate these requirements and build upon them as value-added components. However, the industry is dependent on the customer to indicate what strategic directions are being pursued. To obtain this customer focus, organizations need to identify where client interests are documented. Although this may seem like a difficult task, in reality, customer organizations are actually quite liberal with this information. Specifically, customer interests are reflected in the public statements made by executives.

To illustrate, the current trend of environmental awareness did not appear spontaneously. Rather, as organizations witnessed other organizations gaining attention by making environmental statements, a bandwagon was created that built an initial interest into a corporate responsibility. However, this bandwagon was not a quiet development. Industry and government leaders such as Ted Turner and Al Gore are making public statements about the need for corporate America to address growing environmental problems. When individuals at this level make statements such as this, it is guaranteed that their organizations are going to make an effort to respond to the issue. Similarly, as the initial calls for action grow into a corporate bandwagon, many other organizations are forced by their customers to take a stand on the issue. In this expansion of interest lies a market indicator that service disciplines such as civil engineering will be witnessing new customer requirements for developing projects. In the case of the environmental trend, the early management statements were clear indicators that new services were going to be demanded.

Civil engineering organizations must make the effort to follow these statements in the forums where they are produced. Organizations such as the Business Roundtable, publications such as *Fortune* and *CEO,* and news forums such as the *Wall Street Journal,* provide corporate managers with the publicity outlets that serve as indicators for their strategic directions. The civil engineering organization that identifies these statements prior to the interest becoming a bandwagon has the opportunity to extend services into a strategic advantage.

Public Clients: Opportunities by the Numbers

In developing any report or text on the civil engineering industry, it is difficult to bypass the fact that over 25% of construction is focused on public sector projects. Of greater importance, many of the large firms in the industry as reflected by revenue statements are focused on public or quasipublic infrastructure projects such as roads, wastewater treatment plants, and utility development. With this focus as a primary component, any analysis of strategic market services is incomplete without addressing the public sector. However, at the same time, the argument by some firms that the public sector is a reactive environment that requires organizations to postpone strategic decisions until public agencies have established agendas is unfairly restrictive to proactively focused organizations. In reality, public-sector-focused organizations have a greater amount of market indicators than are available to private sector organizations. Based on the fact that public projects must be financed by local, state, or Federal bodies in advance of the project being developed, published data are available from which civil engineering organizations can establish strategic initiatives. Additionally, the census data that often support these budgetary moves are public information. In summary, civil engineering organizations that plan to focus on public-sector projects have the indicators available to make proactive market maneuvers.

Census Data

For civil engineering organizations pursuing opportunities in the public sector, there exists an information resource that is unparalleled in the private sector: census data. Primarily free of charge and mandated by the Constitution, census data provide civil engineering organizations with a project indicator chain that is unwavering in its ability to predict building trends in localized geographic regions. As illustrated in Figure 8-6, the indicator chain begins with data on population movements, housing starts, and business starts. These foundational data are the core of public sector development and the cornerstone for proactive market segment expansion. The relationship between these data and the civil engineering industry begins with the second component of the chain, public infrastructure demands. With population and business moving freely between geographic regions, demands for support services follow closely behind migrating populations. As these individuals and businesses make significant investments to transfer locations, these same entities expect the local regions to make commensurate investments in public infrastructure to support their relocation. While these entities are quick to enter into purchase or lease agreements with developers in areas that were previously undeveloped to acquire lower property rates, these same entities will not accept reduced infrastructure support. As such, demands for infrastructure development will slowly increase until the local government either ac-

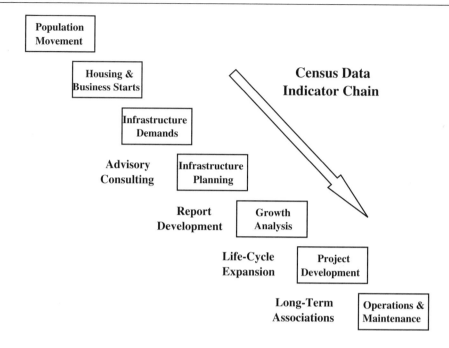

Figure 8-6 Census data start a chain of responses that ultimately result in capital facility expenditures.

knowledges the demands willingly or is forced to pay attention through threats of bad publicity or a turn in the migratory trend.

The acknowledgment of public infrastructure requirements leads to the first market expansion opportunity. Specifically, the demand for public projects such as roads, schools, water treatment plants, and utilities requires specialized civil engineering expertise that is often limited within the public sector. While government entities have engineers on staff, the diversity of projects that fall within the infrastructure category are too diverse for a single staff to address competently. As such, the need for outside advisory consultants grows commensurately with the increase in infrastructure demands. Answering this need for advisors is a market expansion opportunity for civil engineering organizations focusing on upstream project responsibilities. By recognizing the need for these advisory services, a civil engineering organization can focus on a specific entity type such as schools or roads and establish a consulting arm that provides government entities with recommendations on infrastructure development. The consulting component represents a minimum extension of existing services since the knowledge exists within the organization. The only extension required is a proactive move to establish relationships with government officials. However, since the organization is already engaged in activities related to these infrastructure components, emphasizing the need to extend existing relationships to demonstrate additional professional capabilities further focuses the task.

Progressing beyond the advisory consulting role is the opportunity to extend services into report development and project development services. Focusing on the significant regulatory requirements for public infrastructure projects, civil engineering organizations have the opportunity to extend their traditional focal points on design and construction to include environmental, economic, and regional growth analysis. Since every infrastructure project requires regulatory approvals, a builtin demand is created for project development consulting services. In the fragmented approach of the industry, project development specialists have dominated this early project phase. However, there is no inherent limitation that prevents proactive organizations from entering this market segment. Although this expansion may require organizations to acquire personnel with knowledge that extends beyond existing abilities, the appropriate census numbers can justify this expansion. Given census numbers that indicate continuous growth patterns, the market is opened to those organizations that have the vision to capture the additional project development requirements. For example, where an environmental firm once limited itself to physical analysis of project areas, a strong growth indicator provides the basis to expand these services to include economic and growth analysis. When combined into a single professional presentation, the expanded focus presents public clients with a comprehensive analysis ability that addresses a broad range of required analysis. From a client perspective, this represents a budget and schedule savings that cannot be matched by multiple organizations each performing a component of the regulatory requirements.

The final expansion component dictated by the census numbers is the long-term perspective. As documented in current professional and general publications, the urban infrastructure is decaying (Novick 1990). This decay is a direct result of public entities failing to address adequately the operations and maintenance component of physical entities. With design lifespans approaching 50 years, the maintenance and operations of a facility to ensure this lifespan becomes increasingly important as the projected lifespan is extended. Specifically, if an entity such as a school is designed to be operationally effective for 50 years, the investment in maintenance of the facility throughout its existence will have a direct bearing on its actual ability to meet the projected lifespan. A facility that is inadequately maintained either faces early demolition or an expensive maintenance requirement later in its lifespan. Government entities are rapidly beginning to understand this relationship as existing infrastructure begins to demand increasing investments in rehabilitation. This awareness represents a market opportunity for civil engineering organizations. Building on the knowledge required originally to design and build the facility, organizations can propose periodic facility examination and maintenance recommendations as a component of facility development. The knowledge to perform these tasks exists within the organization. It is the repackaging of the knowledge that is unique. Rather than focusing exclusively on pre-construction and construction activities, the long-term perspective provides an organization with the ability to apply existing knowledge to the full lifecycle.

In summary, census data are a strong indicator of the potential moves that government entities will pursue. Census data are equivalent to political power for government agencies. The geographic region with the strongest growth patterns receives the greatest political attention. Retaining this political attention is a continuous objective for public agencies. The civil engineering organization that can demonstrate the knowledge to enhance the agency's ability to retain this attention will have a commensurate opportunity to extend traditional services.

Budgetary Numbers

Complementing the focus on census data is a focus on government budgets. For public-sector-oriented organizations, the final determination of profit potential lies with the government's likelihood to appropriate funds to a particular agency or project. In contrast to private projects that are conceived and acted upon by individuals or small groups, public projects are a result of long budgetary negotiations. Whether the debates occur at a Federal, state, or local level, the underlying premise remains consistent. For a public-sector project to be developed, the funding agency must appropriate the funds as a part of the official budget process.

For civil engineering organizations that specialize in public-sector projects, this fact is nothing new. Organizations in this sector are well aware of the reliance on agencies such as state Departments of Transportation receiving appropriations from budgetary committees. For example, the State of Georgia's expenditures on transportation for fiscal year 1998 were $542 million ("Georgia"). These expenditures are a result of many factors including political favors between state representatives, infrastructure priorities, the power base of senior officials, and influence from the Federal government. From this mix emerges a combination of high- and low-budget projects spanning both urban and rural sectors. From highways to rural bridges, the appropriations committees address political needs and favors from a diverse body of individuals and groups. The issue for civil engineering organizations is how to leverage these diverse influences into market positions by anticipating budget approvals.

The direct response to this issue is to be more active in analyzing the budgetary process. Unfortunately, for most small and midsized organizations the extent of government interaction that is traditionally incurred is through reports from associations such as the Associated General Contractors (AGC), the American Society of Civil Engineers (ASCE), and the Institute of Transportation Engineers (ITE). These societies spend significant resources lobbying at all government levels in the interests of their members. Focal points of this lobbying include the need for project funding, changes in contract requirements, and the reduction of permitting requirements. However, since these are societies that represent hundreds or thousands of members, their focus is not on an individual organization. By necessity, their focus is on the needs of the profession as a whole. Therefore, relying on professional societies to provide market advantages is a questionable strategy.

If a civil engineering organization desires to expand into a new market segment based on anticipated government appropriations, then the organization must become personally aware of the budgetary process. Specifically, the organization must monitor the negotiations, hearings, and discussions that are developing for the upcoming fiscal budget. Individuals must attend budget hearings and legislative sessions, read budget proposals, and meet government staff members. Each of these tasks focuses on obtaining insights into the anticipated spending patterns that government bodies will adopt in the upcoming budgetary cycle. For example, a project management firm, Projects, Inc., which specializes in construction management services for school facilities, desires to extend its services to the preconstruction stage of school facilities. Understanding that the organization's projects will originate from its local school board, the organization has two options. It can either wait for requests for proposals to be developed by the school board and react accordingly, or it can obtain knowledge of what projects are of interest to the board and plan accordingly to win these projects. The former requires no effort by the organization. It follows standard operating procedures, wins a percentage of the contracts, and continues in its current state of profit margins. The latter requires effort and planning. As illustrated in Figure 8-7, a typical school project must pass through several advisory committees and staff members prior to it being recommended for approval by the elected board members. For Projects, Inc., these background discussions are the key to increasing market share. By meeting the members responsible for making decisions on these projects, and attending the meetings where these projects are discussed, Projects, Inc., acquires a valuable asset: knowledge. Understanding why a project is important, who is promoting the project, and what are the key elements of the project provides Projects, Inc., with a knowledge base that reactive competitors fail to obtain. Since predesign and design services often are exempted from the low-bid process, the organization demonstrating the greatest knowledge of the project, its objectives, and its supporters obtains the greatest opportunity to expand its traditional services.

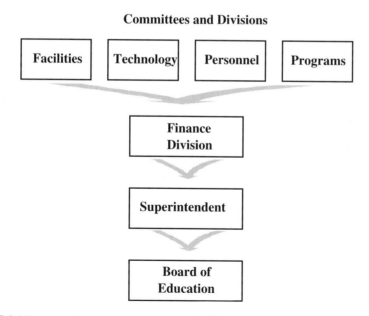

Figure 8-7 Public agencies have a set procedure for authorizing expenditures that often includes several layers of committees and public hearings prior to final authorization.

The school facility example is no different from any other public project. The need for public-sector projects is only a small part of the equation. Politics and public image are a much stronger component of the process. Getting inside this process is the avenue to expanding market opportunities. Professional societies provide lobbying services, but individual organizations are responsible for their market-segment future.

■ The International Market

The United States engineering–construction industry presents a sizeable portion of the world's building market, approximately 20%. However, this number leaves another 80% of the world's market outside the shores of the United States (McClenahen 1997). This substantial market serves as an impetus to many organizations to pursue the dream of expanding from a domestic to an international market. Similarly, the 20% of the world's construction market found in the United States creates a market pull for many international organizations. This international pull is increasingly being demonstrated by the willingness of large European engineering and construction organizations to purchase domestic firms as an entry point to the United States market. Companies such as P&O, the British-based owner of Bovis, Skanska, the Swedish-based owner of Skanska USA, Philipp Holzmann, the German-based owner of Lockwood Greene, and J. A. Jones are demonstrating the willingness of international organizations to invest in the U.S. civil engineering industry. With this level of international expansion, and the 80% of the world market existing outside the United States, the thought of international divisions can be appealing to many organizations.

Although this expansion prospect is one that deserves ample consideration, organizations must be prepared for the challenges of the international marketplace. Specifically, international expansion requires an organization to address two primary issues—control and cultural understanding. The first of these issues, control, is one that easily separates those organizations that will succeed on a global level from those that will never generate the momentum to surge past the initial hurdles. Defined, control is the willingness of the organization to turn over local operations to foreign-born individuals. This concept is arguably the biggest point of management consideration for companies planning a strategic move into foreign markets. Although supporting and contradictory approaches have been tried in respect to control of foreign divisions, the current trend is significantly moving toward local control at very high levels. Although total control may not be turned over, the majority of senior personnel are being recruited from local countries. In this manner, organizations are responding to the need to understand the work processes and rules that dictate operations in an individual location. Increasingly demonstrated by both Asian and European organizations, local control can significantly reduce the learning curve associated with foreign market entry. However, this relinquishment of local control could challenge many management beliefs held by the parent organization. As such, organizations contemplating international expansion should ask themselves if they are ready to make this transfer of control, or are willing to accept the learning curve that accompanies international expansion.

The second issue for consideration is the issue of cultural understanding. As stated by Kangari and Lucas (1997), "The importance of cross-cultural relations should be well understood before one undertakes the management of a foreign operation." This warning cannot be overstated. Local customs and cultural differences can plague an organization during strategic expansion. Everyday issues such as work hours, technology integration, and management–worker relations can lead to delays in operations when a foreign organization does not understand the local culture. As demonstrated over the years, no foreign organization can successfully impose foreign customs over an extended period of time. Yes, a short-term change is possible, but long-term movement against the local customs is a sure path to expansion failure. Organizations contemplating such a move must put time and effort into learning a culture prior to making a final expansion decision. Issues such as compatible financial, labor, housing, and work processes must be addressed, as well as the willingness of the organization to invest the resources to learn the local culture.

The international marketplace represents a vast array of strategic expansion options. However, for every country that an organization selects as a potential expansion site, a thousand details will emerge that need to be learned and translated. International expansion is not equivalent to a foreign vacation. International expansion represents a commitment to education, travel, and financial resource investment.

■ Risk Analysis: The Final Step in Market Expansion

The identification of a market expansion opportunity sets the stage to put in place a strategic initiative to capture that market segment. However, a final barrier exists that separates aggressive and conservative organizations: risk. Every decision encountered during daily operations incorporates an element of risk. Since no civil engineer can fully anticipate the ramifications of a decision, an element of unknown is injected into the decision-making process. Risk is the element of the unknown that an unfavorable outcome will result from the decision (Kerzner 1998). In the context of market-segment expansion, risk is associated with the decision to expand beyond known markets to new market segments. The decision to enter these new market segments carries with it numerous unknowns. Even with the greatest analytic effort, all these unknowns cannot be eliminated. The question that every civil engineering organization must consider is:

What level of risk will the organization accept to enter a new market segment?

Risk Profile

Although market-segment expansion requires organizations to accept risk, the level of risk that an organization chooses to accept will vary. This variance is acceptable and expected. Even the greatest long-term visions must be tempered by the realities of meeting daily expenses and short-term objectives. Weighing these responsibilities against future opportunities is as individual as the organization leaders. As such, each organization must establish its risk profile. This profile establishes the organization's willingness to accept risk, and its willingness to explore market segments that contain potentially severe ramifications if the business exploration fails. Historically, the civil engineering industry has been found to focus on risk minimization rather than risk assessment (Hansen and Tatum 1996). The focus of the following discussion is to emphasize that these same organizations can retain a risk-adverse profile, but move away from a risk-minimization stand to a sensible risk-assessment position. In other words, the organization can make informed and calculated risk decisions rather than retain a risk position with little evidence or rationale to support the position. Figure 8-8 illustrates the three risk profiles typically associated with project and market development (Kerzner 1998).

Figure 8-8a illustrates the risk-aversion profile. In this profile, the willingness of the organization to accept risk (the y axis) reduces as the amount of money at stake (the x axis) increases. In this graph, the curve accelerates steeply at the beginning to signify that the organization is willing to accept risk when a minimum of monetary loss is at stake. However, as the amount of money at stake increases, the slope of the curve reduces rapidly, indicating that the organization is reluctant to risk significant monetary loss in market initiatives. Figure 8-8b represents the next level of risk profile, the risk-neutral profile. In this profile, the organization's willingness to accept risk is in direct proportion to the money that is at stake. The less that is at stake, the less the acceptable risk, and the more that is at stake, the greater the acceptable risk. Finally, Figure 8-8c illustrates the risk-seeker profile. In this profile, the organization aggressively seeks out markets that contain high risk–high return profiles. Although the organization is unwilling to accept risk for low return initiatives, the greater the potential monetary return, the greater the willingness of the organization to accept the risk.

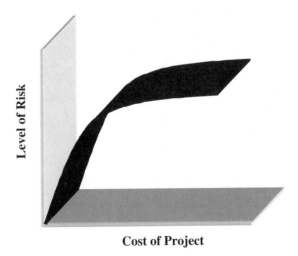

Figure 8-8a The risk-averse profile.

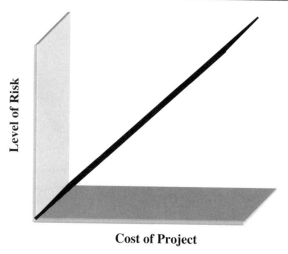

Figure 8-8b The risk-neutral profile.

These profiles represent three foundational characterizations of risk acceptance. The first step an organization must take in examining its willingness to accept the risk associated with entering a new market is to identify itself with such a characterization. This classification will be used to guide risk decisions throughout the following analysis opportunities.

Risk Perspectives

The field of risk management is neither new nor unique to engineering. Business, product development, research organizations, and defense analysts all use risk analysis to determine the potential outcomes of strategic moves (Barnes 1993). What separates many of these users of risk-management techniques from strategic management in civil engineering is the amount of data available for potential outcomes. In terms of product development, years of data have been collected by organizations such as Coca-Cola and General Motors to assist them in evaluating the risk of creating new products. Does this eliminate the risk for these organizations? No. Coca-Cola's ill-fated venture into New Coke is an example of data not reflecting potential risk. In contrast, the civil engineering industry is still in the process of developing such statistics. With very little, if any, data developed on market-segment risk, civil engineering organizations are forced to accept a greater amount of risk, since a greater number of unknowns exist in the marketplace. Concurrently, these organizations must interpret the results from risk models with the same degree of caution. Although the following risk perspectives provide different analytical insights into evaluating the degree of risk associated with a particular market-segment expansion, the models can only deliver results that are as good as the data entered into the model.

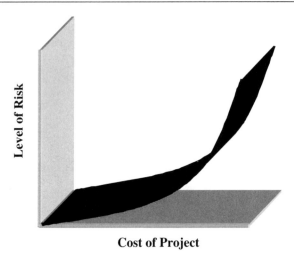

Figure 8-8c The risk-seeker profile.

Objectives Analysis

The first level of risk analysis available for organization leaders is an objectives screening analysis (Souder 1983). Based on decision-making theory, the objectives screening analysis provides a value-oriented measurement related to long-term and short-term objectives. As illustrated in Figure 8-9, an objectives screening analysis form is divided into two segments: an objectives list and a scoring spectrum. The objectives list is further divided into two sets of criteria: short-term objectives and long-term objectives. The example objectives listed in Figure 8-9 represent a sample of the objectives developed by Projects, Inc., in their pursuit of project management work in the predesign phase. These objectives reflect the organization's short-term focus on establishing reputation and name recognition in a new market segment in addition to

Objectives		*Scoring*				
		-2	-1	-0	$+1$	$+2$
Short-Term						
	Reputation	-2	-1	$\boxed{0}$	$+1$	$+2$
	Name recognition	-2	-1	$\boxed{0}$	$+1$	$+2$
	Market Entry	-2	-1	0	$+1$	$\boxed{+2}$
	Revenue	-2	-1	0	$\boxed{+1}$	$+2$
Long-Term						
	PM Organization	-2	-1	0	$\boxed{+1}$	$+2$
	Market Leader	-2	$\boxed{-1}$	0	$+1$	$+2$
	Preferred Provider	-2	-1	0	$+1$	$\boxed{+2}$
Total Score		$+5$				

Figure 8-9 An objectives analysis provides an objective analysis tool for organization to compare positive and negative ramifications of a market decision.

the long-term objectives of establishing a recognized, full-service project management organization. In addition to these objectives, the analysis form includes a scoring spectrum that includes a scale from -2 to $+2$. This scale can be changed to meet the preferences of an individual organization. In this case, the -2 to $+2$ scale is used to provide both a positive and negative indicator of the market initiative's ability to meet the specified objectives. An additional adjustment that may be included is a weighting of the factors. For example, name recognition may be twice as important as any of the other factors. In this case, the organization may elect to double the score of this factor. However, the organization should be careful in utilizing this approach, as it has the potential to bias the score if the organization predetermines a single factor, or small number of factors, that will outweigh all other considerations.

For each objective, the organization makes a determination as to the positive or negative qualities of the new venture in relationship to the stated objective. For example, given the opportunity to extend their project management work through an advisory role for a rural school board, Projects, Inc., gives positive marks to market entry, but neutral marks to name recognition. Although the market entry point gives good experience, it gives little opportunity for recognition in the urban marketplace. However, the name recognition factor is not a negative factor. In contrast to areas such as environmental hazards, accepting the market entry point does not have negative associations. The cumulative effect of evaluating these objectives based on the scoring criterion is a total score characterized as follows:

$$S = \Sigma c_i$$

where, S is the total score and c_i is the score for the market initiative on the ith criterion.

In relationship to the risk profile, an individual organization must determine what value of S is an acceptable level for an individual initiative. The lower the number, the greater the risk of its failing to meet organization objectives.

Boundary Analysis

Closely related to the risk profile curves is the concept of a risk boundary analysis for new market initiatives (Cleland and King 1983). The focus of this analysis is to evaluate market initiatives based on risk–return values. Figure 8-10 illustrates a boundary graph for several market expansion initiatives for Projects, Inc. In this graph, each initiative is plotted on the risk and return scales in relationship to the estimated levels of risk and monetary return that may be achieved. The focus of this exercise is to determine if the proposed initiative exceeds the maximum risk desired, or fails to exceed the minimum return desired. Although this analysis is a rough indicator of market expansion success, it provides a checkpoint to ensure that the organization is retaining a consistent risk profile for each market initiative. Maximum risk and minimum return are concepts that should remain fixed for an individual organization. Changing these boundaries on a constant basis leads to an inconsistent perspective on market expansion.

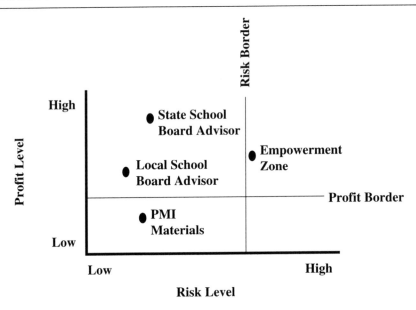

Figure 8-10 A boundary analysis provides an organization with a graphical representation of the risk and return levels associated with a specific project.

Placing this analysis back in the context of Projects, Inc., it can be seen from the graph that an initiative to pursue project management work for Federal Empowerment Zone projects falls outside the maximum risk boundary, while an initiative to develop project management materials for the local Project Management Institute chapter falls below the minimum return boundary. In contrast, two initiatives focusing on local and state school board advisory programs fall within the specified boundaries. In each case, Projects, Inc., has examined the probability of success, the benefits of success versus the ramifications of failure, and the monetary objectives to develop an estimated plot location. The boundaries are not absolute. An initiative falling close to the boundary should not be dismissed one way or another. Rather, the boundaries are an information point that should be evaluated in combination with additional data points.

Probability of Return

At the heart of every business organization is the objective for profit. Organizations cannot succeed and grow unless they demonstrate the capacity to support their operations. Given this need for profit, organizations must focus on this bottom-line number and the risk associated with achieving it. Figure 8-11 illustrates a graphical method for analyzing the different risks associated with multiple market initiatives. In this graph, the local school board advisory initiative of Projects, Inc., is plotted based on anticipated profits (the x axis) and the probability of achieving those profits (the y axis). As seen by the flat bell curve, the most likely scenario for profits sets the profit

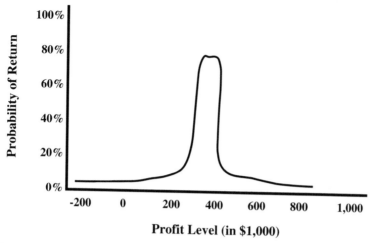

Figure 8-11 A probability of return curve illustrates the potential for profit and loss for an individual project.

range from $300,000 to $400,000. Significantly lower probabilities exist for profits above and below this range.

By itself, the curve in Figure 8-11 tells Projects, Inc., that a 70% chance exists for the organization to realize the $300,000–400,000 profit margin from the proposed market initiative. Additionally, the sharp slope of the curve below and above these amounts indicates that little chance exists for higher or lower profits. In this scenario, the organization is taking little risk to adopt the proposed initiative. In contrast, Figure 8-12 overlays the initiative to provide advisory services to the State School Superintendent's office. This market focus has a grander scale, but is accompanied with

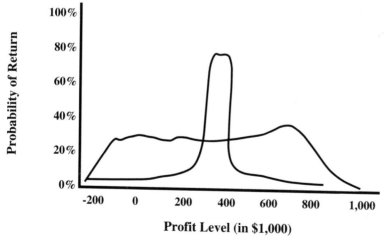

Figure 8-12 An illustration of two projects and their relative opportunities for profit and loss for the organization.

greater probabilities for reduced returns. These risks reduce the probability of meeting the profit opportunities reflected in the second curve. Although the second curve encompasses potential profits of $750,000, it also includes a real potential for a $100,000 loss. This increased probability for loss is referred to as the downside risk potential between the initiatives.

The decision to pursue either initiative is influenced by the organization's risk profile. An organization that has greater risk tolerance may pursue the second curve with the anticipation that higher profits are available, but a higher downside risk accompanies the endeavor. Similarly, an organization with a risk-aversion profile should aggressively pursue the first curve with the knowledge that a high probability exists to achieve the $300,000–400,000 profit range. However, it is important to note that these risk estimates are based on estimated probabilities of success and failure that the organization must develop. Once again the contrast between product and civil engineering service is analyzed. The probabilities for a product rollout are often based on significant data points. In contrast, a civil engineering organization must rely on estimates developed from modeling techniques or experience to complete a probability of return analysis.

Economic Analysis

The final evaluation tool that can be applied to new market initiatives continues with a monetary return emphasis. Anticipating that a new market venture will require several years before it returns a profit, an organization must analyze the value of the return in present-value dollars in comparison to the investment that is being made to enter the market. This analysis is based on the knowledge that a dollar is worth more today than it will be a year from now. Due to inflationary pressures, the value that a dollar represents in terms of purchasing power is greater at the present time than it will be at a later time. This is reflected by the following equation:

$$F = P(1 + k)^n$$

where F is the future value of money, P is the present value of money, k is the inflation rate, and n is the number of years from the current time that is being calculated (Kerzner 1998). Taking a simple example of $100 with a 3% inflation rate, it will take $106.09 in two years to equal the same value of today's $100.

Taking this concept a step further, an economic analysis can be made of a market initiative based on the investment made, the interest rate on the capital borrowed (or the interest rate lost from other investments), and the anticipated financial return over the next several years. In this extension to the present value concept, the previous equation is changed to reflect the perspective of present value rather than future value. In the equation below, the question of worth for $100 is changed to reflect what that $100 obtained in the future is actually worth at present value.

$$P = \frac{F}{(1 + k)^n}$$

Using the same $100 and 2-year time frame, the equation returns that $100 received 2 years from now, with an inflation rate of 3%, is only worth $94.26 in today's money. With this concept as a basis, an analysis of multiple-year returns and initial investment is reflected by the following net present value formula:

$$NP = \sum_{t=1}^{n} \left(\frac{F_t}{(1 + k)^t} \right) - I$$

Where NP represents the net return of the cumulative present values over the anticipated time frame minus the initial investment made in the initiative. Placed in the context of Projects, Inc., the net present value can be calculated for their local school board initiative. In the probability analysis above, Projects, Inc., identified a probable profit of $300,000–400,000. This return is based on a 10-year association with the school board that would result in the anticipated profits after project costs are subtracted shown in Table 8-2.

To achieve these anticipated returns, Projects, Inc., is anticipating a startup cost of $150,000 to cover an additional employee, marketing costs, and new computers for remote data collection. The organization is going to cover these costs by selling rental investments that were returning an 8% yield from lease agreements. Using the profit calculations and the initial $150,000 investment, the Table 8-3 illustrates the net present value on the investment by Projects, Inc.

In stark contrast to the approximately $200,000 that Projects, Inc., anticipated as a return after their $150,000 investment was repaid, the net present value calculation shows that the real return is only $46,448. Placed in the context of percentages, the initial calculation would have resulted in a 57% return on investment, while the net present value return is reduced to a 13% return on investment. Does this mean that Projects, Inc., should not pursue the new market segment? Not necessarily. Rather, it

TABLE 8-2
Anticipated Profits From a 10-year Relationship Between Projects, Inc., and the School Board.

Year	Profit
1	$(15,000)
2	$10,000
3	$20,000
4	$25,000
5	$25,000
6	$35,000
7	$50,000
8	$50,000
9	$75,000
10	$75,000
Total	$350,000

TABLE 8-3
Net Present Value of Profits for the Projects, Inc., Proposal to the School Board.

End of Year	Profit	Present Value
1	$(15,000)	(13,889)
2	$10,000	8,573
3	$20,000	15,876
4	$25,000	18,375
5	$25,000	17,014
6	$35,000	22,055
7	$50,000	29,174
8	$50,000	27,013
9	$75,000	37,518
10	$75,000	34,739
	Present value of profit	196,448
	Less investment	150,000
	Net present value	46,448

provides an additional data point from which to base a strategic decision. Although the venture is not returning the results that were anticipated, the investment is still returning a profit and it is establishing a new market segment on which to achieve a long-term objective. The issue for Projects, Inc., is whether or not the 13% return is worth the risk associated with pursuing a new market-segment endeavor.

Summary

In summarizing risk analysis, the final determination always returns to the issue of threshold and profile. Every organization must determine where its limit exists for tolerating risk. However, to break from the traditional project perspective, an organization must assume an increased level of risk. Entering new markets is inherently risky. There is no magic formula that can eliminate this risk. Every method presented in this section represents an element of educated predictions. The value of the information returned will only be as good as these predictions. In the end, an organization must return to the starting point that began the journey toward a strategic management perspective, the vision. Achieving long-term vision must remain at the forefront of the objectives. If achieving this objective includes adopting risk, then adopting reasonable risk is the price that must be paid to achieve proactive market expansion.

A Summary Checkpoint

Entry into a new market segment requires planning and analysis equal to any project development plan. Financial errors are created when organizations fail to conduct this analysis in a thorough and comprehensive manner. Summarizing the preceding sections into a single series of methodological steps, a checkpoint is presented as a guideline for organizations preparing to enter a new market segment.

1. *Understand why the expansion is being pursued*—Expansion and growth is an essential component for strategic management success. Expanding without a focused reason for expanding is an invitation for financial failure.

2. *Focus the expansion effort*—The redefinition of the civil engineering industry and the project lifecycle in which it operates is creating numerous market opportunities that were previously unavailable. Establishing a proactive market position in a focused area will establish a strategic advantage before competitors have the opportunity to respond.

3. *Develop an expansion objective*—Opportunities exist in both private and public sectors for expansion. Develop an action plan for entering the sector and establishing an initial position.

4. *Analyze the risk*—Every market initiative encompasses risk. Successful organizations will balance this risk against the opportunities presented.

Next Steps

The internal and external focal points have been established to move the organization to a strategic management focus. Now it is time to place the redefined organization into a competitive situation against other organizations pursuing similar objectives. Chapter 9 presents the final strategic management issue, competitive analysis, and its role in developing a long-term basis for strategic success.

DISCUSSION QUESTIONS

1. The civil engineering industry is divided into traditional market sectors and regional boundaries. These divisions provide organizations with established foundations upon which to develop strategic marketing objectives. Given this established structure, why would an established civil engineering organization pursue new market sectors?

2. Strategic differentiation is a concept receiving an increasing degree of attention by organization strategists. Is this concept viable in a service-based industry such as civil engineering, where similar services are provided by each organization within a market sector?

3. Setting a proactive response to industry forces requires organizations to reinforce current positions through market barriers or roadblocks. Given that very few barriers exist to new entrants, what can a civil engineering organization proactively do to protect market share?

4. A civil engineering organization decides to attract new customers from the private sector and identifies Super Grocers as a potential customer. Super Grocers has a strong balance sheet and requirements for expansion. The question for the civil engineering organization is how to attract Super Grocers' business. What can the organization do to get the business before the competition?

5. Risk profiles provide guidelines for civil engineering organizations to identify desired levels of risk to accept in given projects. How can a civil engineering organization balance desired risk levels against the need to retain a project backlog?

6. At what point does a risk level exceed the potential for return in selecting projects? Under what circumstances should this risk level be exceeded?

Case Study	**Expanding beyond an Established Market**

The establishment of a core market is a fundamental component of organization success. Building a successful strategy upon that core market is the first step along the path to achieving long-term goals. As an organization matures, the level to which the organization controls an individual sector often determines the ability of the organization to achieve its stated objectives. The greater the control that an individual organization has in an individual market, the greater the opportunity that organization has to control its progress toward strategic goals. Organizations at the top of the market sector have the greatest opportunity to continue that dominance based on name recognition and experience. However, at the same time, these organizations can also fall into a comfort zone that focuses on staying within that market sector and consolidating strength rather than accepting the risk of a new venture. The challenge for organizations is to determine whether it is acceptable to stay within that comfort zone or the risk is acceptable to broaden current strengths into new areas.

An example of this challenge is exemplified by one of the largest civil engineering design firms in the United States, Parsons Brinkerhoff. The Parsons Brinkerhoff group of companies (PB) has been the number-one engineering organization in the transportation sector for the past 10 years and one of the largest civil engineering consulting organizations overall. Founded in 1885, PB is one of the oldest continually operating engineering consulting firms in the world. Starting with the building of the New York City subway, PB has expanded to over 5,500 employees worldwide in over 150 offices. With revenues of almost $600 million, PB has firmly established itself as a dominant force in the civil design market overall and the transportation/infrastructure sector in particular. Building on this foundation, PB has its sights set on becoming the number-one provider of engineering services in the world in the next 5 years while doubling the size of the organization.

Market Opportunities

The traditional transportation sector continues to provide significant business opportunities for organizations of all sizes. From individual consultants to international conglomerates, opportunities exist throughout the international community. However, while these opportunities continue to exist, new markets previously unknown to the traditional civil engineering industry are providing expansion opportunities. A specific example of this market emergence is the expansion of the global communications market. Whereas the traditional communications sector was defined by telephone companies and copper wire, today's telecommunications sector is highlighted by fiber-optic lines transferring billions of bits of digital information including voice, data, video, and still to be defined items. With expenditures in the billions of dollars, the telecommunications market is combining the interests of the computing, telephone, entertainment, and consumer goods communities for the development of a single delivery mechanism.

With this common focus, research, development, and implementation are being conducted by private and public organizations and by government and industry organizations, financed by government, business, and venture capitalists. This infusion of finance and development has led to one of the fastest-growing business sectors in the United States. As part of this growth, the industry is now progressing from the research and development phase to the implementation phase. Throughout the world, advanced communication services are being put in place to support the expansion of digital communications. With this expansion has emerged a need for consultants in all areas of telecommunications design, construction, and operations. However, in contrast to expansions in areas such as commercial real estate or water treatment facilities that focus on establishing locations in a single geographic locale, telecommunications represents opportunities to develop projects at national and international scales.

The Decision

The expansion of the telecommunications market has not gone unnoticed by the PB organization. With its international reputation and global work force, PB has invested time and money to identify emerging market opportunities. A central result of this analysis has been the identification of telecommunications infrastructure as a primary component of the twenty-first-century engineering market. However, rather than remaining in a market identification stage, PB has made the commitment to interact with the public and private entities that are emerging as the driving forces in the industry. A direct result of this proactive action has been the opportunity to participate in one of the largest privately financed fiber-optic development projects undertaken in the United States, Fiber 2000. Encompassing a nationwide scale, the Fiber 2000 project is being financed by a consortium of private organizations both inside and outside the telecommunications industry.

Anticipating the twenty-first-century competition in the telecommunications industry, the private consortium is making an aggressive move into the fiber-optic sector. Presenting the opportunity to PB to participate in this venture as the engineering consultant, the consortium is opening a door for PB to enter a new market that has the *potential* to be a leading source of revenue in the next century. To put the Fiber 2000 project in perspective, the development of this first project will create a 10% change in the public–private ratio of projects awarded to PB nationwide. Very few projects of this magnitude are made available to civil engineering organizations on a regular basis. However, with this expansion opportunity also come organization decisions. Specifically, the transportation and traditional infrastructure expertise of the PB organization does not specifically coincide with the telecommunications opportunity. Although the fundamentals remain the same, specific knowledge of fiber optics is different from knowledge of roads, bridges, and airports.

With the difference between infrastructure and fiber optics, PB is facing the prospect of investing in human and technical resources that are currently outside the corporate family. From project managers to technical specialists, PB is facing an investment in resources that will create expertise in a new market segment. Although the initial project represents an excellent opportunity to jump start this market segment, no guarantees exist that the market segment will continue to support the investment in resources. PB will be relying on the market projections that fiber optics and telecommunications represent a continued need for U.S. and global organizations. However, as with any technology, the stability of that market is a dynamic premise. In contrast to the long-term need for transportation facilities, today's telecommunications infrastructure requirements may or may not be a long-lasting market sector. This risk represents the market question for the PB organization.

The Decision Maker

Although the decision to undertake the fiber-optic project and enter the telecommunications market will be made by a management committee, one individual at the center of this decision-making process is Susan Kelly. Kelly is currently an Area Manager for PB with the responsibility to oversee PB operations in four states. A civil engineer by education and training, Kelly has been with PB for 11 years and worked her way up the career ladder from project manager to engineering manager and finally to oversight responsibilities of a four-state region. Viewed as a progressive individual with an emphasis on long-term planning, she has built a reputation within PB as a senior manager who is preparing the organization for the competitive playing field of the next decade.

Since arriving at PB, Kelly has been committed to achieving the long-term PB vision of becoming the preferred provider of infrastructure services worldwide. Emphasizing education of employees within her region and the development of young engineers for positions of responsibility and project development, she has firmly established long-term vision as the emphasis of her business planning. Concurrent with this education focus has been a focus on new markets and economic stability. Kelly continually emphasizes to her project offices that the future for PB is in the identification of changing infrastructure requirements and customers. A central component of this analysis is the emergence of the private customer as an infrastructure client. She believes that infrastructure projects are moving away from a pure public agency process to a combination of public–private partnerships.

This focus on future development has materialized in the Fiber 2000 project. Since it originated out of her office, Kelly has been placed at the center of the PB decision-making process. Specifically, it will be her project managers and primary technical experts who will lead the national engineering team. Dispersed geographically throughout the United States, these professionals will oversee an engineering effort on a scale that has not been previously seen in the private fiber-optic sector. However, Kelly also realizes that she does not currently have the complete range of expertise on staff to complete this project. As such, PB will be forced to invest in experts to form the core of a new telecommunications market group. The question for Kelly and the remaining PB management team is whether this investment will pay off over the long term, or will this investment represent a single project that diverges PB's focus from its core business sector?

QUESTIONS

1. Placed in Susan Kelly's position, how would you evaluate the risks associated with the Fiber 2000 project?

2. Placed in Susan Kelly's position, what recommendation would you make to the overall PB management committee, and how would you defend this decision?

3. Placed in an employee stockholder's position, what questions would you want PB to answer prior to making a decision as to whether or not to endorse the PB investment?

4. Placed in the position of a Fiber 2000 investor, what attributes would you look for, and what commitments would you expect from, the engineering consultant on the project?

9 COMPETITION

Battle in the Civil Engineering Arena

What positions are established to protect against new competitors and leverage the organization into new competitive positions?

Military and business leaders throughout the ages have learned that a central component of strategic management and planning is obtaining an advantage over competitors by anticipating their moves. Through this anticipation, operational plans can be established that combine defensive positioning with offensive movements in a comprehensive package. This focus on a comprehensive approach to achieving victory provides valuable insights for setting strategic goals within the civil engineering industry. In an industry characterized by intense but congenial rivalry, the role of competitive positioning has never been more essential. As such, the final component of strategic management for civil engineering, competitive positioning, is the ultimate implementation test. In contrast to military objectives, competitive positioning for civil engineering organizations does not emphasize weapons or surrender. However, similar to the original military objectives, competitive positioning emphasizes preparation for battle, protection of territory, entrance into new territory, and strategic analysis of competitors.

With a traditional project management emphasis, competitive positioning has often taken a backseat to the needs of individual project managers. However, this emphasis proves insufficient when elevated to the long-term goals of strategic management. In this broader perspective, the organization emphasizes expanded revenue streams, profit margins, customer bases, and service opportunities. With the continued appearance of new competitors and markets, the need to establish strategies that defend or attack market positions gains increasing importance. The era of casually competing with a limited number of similarly positioned competitors is drawing to a close as both international and domestic conglomerates enter the domain of the traditional civil engineering organization. Companies such as Korean giant Hyundai Engineering and Construction and domestic consulting giant Anderson Consulting are actively looking for opportunities to expand their presence in the U.S. civil engineering market. Through this challenge, civil engineering organizations face the difficult question of whether to face the competition on an equal footing or fold due to insufficient resources that can be used to defend long-held market positions.

In this chapter, this question is addressed through the examination of strategic positioning from both a leader and challenger perspective. Subsequently, the development of an expanded marketing strategy is examined in the context of market positioning. This chapter emphasizes the need to develop external marketing objectives as the basis for long-term, strategic positioning. Advocating a proactive, aggressive approach to challenging and defending markets, the chapter examines the technological impact of innovative marketing as well as the opportunities available to organizations that quickly pursue the window of opportunity that has temporarily opened. Establishing this aggressive approach to strategic positioning sets the challenge for this chapter—*to identify existing barriers to strategic positioning and establish a vision aggressively to pursue and defend market opportunities.*

■ Putting Competition into a Strategic Advantage Perspective

As discussed by Porter (1979), the focus of developing strategic positioning is to find a position in the industry where the organization can best defend itself from competitive forces or influence them in its favor. Given this position of strength, the organization obtains a strategic advantage. Placed in the civil engineering context, this strategic advantage is the ability of an organization to establish a market advantage that is applicable to the entire business and not just a single project. For example, the organization that develops an education program to provide members with a greater understanding of market changes will have a strategic advantage if the knowledge can be leveraged into results such as early entry into emerging market sectors. Similar types of strategic advantages are reflected by organizations entering sectors such as enterprise management services, environmental sustainability, and collaborative technologies prior to the competition. In each case, entering the market sector early in its evolution provides opportunity for name recognition and experience that competitors will be lacking. Similar advantages can be developed through subcontractor relationships, client relationships, and technology advances. In essence, any market or operational advantage that enables an organization to develop an edge over competitors may be considered strategic. In contrast, improvements in project-specific endeavors are not strategic, since they do not address overall organization objectives.

Although many business writers place an analysis of the competition at the beginning of the strategic management or planning process, the topic of competition is moved to the end in this context, since it emphasizes the culmination of all previous analysis efforts. At the risk of entering the overdiscussed topic of how to analyze business efforts, it is proposed that civil engineering organizations can expand the traditional project focus to a strategic focus by following three specific steps:

1. Understand the position of the organization in the business environment
2. Understand the position of the competition in the business environment
3. Establish a market position with a strategic advantage

With this simplified three-step process as a model, organization-level efforts are placed in the context of providing organizations with strategic advantages in the marketplace. Without these advantages, the civil engineering organization is just another player in a crowded field competing for limited resources. The successful organization stands above this fray to focus on long-term goals and objectives. However, to obtain this position, the organization must be able to answer the question of where the organization stands in the current or projected marketplace. Developing an answer to this question is the emphasis of each of the preceding chapters. Building on concepts such as vision, competencies, and finance, the organization establishes a self-developed understanding of how their market sector operates, the barriers that exist within the sector, and the challenges to expanding beyond its boundaries. The proper development of a response and action plan requires preparation. The previous chapters within the text established the boundary for where the organization currently stands. Strategic advantage completes this analysis by building on the current market position and ensuring a profitable future.

Knowing Your Position

The first step in building a strategic advantage focus is ensuring that the organization has a firm understanding of its place in the competitive market. This standing is composed of two components: a market-share analysis and an influence analysis. The former element is a quantitative look at the percentage of a market the organization currently holds. For civil engineering organizations, this basis is generally made on the calculation of revenue. Although it can be argued that revenue data may be misleading, the fact remains that the current focus on revenue within the industry places this figure at the center of market-share analysis. In conducting this analysis, the focus should be on several factors, including the current share the organization holds, the projected increase desired, and the realistic opportunity to achieve this increase.

The analysis of current market share is straightforward. The equation divides current revenue by overall revenue in the markets in which an organization operates; in most cases, this number will be relatively small. For example, looking at the case of Fluor Daniel, Inc., one of the largest engineering–construction organizations in the United States in 1996, it is seen that Fluor-Daniel held less than 2% of the market (ENR 1996a). This is in contrast to engineering domains such as automotive or aerospace, where leading organizations often hold double-digit market shares. The point of this comparison is to bring expectations of civil engineering organization leaders into a reasonable perspective. The civil engineering industry is not going to produce a dominant player such as those seen in other engineering disciplines. However, the industry does produce local, regional, and national leaders. The focus of an individual organization must be on obtaining one of these leadership positions, and not necessarily on becoming the overwhelming dominant force among the competition.

Continuing with this line of thought, the projected increase in market share must be put in perspective to the market share held by competitors. In many cases, a 1% increase in market share may be a significant increase. Rather than viewing this from the perspective of pure market-share numbers, the organization should analyze the impact of the potential increase on the organization. For example, if the organization currently has revenues of $50 million dollars, which translates to a 5% local market share, then a projected 3% increase would translate into an additional $30 million in revenue. The organization that desires to achieve this increase must analyze whether it is in the position to handle this increase with current resources. If not, can the organization afford to invest in the resources required to absorb the additional business? If the answer is not a definitively positive response, then the organization should reexamine the market-share goals. Not every organization desires to assume this increase in resource investment. Therefore, market-share increase should be in line with the ability to assume this increase, and not just a projected number that would make investors or financiers satisfied.

This market-share analysis provides the starting point to consider the realistic opportunity for an organization to achieve market-share increases. In contrast to the previous two considerations, developing a realistic picture of market advancement cannot be accomplished through direct data analysis. Rather, developing this picture requires an organization to analyze internal strengths and the relationship of these strengths to the competition. Specifically, the organization must examine its core competencies to determine if the strengths exist to threaten other players within the same market sector. To reach this decision, the emphasis moves from knowing the position of the organization to knowing the position of the competition.

Knowing the Competitor's Position

The second step in building a strategic advantage is to develop an understanding of the positions that competitors hold in the marketplace. Through this understanding, an organization determines the strengths, directions, weaknesses, and practices that characterize key competitors. However, since civil engineering organizations have a long-held belief that each performs a similar task on a given project, the need to examine competitors often takes a secondary role to developing greater operational effectiveness. In this latter focus, the organization develops advantages by establishing project procedures that are superior to the competition. The problem with this approach is that it decreases the emphasis on establishing long-term advantages. Although project advantages are important to creating greater profit opportunities on given projects, these advantages do not address the need to develop strategic advantages such as entering markets prior to the competition, establishing positions that reduce competition, and developing services that command a greater profit margin. These issues cannot be addressed at a project level; rather, they need to be addressed by organization leaders who have an understanding of where the competition is attempting to strengthen its position in the marketplace.

A large auto manufacturer illustrates how other engineering sectors have adopted this competitive analysis as a key component of their operations. During an interview with a manager at the central research and design facility, one of the authors spotted a large number of new model year cars from competing manufacturers in a parking lot next to the testing lab. When asked why competitor vehicles were in the parking lot, the manager responded that this was part of the design process. Although the manufacturer was aware of the strengths and weaknesses in its own designs, it also studies competitors' vehicles to understand where improvements must be made on its own cars to stay competitive. As such, each of the competitors' vehicles was disassembled to gather data on competitive advances and weaknesses. Through this method, the manufacturer obtained a greater strategic understanding of where each competitor stood in the automobile marketplace.

Although it is not practical to disassemble a structure to analyze a competitor's final product, the need to develop an understanding of the competition's strengths and weaknesses is as much a foundational element for civil engineering organizations as it is for automotive manufacturers. In fact, this need may be greater for civil engineering organizations, since there are a greater number of competitors offering similar services. However, addressing this issue requires civil engineering organizations to adopt a mindset that is more aggressive than typically found in the industry. Although the civil engineering industry is characterized by strong competition as a result of fragmentation and specialization, competition within the industry often remains focused at a project bid level. Specifically, organizations focus on developing an edge with subcontractors and suppliers to obtain better quotes for services or materials that can be translated into lower bids. The belief in this process is that the ultimate goal is to develop strategies for generating lower bids for the customers. While this perspective is true in certain respects, the previous chapters have argued that a broader perspective is required for the next generation of competition. Although winning projects may remain individual battles, new market sectors and services will characterize the overall battleground.

The previous statement may sound out of context in the civil engineering industry, but it represents the necessary focus for organizations that intend to develop competitive advantages in the increasingly complex civil engineering industry. Organizations that desire to achieve competitive positions need to adopt a competitive attitude. Although a competitor may be a partner in the future, until that time arrives that organization is a competitor and must be analyzed as one.

Analyzing the Competition

In developing a process to analyze competitors, the question that must be asked by the organization is what are the boundaries for obtaining knowledge about competitors. The opportunities to gain knowledge are numerous and varied. The options and ideas listed below are a sampling of the opportunities that are available to the civil engineering organization. The decision as to which methods to employ is based on the preferences of the organization and the competitiveness of the market in which the organization operates.

1. *Meetings and society functions*—One of the most overlooked opportunities to obtain information on competitors is the regularly scheduled meetings and functions held by professional societies. Local chapters of ASCE, AGC, and other organizations have regularly scheduled meetings open to all members. Many individuals do not attend these meetings due to a perception that doing so takes time out of busy schedules or provides minimal information. However, these meetings serve an important purpose for knowledge gathering. In a relaxed setting, members are more likely to talk about current projects and other business-related items. These conversations provide extensive information when analyzed. For example, a transportation design firm known to do a majority of its work in Boston, when heard casually talking about the difficulties of designing for traffic of New York suburbs, has just revealed that it is expanding its firm from local to regional customers. Similarly, a construction firm that has built a reputation in the commercial sector, when heard discussing the difficulties of working with doctors, is stating that they may be developing a proposal for a medical center known to be in the planning stages, an expansion of their core market focus. Similar statements concerning overtime, low employee morale, or business trips to specific cities reveal important information.

2. *Marketing statements*—One of the easiest things that an organization can do to acquire information on competitors is to obtain marketing information. In today's age of electronic communications, this task is made easier through information that is available on-line. Every organization should obtain as much marketing information generated by competitors as possible. This means spending time searching for on-line information, gathering printed materials, and locating articles that have been written about the competition in newspapers and professional periodicals. These materials provide two principal benefits: (1) They provide a picture of how the competitor views their position in the market, and (2) they provide insights into the competitive direction the competition is pursuing. In particular, printed materials and articles provide knowledge-gathering opportunities, since these media are expensive to update and, in the case of articles, may be a one-time opportunity for exposure. In either case, the competitor uses these materials to portray direction and accomplishments to the customer. For example, an article about a structural design firm that highlights the firm's work in historic preservation is telling the market that this organization is establishing a competitive position in this sector and is attempting to create a barrier to others who want to enter.

3. *Customer analysis*—Focused primarily on the public and industrial sectors, customer analysis is the pursuit of information through public meetings or published data. In the case of public customers such as school districts and state agencies, regularly scheduled meetings provide an excellent opportunity to learn how competitors are performing tasks. In the case of school districts, every meeting that emphasizes the delay of projects due to problems with HVAC equipment or cost overruns is an opportunity to gather additional evidence on the difficulties a competitor is encountering. In the industrial sector, reading industry publications such as *Plant Engineering* or the annual reports of individual companies provides similar information. In the case of the former, large projects are featured upon completion. These articles spotlight key areas of interest and success in the project. This information provides the organization with ideas for future emphasis in similar projects. Conversely, annual reports spotlight facility expenditures and delays in getting facilities in operational order. These financial notes represent opportunities for competitors to promote better service in future projects. In each of these public and private scenarios, a common theme emerges: The focus is not to disparage competitors, but rather to obtain information on potential areas of dissatisfaction by the customer.

4. *Recruitment*—The final analysis opportunity is one of the most aggressive an organization can take. If an organization is committed to challenging a competitor, then the most powerful knowledge-gathering tool is the recruitment of key personnel from competitors. Often discussed in the domain of entrepreneurial enterprises such as software development, the recruitment of a competitor's personnel is a viable method for obtaining knowledge at all levels (Taylor 1998; Donkin 1997). The central issue here is what is defined as key personnel. Ideally, the organization should target midlevel managers at the senior project manager level. The reason for this focus is twofold. First, lower-level personnel are typically not worth the recruitment investment. Even though there may be an up-and-coming junior-level person, the odds of this person making a strategic impact in a new organization are negligible. An organization is better served recruiting a new individual out of school who shows promise. Second, senior-level personnel may require greater incentive to change organizations. The demands associated with this incentive may be too great to justify unless the individual is a recognized leader in the industry. This decision must be made on an individual basis, but in general, the cost will outweigh the benefit. This leaves the midlevel personnel. These individuals are a central focus due to their knowledge of the competitor's processes and their desire for career advancement. A committed organization can provide the latter while gathering the former. What this process emphasizes is that civil engineering is about service, and service originates and concludes with personnel. Changing and augmenting personnel is essential to obtaining competitive advantage.

The above represent only a fraction of the opportunities that exist for organizations to obtain information on competitors in a given market segment. Additional opportunities include attending society meetings of potential customers, looking at court records for cases that may involve competitors, and asking potential new hires what they thought about the competition in their interviews. However, these opportunities will not emerge by chance. The organization must make a commitment to focusing on competition. If an organization is not aggressive in pursuing an offensive position, then the competition will gladly accept the opportunity. Knowledge of the competition is a competitive asset. The organization that can acquire the greatest amount of this knowledge is positioning itself for strategic gain.

■ Competitive Perspectives

Obtaining knowledge of the competition is not an end unto itself. While the accomplishment of the task is a vital component of competitive positioning, it does not represent the end result. Rather, the goal of this process remains developing competitive advantages. With this as an underlying purpose, knowledge related to competitors can be placed in the context of a competitive perspective. In this perspective, competition is analyzed as either objectives that need to be fulfilled or positions that need to be guarded. In either case, an organization must prepare and execute plans to enhance its position within a market sector. Put into the context of the market, competitors can be divided into three categories: those that are ahead, those that are behind, and those that are equal (Fig. 9-1).

Competitors that are ahead are those that have a superior position in the market and are targets for market-share acquisition. Competitors that are behind are those that have a weaker market position but must be assumed to be eyeing the organization as an opportunity for their own acquisition goals. Finally, competitors that are equal are those that have a relatively similar position in the market. However, it must be assumed that these equal organizations are also attempting to secure similar strategic goals, and thus they will eventually fall into one of the preceding two categories. As such, organizations that are currently in an equal market position will be assumed to be in a leadership position unless they fall into a follower position at a later time. Given this division into market leaders and market followers, the organization must look at both perspectives to establish a complete competitive perspective.

Competition from a Market Leader's Perspective

The first of the two competitive perspectives is that of the market leader. The market leader is in a position of strength. The leader has the advantage of resources and

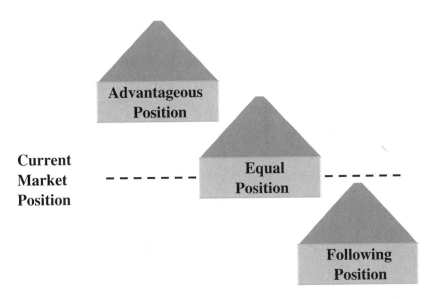

Figure 9-1 A competitor analysis places the competition into one of three positions: ahead, behind, or equal. These positions determine competitive strategies for market positioning.

existing market share that a follower has not obtained. The objective for the leader is to leverage this advantage into both a retention of the existing position and a drive for a stronger leadership position. This objective has been the focus of management writers for over two decades. Today, this research extends beyond the original concept of strategic advantage to incorporate concepts of sustainable competitive advantage, technology advantage, and people as strategic advantage (Pfeffer et al. 1995; Price 1996). However, each of these topics is a deviation from the same central theme. Placing an organization in a position of market leader requires foresight, planning, and aggressive operational implementation. Consolidated into a series of steps that can be implemented within the civil engineering industry, retaining a competitive advantage can be characterized as requiring three components: building on core competencies, investing in the future, and developing a corporate perspective to stay ahead of the competition.

Building on Core Competencies

The first of the strategic components, building on core competencies, returns to one of the basic tenets of strategic management in civil engineering. Core competencies are the foundation of the success that an organization has achieved. The objective of a market leader is to use these core competencies as an advantage against potential competitors. Put in a different perspective, an organization with strength in its core competencies needs to erect market barriers against competitors that are prohibitively expensive for a follower to overcome.

Illustrating this concept is Energy Engineering, a fictional large industrial-sector engineering–constructor based in Los Angeles. Energy has been designing and building oil and gas refineries for 20 years and has established itself as a leader in the field. A key component of Energy's rise to the leadership position has been its commitment to build on a core competency of computing technology application. Founded by a partnership that believed in the use of technology to assist in the design, visualization, coordination, construction, and maintenance of refineries, Energy invested heavily in both personnel and computing resources. Additionally, Energy invested in the development of proprietary software that allows Energy personnel to monitor customer plants for maintenance requirements. With this commitment, Energy has become a strong force in the pursuit of any projects that are proposed within the Western United States. The advantage that Energy relies upon is its ability to give a computer-based presentation that walks a customer through every aspect of the project design, construction, and development process. As a final marketing pitch, Energy presents its real-time monitoring capabilities as unique within the industry.

Energy Engineering's approach to obtaining customers and building upon its success is a continued commitment to its core computing competency. Energy does not hesitate to sell this capability to customers, nor does it hesitate to emphasize that competitors do not have this technology. In this manner, Energy is establishing a barrier for market followers. The followers cannot compete on equal terms with Energy unless they have a response to the technology that Energy is presenting. The success of this barrier is dependent on Energy's commitment to building on this core competency. Given the relative equality in experience that many organizations can project to customers, differentiation will continue to be a cornerstone for Energy's future success. However, this differentiation based on technology will be short-lived if Energy does not continue its investment into the development and integration of technology into its engineering and business practices. Since the technology industry moves at such a rapid pace, Energy must retain its commitment to its core competency or risk the loss of the created market barrier.

The lesson to be learned from the Energy example is that an organization's strength lies in its core competencies. Diverging from these competencies dilutes these foundational elements. Therefore, if an organization is in a market leadership position, retaining a focus on core competencies is equivalent to building on strength rather than weakness.

Investing in the Future

The second strategic element of a market leader's perspective is the need to invest in the future. Obtaining the position of a market leader can be a tenuous stay at best if an organization does not focus on retaining that position. This focus must emphasize the needs of the customers in future initiatives. As demonstrated by the U.S. automotive manufacturers in the 1970s, resting on past achievements and ignoring changing trends will have a direct result in loss of market share. If the automakers had paid greater attention to the consumer desire for greater fuel efficiency and quality, they may have headed off the Japanese attack on their markets. Instead, the Big Three believed they knew what the customers wanted better than the customers did. The result? Near bankruptcy for Chrysler, and significant market loss for all three manufacturers. Although still market leaders, their world dominance was probably lost forever.

For the civil engineering industry, this example is a warning for smaller-scale markets. It does not matter how large a market an organization possesses or how strong past performance has been. The only factor that counts in the business perspective is how the organization is focusing on the next challenge. For a market leader, this translates to leveraging existing resources and strengthening market position in future endeavors. Specifically, the market leader needs to address the question of what business factors are changing that may threaten the current leadership position. The answer could come in a number of forms. Demographics may be changing the marketplace, technology may be changing processes, legal rulings may be changing risk and liability factors, or international buyouts may be changing the nature of the competition. In either case, the future battleground is going to be different from the one that currently exists. As demonstrated by Alexander the Great 2,000 years ago, anticipating changes, preparing for changes, and leveraging expertise are the elements that result in victories. Two thousand years later, these principles remain true in the business market.

Continuing with the Energy Engineering illustration, the focus on future requirements can be placed in terms of Energy's technology commitment. Whereas Energy built a reputation for using advanced technologies to solve the engineering and construction problems associated with refineries, the rapid decrease in technology costs are leading to a very different future. The reduction in costs is providing competitors with the ability to purchase much of the technology that Energy has previously held as a competitive advantage. Thus Energy must focus on augmenting this advantage in future competition. Energy can respond to this challenge in several ways, including the development of new technology, the application of the technology to new customer needs, or holding and waiting for the market to react. Although this last option is the weakest of the three, the civil engineering industry has more often than not selected this as a preferred position. Emphasizing liability and cost concerns, the industry has increasingly taken a conservative wait-and-see approach to the future.

The challenge for Energy and all other market leaders is to avoid the wait-and-see trap. A competitive battlefield never remains static; it is a dynamically flowing entity that sees leaders and challengers change positions on a regular basis prior to the conclusion of the battle. The side that emerges victorious is more often than not the side that was able to anticipate the other's moves and prevent the challenge. The same is true in the marketplace. The position of market leader is a temporal concept; over time, the position changes. The goal for the market leader at any given time is to retain a focus on the future. Investing the resources to respond to that future is the greatest advantage in preventing a challenger from being the first successfully to address future needs.

Developing a Corporate Perspective

Complacency is the enemy of market leaders in any industry. Achieving an objective can be more damaging than failing to reach a goal. The reason for this is that once an organization reaches a goal, such as becoming a market leader, the desire to continue to push forward may be lost (Collins and Porras 1996). As the last component of the market leader perspective, establishing a corporate perspective consolidates the leader's focus into an organization pursuit. The difficulty with many strategic planning efforts is that often the only individual who understands what is being developed is the head of the organization; everyone else is either waiting for instructions or focusing on the strategic aspects of a minor organization component. In contrast, a corporate focus building on the current leadership position emphasizes that organization members analyze how the individual organization components can continue to combine to retain this position.

This focus should not be confused with ideas such as leadership circles, quality circles, or any other participatory event. Rather, committing the organization to competitive success means committing every individual to that end. The organization must examine the role of each department and division in the pursuit of developing further advantages. Each of these components must be coordinated to create a greater challenge for competitors. Allowed to pursue independent goals, the divisions lose the strength of the overall organization. Collaborating, the separate areas can leverage the organization knowledge base and present a stronger organization. It is important to remember that a significant component of a battle is psychological. If the challenger believes that the battle will be difficult to win, then the leader has an immediate psychological advantage. However, if the challenger believes that the leader is disorganized and can be defeated by dividing the leader into several defeatable components, then the challenger receives a psychological lift (Hart 1954). This concept is directly applicable to the marketplace. If a challenger believes that the leader is a large, coherent organization focusing all its efforts on succeeding in the marketplace, then the challenger is placed at a psychological disadvantage from the beginning. In contrast, if the challenger sees a leader with multiple divisions working independently, without the support of a coordinated effort, then the challenger can target independent divisions for weakness in the marketplace.

The corporate focus emphasis is independent of organization size. For a large organization, multiple engineering and construction divisions can either collaborate and demonstrate a strong corporate force or pursue divergent goals. In the latter case, the divergent goals present an opening for competitors to exploit. Corporations from divergent industries have illustrated both sides of this perspective. Corporations such as Coca-Cola, 3M, and Microsoft have developed leadership positions by presenting strong corporate fronts to the competition. These organizations have essentially said to competitors that if they intend to challenge for market leadership, then they will be forced to challenge the entire organization. In contrast, organizations such as IBM, General Motors, and Sears each believed they held unassailable market positions (Moore 1993). However, each allowed competitors to chip away at their positions based on their failure to place the corporate focus on new developments. Civil engineering organizations can similarly experience either side of the corporate experience. Although corporate focus will not ensure market success by itself, an absence of the focus will significantly enhance the chance of failure.

Summary

In summary, the market leader is faced with a dual challenge: Prepare for the future while protecting existing gains. The successful organization will spend a greater amount of effort focusing on the former. To assist in this process, the market leader needs to remember that it is a market leader and must project this position to challengers. The psychological component of battle cannot be lost in the business environment. If a challenger believes that the challenge is beyond their capacity, then the challenger will be smart to find an easier target. A challenger that perceives weakness will redouble efforts to dethrone the leader.

Competition from a Market Challenger's Perspective

In contrast to the dominant position of the market leader, the market challenger is faced with the task of either penetrating a defensive position or determining a path around the defense. The challenger may not have the luxury of relying on existing competitive advantages to establish barriers to competition or name recognition as an automatic entry point. Rather, the challenger must focus on innovation, flexibility, and creativity as tools to achieve stronger market positions. Although this may seem daunting to the structured problem-solving demeanor of the civil engineer, the market challenge has many of the elements found in the engineering design process. Developing a market challenge requires the organization to develop objectives, develop a solution, and implement the solution. This process is similar to the engineering process that forms the core of the industry business. From this perspective, developing a market challenge closely follows the training given to civil engineers. The key is to transfer the problem-solving experience of the engineer to the business side of the industry.

The development of competitive challenge strategies is a popular topic in the business press and research journals (Oliva, Day, and DeSarbo 1987; Tersine and Hummingbird 1995; Nord and Nord 1995). This is understandable considering the number of challengers that exist in any given market compared to the number of leaders. Although much is written about the changing list of Fortune 100 companies over time, the fact remains that although an organization may slip in the rankings, it is more common for the organization to remain near the top in its market rather than disappear from existence. In contrast, the appearance of a challenger in this elite rank is becoming less and less frequent as larger organizations consolidate their positions. While the typical civil engineering market sector does not incorporate organizations that will appear on the Fortune 100 list, the stakes and maneuvers remain the same on a smaller scale. Thus the guidelines for developing a competitive challenge remain the same whether the organization is planning an international marketing campaign or a local market challenge. In each scenario, an organization is required to evaluate market opportunities, change the leader's strength into a weakness, and invest in future opportunities.

Evaluating Market Opportunities

The first component in the development of a competitive challenge is to determine the market opportunities that exist. This determination can occur in several different areas, including the core business area, emerging sector areas, and created need (Fig. 9-2). The first of these areas, the core business area, represents the heart of the leader's market strength. The core business area represents the main body of customers that the leader services within the market segment. For example, the core customers of a transportation engineering firm would be the public agencies that contract for local or regional design efforts. From a battle perspective, this core segment represents the frontal defensive position of the leader. This core area is to be protected from all challenges, since a breach in the defense could lead to an eventual collapse in the entire business. As such, the leader places an extraordinary focus on retaining these customers and establishing barriers that are difficult for challengers to overcome.

Given the knowledge that the leader is emphasizing the defense of a core business area, the challenger must decide if it is worth a direct attack or more prudent to approach from a different direction. It is up to the challenger to find and exploit these

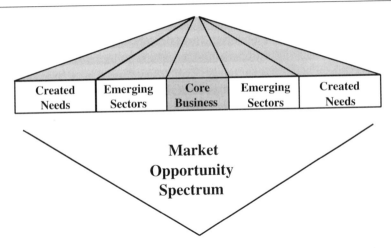

| Created Needs | Emerging Sectors | Core Business | Emerging Sectors | Created Needs |

Market Opportunity Spectrum

Figure 9-2 A market challenger has the option of attacking a leader at the core business, emerging sectors, or created needs.

weaknesses. With this as a basis, the decision to challenge the leader directly in its core market can be accomplished with a quantity of ingenuity and knowledge gathering. Specifically, developing a strategic challenge requires the challenger to find the weakness in the service that the leader is providing. Following the methods for analyzing the competition outlined above, a challenger can develop insights into the weakness of leader by focusing on knowledge gathering. In this knowledge-gathering effort, the challenger focuses on the areas where customers may be unhappy with services, but are unaware that an alternative exists. For example, every year newspapers report the start of the school year with stories of schools that are not ready for students due to construction delays. In most cases, the school board accepts this situation, since it believes that is the risk of undertaking construction projects. Since school boards often do not have sufficient staff knowledgeable in the area, the problems persist on an annual basis. This scenario is the perfect setting for a direct challenge to a core constituency. A focused effort to meet with the school board staff to discuss their problems and present potential alternatives can be developed by the challenger. It may take a year or more for the effort to result in a new contract, but the weakness demonstrated by the market leader is an opportunity that can be exploited.

This example is representative of the elements in direct market challenges. Collecting knowledge of the competitor's work is the strength of the challenger. In a challenger's role, the advantage that the challenger possesses is the complacency that the leader develops towards customers. By making the assumption that name recognition and familiarity will retain customers, the leader begins to let down its guard in some areas of the core defense. The role for the challenger is to develop the knowledge required to identify these lapses and then pursue the weaknesses with aggressiveness.

Emerging Market Sectors The decision to attack a market leader at the point of their strength may not always be a feasible alternative due to relative strength, resources, or local politics. However, this does not eliminate the opportunity to threaten a market leader. Rather, it requires the challenger to adopt more creative approaches to the strategic challenge. The basis for this approach is seen throughout history, where a perceived leader in domains as diverse politics, military, and industry has been overtaken by challengers who had the ability to find opportunities in situations where others perceived only barriers. As discussed at the beginning of the book, the greatest military leaders in history built their reputations on the ability to attack an enemy with fewer resources and perceived losing positions. Similarly, some of the greatest industrial stories of the twentieth century emerged from industry startups threatening established leaders by bypassing the direct assault and adopting an emerging sector approach. Perhaps the greatest example of this is the evolution of Microsoft. Adopting an approach that emphasized software over hardware, Microsoft placed itself in a technology leadership position by bypassing the strength of established hardware manufacturers such as IBM and attacking the emerging market of personal computer software.

Civil engineering organizations can adopt a similar approach to competitive market challenges. In the case of the Energy Engineering example, the organization has developed into a strong regional player, but has yet to be able to move into the international market. The barrier perceived by Energy is that engineering firms with strong political ties at the international level have established barriers that Energy cannot directly assault. However, at a strategic planning session, Energy executives identified that real-time maintenance systems were emerging as an important issue in process facility design. Moving from the research lab to design objectives, these intelligent systems require strong backgrounds in technology together with process plant design experience. Energy personnel identified this trend as an opportunity for their organization. Rather than battling established leaders on a pure design basis, Energy would bypass this confrontation and establish a challenge on their own terms. Specifically, Energy would sell their technology experience as a benefit the traditional organizations could not match.

The lesson to be drawn from this example is that market challenges do not have to occur solely on the terms outlined by the market leader. If a weakness is not obvious to the challenger, then alternative focal points must be identified. Emerging market sectors provide this alternative. The key is initially to pursue these opportunities quietly until a foothold can be developed before the market leader detects the presence of a challenge.

Created Needs The final approach to analyzing market opportunities is the focus on new needs created within the civil engineering sectors. The technology, knowledge, and corporate emphasis that created traditional industry are slowly eroding. Today, civil engineers are speaking of concepts such as systems engineering, concurrent engineering, and supply chain management (Ballard and Howell 1998). These concepts have one element in common, an interdisciplinary emphasis. Rather than building on traditional fragmentation, the emerging concepts emphasize the destruction of these barriers and the development of collaborative teaming arrangements. Looking closer, the new concepts each have an emphasis on satisfying the emerging needs of customers.

The establishment of new customer needs is a direct reflection of the changes that customers are addressing in their own business sectors. The focus on lean manufacturing, enterprise management, and process integration is being reflected in customer demands to work with organizations that have similar business perspectives. This demand represents opportunity for a market challenger. Specifically, the entrenchment that often accompanies corporate success can result in the market leader placing additional emphasis on consolidating the leadership position through further focus on the areas that initially led to the leadership position. It is less common to see an organization remark that once they achieve a leadership position they will then be ready to introduce new services. This entrenchment philosophy represents an opportunity for organizations developing a strategic challenge.

In the Energy Engineering example this concept is illustrated by Energy examining the opportunity to utilize its knowledge of advanced technologies as a central component of its market challenge. However, Energy can take an additional step that emphasizes customer needs. The emergence of real-time computing technologies not only creates a demand for organizations that can develop design solutions that incorporate the technology, but also creates a demand for organizations that can assist customers in evaluating the technologies. This evaluation process is an additional need that emerges from changes in the existing market. Energy can build on this change by presenting customers with a technology evaluation and recommendation service. This service does not diffuse the core strength of Energy's business, but rather it exploits the complacency displayed by its competitors. By focusing on a new need within an existing sector, Energy can challenge the market leader without mounting a direct assault on the leader's position.

As with the emerging sectors, analyzing customer needs provides a lesson for organizations in all phases of sector competition. Market sectors are in a constant state of flux. What was considered advanced 2 years ago may now be standard procedures. Similarly, the customer perspective changes along with the sector. What satisfied a customer 2 years ago might now be considered adequate at best. The organization that is able to identify these changes prior to the competition establishes a strategic advantage. If this advantage provides an opportunity to bypass the core strength of the market leader, then the challenger is in a position to challenge the leader from an unexpected position.

Changing Strength into Weakness

Evaluating market opportunities should be considered a preferred method to establishing a competitive challenge. The identification of a weakness in service, the complacency of an organization or the emergence of new customer requirements are all opportunities for the market challenger. However, this identification process may not provide the foundation an organization requires to pursue the challenge. The perceived strength of a market leader may be intimidating enough that the challenger is reluctant to pursue the battle and instead accepts a secondary market position. Adopting this position can be a fatal mistake. The changing factors within the civil engineering business environment are eliminating the opportunity for organizations to be complacent in a second-tier position. The influx of international financing is increasing the gap between challengers and leaders. Therefore, market challengers must change their perception of the market strength and leadership. Specifically, the challenger must turn the leader's strengths into weaknesses.

Every market leader believes that his or her position is strong due to their market positioning. Whether the positioning is based on customer contacts, resources, or tradition, the market leader establishes a position of strength that is assumed to be unassailable. The fallacy in this argument is that no position is unassailable. Rather, the challenger has to think of a unique solution on which to mount the challenge. In the business scenario, this reversal of strength focuses on the inability of leaders to respond to marketplace changes as quickly as challengers. This concept of market response is central to the strength reversal philosophy. Whereas the market leader is focusing on building on existing strength, the challenger can focus on meeting customer needs. Since the challenger is not protecting a leadership position, the challenger does not have to focus on blocking market movement. This gives the challenger flexibility of response. As documented by many management writers, flexibility of response is a primary key to twenty-first-century business success. The civil engineering industry is no different in this regard. The ability of an organization to respond to a market change prior to the market leader is a competitive advantage. When combined with the ability to portray the market leader as slow to respond, the challenger renders a blow to the leader's image, a primary source of strength.

To illustrate this concept, assume that Energy Engineering is challenging a well-known, international firm for an overseas refinery expansion program. The multi-billion-dollar expansion is necessary due to a recently discovered oil reserve that the foreign country wishes to explore. On the surface, Energy appears to be in a difficult position due to a much lower degree of name recognition. However, Energy decides to turn this disadvantage into an advantage over its competitor. Energy accomplishes this task through a two-pronged approach. First, Energy creates a marketing effort that emphasizes its strength in new technology and its ability to respond to refinery automation requirements. Portraying itself as the choice for innovation and future technologies, Energy attempts to change the external perception of the market leader. Specifically, Energy attempts to make the leader appear to be a guardian of old technology and processes. For a customer that demands the newest in technology, this characterization is an immediate attack on the leader's position. Second, Energy emphasizes how the size of the leader is a detriment to project execution and customer interaction. Energy emphasizes that they can provide direct contact with organization leaders at any time, whereas the size of the market leader and its geographic distribution makes it difficult for the same personal touch to be provided. Similarly, Energy emphasizes that decision-making authority is placed in the hands of its project managers. This is in contrast to the long-held procedures of the market leader that demand formal approval procedures for all decisions. Once again, Energy is portraying itself to the customer as the agent of response and future innovation versus the market leader, which is an agent of processes and self-promotion.

In this scenario, Energy has not changed its procedures, nor has it tried to compete directly with the stronger organization. Rather, Energy has turned the perception of size and tradition against the market leader. By making size and tradition sound like bureaucracy and customer indifference, Energy has opened the door for a challenge that was unexpected by the leader. Note, this challenge will not be successful if Energy does not have the technical skill to back up its claims. However, given the technical ability to perform the work, Energy has damaged the previously perceived unassailable position of the leader. This process can be transferred to any civil engineering organization preparing a market challenge. Similar opportunities for changing strength into weakness include:

- Portraying a local market leader as unable to meet the demands of a regional or national customer
- Portraying a leader who emphasizes having members with long organization tenures and experience as a guardian of out-of-date procedures and protocols
- Portraying a partnership avoiding risk and as such not able to pursue new project delivery alternatives
- Portraying an organization with few executives trained in technology or business as ill suited to speak the language of the customer and address the customer's strategic requirements

These examples are only a few of the possible opportunities to change strength into weakness. While these examples may seem harsh to the traditionalist, it cannot be overemphasized that market forces are creating a marketplace that demands aggressive tactics. A market leader is only as strong as the challenger permits the leader to be. If the challenger believes that the leader is unassailable, then the leader will remain in that position. However, if the challenger can see past the perceived strength to turn it into weakness, then what was once unassailable can become a fortress for defeat.

Investment as a Tool for Opportunity

The final area of focus for market challengers is similar to that of the market leader. An organization preparing to challenge for market leadership needs to invest in the future. This investment may be in personnel, technology, marketing, or any number of other organization requirements. The only requirement for the investment is that it assists the organization in reaching its strategic objectives. Without restating the arguments put forth for the market leadership position, it is sufficient to state that an organization that fails to invest in the future is not building a foundation on which to attack the market leader. Where the challenger must be wary is in the investment of critical resources in speculative areas. For example, the organization that is considering purchasing powerful visualization technology in the hope that it may provide a competitive edge is taking a significant risk if that technology cannot be directly linked to a market challenge. Similarly, the investment in new personnel on the thought that additional skills might be the key to expanded opportunities is making a commitment that may be unfulfilled if the market does not match these skills. In both scenarios, the mistake being made is the thought that investment for the sake of investment will result in market opportunity.

An example of this misplaced investment was the investment in artificial intelligence (AI) technologies during the 1980s. Many large manufacturers believed the advertisements of commercial AI organizations that investment in the technology would enable the manufacturers to automate even their most knowledge-intensive tasks. Millions of dollars later, the customers found that these claims were just that, unfulfilled claims. In one extreme example, an aerospace company spent 3 years on a project and had to abandon it after realizing that if the task was extremely difficult for a human engineer to accomplish, it was almost impossible to explain it to a computer. This is not intended to say that AI technologies were not worth the investment. Rather, the organizations making the investment failed to understand fully the context of the investment. Without a clear understanding of where the investment is leading, the investment cannot be relied upon to serve as a basis for a strategic challenge.

Summary

In summary, the market challenger is faced with a fundamentally different problem than the market leader. For the challenger, the objective is to utilize ingenuity, creativity, and rapid response to outmaneuver the market leader. Similar to the battle plans of Alexander and Napoleon, the market challenger must turn the strength of the leader into weaknesses that can be exploited. This change will not be achieved by organizations focused on winning the battle through head-to-head project competition. Rather, the challenger must be aggressive and find the cracks in the defensive position. A toe-to-toe battle rarely results in a victory for a weaker challenger. However, a deceptive blow can defeat even the strongest defender.

■ Remembering the Personal Relationship

With a competitive plan in place and a market perspective that sets the organization on a challenger's or leader's path appropriately, the organization is set to battle in the marketing domain. However, before the discussion moves from making strategic advances to implementing marketing plans, a moment should be taken to remember a second component of the competitive battle field: the personal relationships. Although this text has emphasized the need for greater competitive perspectives by civil engineering organizations, it would be a mistake for an organization to abandon all policies related to the strength of personal relationships. For decades, it has been these personal relationships between all stakeholders in the A/E/C industry that has helped to overcome the fragmentation that is evident in the industry. Specifically, the personal and professional relationships between organization members has permitted organizations to be competitors one day and partners the next with a minimum of difficulty.

The concepts of strategic advantage and strategic position do not necessarily conflict with the concept of personal and professional relationships. Rather, an organization can retain these relationships while continuing to pursue an advantageous position. Specifically, an organization should continue to foster existing relationships to enhance opportunities related to large projects, but concurrently, the organization should pursue strategic differentiation to place the organization in a position to win projects in head-to-head competition. To illustrate, take the case of two national A/E/C firms, TWG and Howard International (not the real names). TVG and Howard are both full-service firms that were founded in the 1960s. Both organizations started with an architectural focus, but soon added engineering and construction services to provide a full-service option for customers. Both organizations expanded from regional bases to emerge as national leaders in both design and engineering services. Finally, the organizations share a common educational background with the founders and current employees primarily originating from one of three universities. Given this similarity, TWG and Howard have battled for years in the A/E/C marketplace. However, while continuously attempting to project a competitive advantage, TWG and Howard understand the strength that the two organizations portray when they join forces. As such, the two organizations retain and foster a close professional relationship for the specific purpose of creating joint ventures to enhance the position of both organizations when appropriate. Building on this relationship, the two organizations have successfully combined forces to win several stadium and convention center contracts as a single entity. Opening joint project offices, the two organizations create a unified team that appears as a single entity to the customer, while the main offices continue to battle for smaller projects.

Although this arrangement may seem strange to outside industries, it is a way of business for the civil engineering industry, and will continue to be as long as fragmentation exists in the A/E/C domain. Given this fact, the civil engineering organization should heed a caution prior to putting a competitive plan in place; gaining a strategic advantage should emphasize the projection of organization strength and not an emphasis on negativity related to the competition. Building bridges should always remain an underlying strategic goal, not burning them.

■ Putting the Battle Plan in Place: Technological Marketing in the Twenty-First Century

The battlefield has been set and the strategy developed; now it is time to project that strategy outward to the customer. This link to the customer is the role of marketing. As defined by Stanton, marketing is a system of business activities designed to plan, price, promote, and distribute goods and services to the benefit of the market (1984). The key ingredient of food companies, car companies, and many other products and services, marketing has often taken a back seat in the civil engineering industry. However, the launch of the *Construction Marketing Journal,* conferences on marketing civil engineering services, and research originating on the topic are indicators that civil engineering organizations have latched onto marketing as a necessary component of strategic management. However, while these advances indicate a move toward accepting marketing as a greater component of strategic plans, it is the development of a specific type of marketing, technological marketing, that is providing organizations of all sizes with the opportunity to utilize marketing as a strategic tool.

The differentiation here of marketing as a general tool and technological marketing as a tool for all civil engineering organizations is significant. Traditionally, only large civil engineering organizations could afford to employ marketing specialists as full-time organization members. These specialists were reserved for organizations such as Bechtel, Stone & Webster, and Fluor. Organizations such as these could afford the overhead of marketing designers, copywriters, and graphic artists to develop glossy marketing brochures and corporate newsletters. For the remainder of the industry, marketing was often limited to the publishing of a brochure every few years and a news release when significant corporate events occurred, the reason being that the cost of producing marketing materials is extremely high and must be covered by overhead charges. For an industry that faces continual pressure to reduce costs and trim overhead, investing in marketing materials is often placed as a secondary priority. For smaller firms, this emphasis is even less, with organization leaders preferring to rely on personal connections to a greater extent than printed materials.

Today, the introduction of inexpensive electronic media is changing the playing field. No longer are marketing materials restricted to the domain of the large organization. Today, an organization with a personal computer can develop electronic materials that are dynamic and eye-catching. Of greater importance, the use of electronic tools is overcoming the marketing barrier that has existed between large and small organizations. Two components are creating this opportunity: graphics programs and the Internet. In terms of the former, the introduction of sophisticated graphics programs such as Microsoft PowerPoint and WordPerfect Presentations allows individuals with a minimum of graphic abilities to produce impressive presentations. No longer does an individual need extensive knowledge of color theory and artistic technique. Now, the programs provide the templates and presentation layouts that address these concerns automatically. In this manner, the presentation programs are leveling the playing field. Any organization can purchase one of these programs for minimal cost and have in-house personnel develop presentations, as they are needed.

Concurrently, the introduction of the Internet and the World Wide Web has changed the face of marketing. In the past, marketing materials were delivered by hand or through the mail. An impressive set of glossy documents in a bound folder was the accepted manner of projecting an organization image. Today, the Internet has changed that perception. With the introduction of World Wide Web development tools, any organization can develop on-line marketing materials that are available for any person to view anywhere in the world. With the acceptance of the World Wide Web as a greater force in information distribution, this mode of information transfer is becoming the first introduction a customer may have to a civil engineering organization. With this as a basis, the battlefield for civil engineering marketing will slowly move from the traditional glossy brochure to the technological arena.

The Presentation as a Deliverable

The first area for civil engineering organizations to explore is the use of presentation graphics to create on-demand marketing materials. Rather than relying on printed materials that can often take months to design and print, and which ultimately become out of date almost as quickly as they are published, electronic presentations provide the flexibility required in a competitive marketplace. The electronic presentation can include many different media elements including video, sound, pictures, graphics, and text. In combination, these elements tell the organization story to potential customers in multimedia action. Using the multiple presentation elements, the electronic presentation can be tailored to a customer segment or individual customer with a minimum of time. As illustrated in Figures 9-3a and 9-3b, the same presentation can assume a different emphasis with the slightest change in a title slide. As illustrated in Figure 9-3, a change in the background and the graphics can change a presentation from an international business context to an industrial context. Similarly, the qualifications of the organization stated in the presentation can be altered to meet the needs of a specific project or customer.

Although this flexibility is a significant benefit, the greater advantage to organizations is the reduced cost in developing and modifying the materials. For a small or

Figure 9-3a An example of an electronic image for an Energy Engineering presentation.

Figure 9-3b An alternative look to the presentation can rapidly create a different perspective on the Energy Engineering organization.

medium organization, the deliverable presentation provides an opportunity to project an image that is closer to that of a large organization. By focusing on the multimedia aspects of the presentation, a small or medium organization can develop a presentation that includes all the technological wizardry of a customized show developed by an ad agency. With the purchase of inexpensive equipment such as digital cameras and microphones, the organization can turn a static presentation into a showcase of the organization's accomplishments. As illustrated in Figures 9-4 to 9-6, the presentation can capture the range of projects performed by the organization as well as introducing organization personnel.

The Internet as Delivery

The second method of electronic marketing that is changing the competitive battlefield is the Internet and the World Wide Web. Similar to the development of presentation graphics, the development of software tools for building Web pages has brought the Web into the realm of every civil engineering organization. However, taking the electronic marketing paradigm a step further, the World Wide Web has created the opportunity for every organization to have an international presence. With a minimum of development, even the smallest civil engineering office can create a Home Page and notify the world that they are in business. In this manner, the Web provides

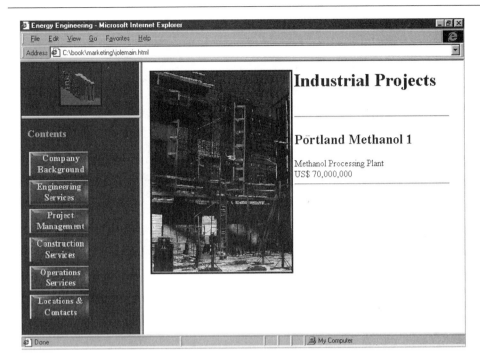

Figure 9-4 An example of a project page in a Web-based marketing presentation.

an equalizer for organizations. The Web allows any organization to portray an image that is specific to their target audience. Whether the image is of a traditional engineering organization or a forward-looking organization taking advantage of new technologies, the Web allows organizations to create the image that best supports their challenge or defensive positions.

Similar to the advantages presented with presentation graphics, the Web provides an organization with marketing flexibility. Personnel, projects, and objectives can be changed at a moment's notice. An organization can place real-time images of projects under development to inject a sense of dynamism to the pages (Fig. 9-7). Additionally, project information including drawings, visualizations, and economic numbers can be added to provide potential customers with a feeling of the interaction the organization will create with each project. Soon, it may even be possible for a potential customer to take a real-time, virtual-reality walk through the organization's offices. While difficult to imagine only a year ago, the advancements in Web technology are occurring at a rate that will enable aggressive organizations to employ possibilities in the near future that cannot even be imagined today. Each of these items is an example of projections that are difficult, if not impossible, to create in traditional print media.

Placed in the context of the Energy Engineering example, electronic marketing can transform a strategic plan for a market challenge into a multimedia assault on the market leader. As stated earlier, Energy is attempting to expand from its regional position to an international organization. Energy has identified its background in tech-

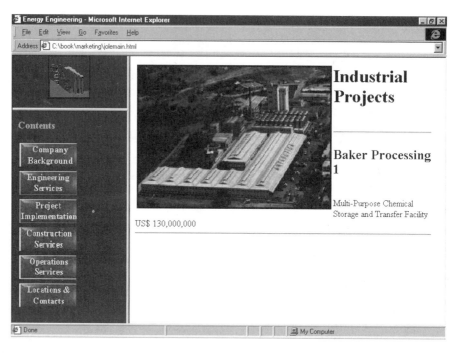

Figure 9-5 An example of a project page in a Web-based marketing presentation.

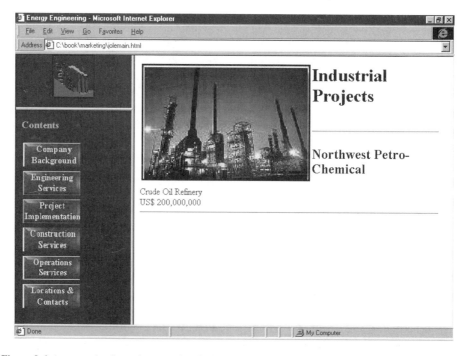

Figure 9-6 An example of a project page in a Web-based marketing presentation.

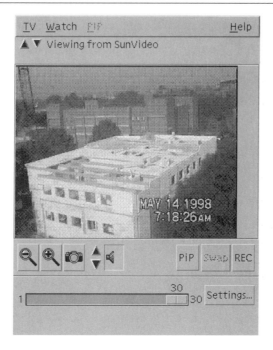

TV   Watch   PIP   Help
▲ ▼ Viewing from SunVideo

MAY 14 1998
7:18:26 AM

🔍⁻ 🔍⁺ 📷 ▲▼ 🔇   PiP Swap REC

                                    30
1 [▭▭▭▭▭▭▭▭▭▭▭▭▭] 30 Settings...

Figure 9-7 Live video can enhance a Web presentation by bringing the viewer directly to the project site.

nology and its knowledge of real-time systems as an emerging market area that will permit it to attack the market leader from an unexpected direction. However, Energy has found that identifying a market niche and portraying its strength in that niche are two distinct items. Energy needs an aggressive marketing medium to match its aggressive battle stance; electronic marketing provides this opportunity. Energy decides to create a Web site that spotlights three items, its knowledge of real-time systems, its technology history, and its dedication to customer interaction. For the first item, Energy develops a Web page that links to all of the major real-time work that would be of interest to potential customers (Fig. 9-8). For the second, Energy creates a Web page that describes its technology developments and links to projects where these technologies were implemented (Fig. 9-9). And, finally, for the third item, Energy places examples of reports and project status information that it makes available to its customers (Fig. 9-10). Given this creation, Energy personnel arrange to have a qualification meeting with the international government of interest. Rather than arriving with only printed material, Energy arrives with a combination of printed material and a laptop computer with a wireless modem connection. In the presentation, Energy uses its Web site as a marketing tool, using Web pages as support materials for presentation topics. However, of greater impact is the message that remains after the presentation. Rather than leaving the customer with a pile of static materials, the customer has an electronic entry point into Energy's organization. The Web pages remain as a dynamic marketing tool that is constantly updated to present an aggressive perspective for the customer to judge against the market leader.

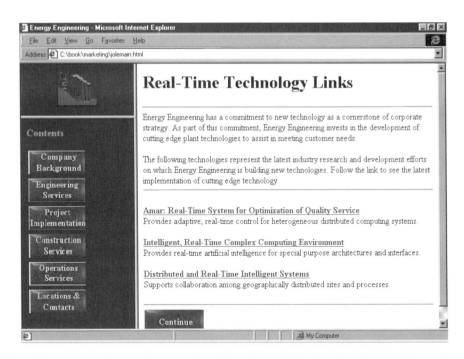

Figure 9-8 An organization can demonstrate knowledge of a specialty by including electronic links to relevant Web sites.

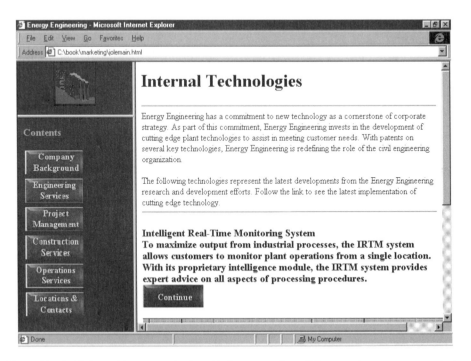

Figure 9-9 The Web provides an organization with the opportunity to link past projects to current marketing materials.

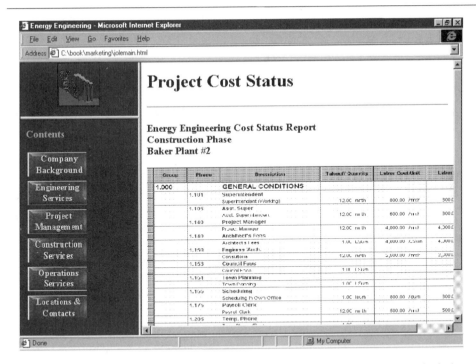

Figure 9-10 Example project reports and designs can be displayed as representative of an organization's capabilities.

The Energy example is not a unique situation. The introduction of the Internet has the potential to revolutionize marketing within the civil engineering industry. No longer divided into the organizations that can afford advertising agencies and the organizations that have to find alternatives, the Internet changes the marketing battlefield. Now, with a component of creativity, a component of technological awareness, and a component of aggressiveness, any organization can develop a focused marketing tool to support its competitive challenge or defense.

The Presentation Components: The Marketing Forces

Although no right or wrong method exists for developing electronic marketing materials, the importance of the materials to the overall competition demands that an organization put sufficient planning into the development process. Where organizations can make a mistake is in confusing the ease and flexibility of developing an electronic presence with the need to plan and design the materials. Whereas organizations clearly understand the ramifications of developing a poor paper-based brochure, most notably the cost of changing the materials, the same realization has not as yet been equated with electronic materials. The responsibility for developing these materials is often given to a new employee as a special project or given to the information technology specialist to create, as they deem appropriate; this can be a costly mistake. Electronic marketing materials are a projection of the organization that will be seen worldwide. The fact that they are not as permanent as the traditional paper materials should not be confused with a reduction in impact. The materials still represent the organization and as such must have a commensurate development effort. Where organizations can reduce the development effort is in the concern for perfection. Additional materials can always be added to the electronic presentation. Therefore, the organization does not have to be concerned with including every item of importance in the initial material release.

Given the ability to add and change materials, the focus of the development process focuses on establishing a strong base from which to build. Although each organization will have different requirements, a core set of elements should be included in all electronic materials. The following descriptions provide an introduction to these elements together with examples, as they would be created for Energy Engineering.

The Introduction

In the world of Web-based electronic marketing, the importance of the Home Page can not be overemphasized. Just as people develop opinions of others in the first 30 seconds of meeting, individuals rapidly develop opinions of an organization based on their Home Page. This introductory element is the image that the organization decides to portray to the world. For an organization locked in a heated competition, every advantage is important. Wasting an opportunity by creating an introductory page that does not reflect the image of the organization can be a fatal mistake. The introductory page must reflect the professionalism, demeanor, and atmosphere of the organization. A conservative organization should not include things such as cartoon clip art on their introductory page. Similarly, an aggressive organization with strong technology abilities should include interesting graphic effects to portray their strength in technology. In this manner, the introductory page is the initial portrayal of the organization and should receive appropriate consideration. Figure 9-11 illustrates a possible Home Page for Energy Engineering. Reflecting Energy's focus on technology and process plant design, the page integrates the themes into a single organization image.

Company Background

The second component in the presentation is the company background. Given the potential for customers to view these materials from locations around the world, an organization must provide a context for the customer in terms of where the company originated. In many instances, the customer may find the electronic materials prior to ever speaking with an organization representative. In these cases, the electronic materials must tell the organization story without the benefit of a business development employee guiding the customer through the history. Customers want to know the organization history. A 5-year-old startup has a very different background than a 100-year-old organization. One is not better then the other, but customers want to know

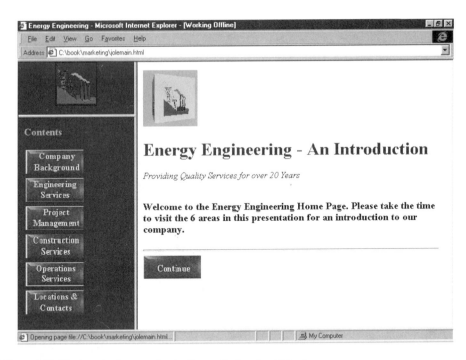

Figure 9-11 The introduction page for the Energy Engineering Web site.

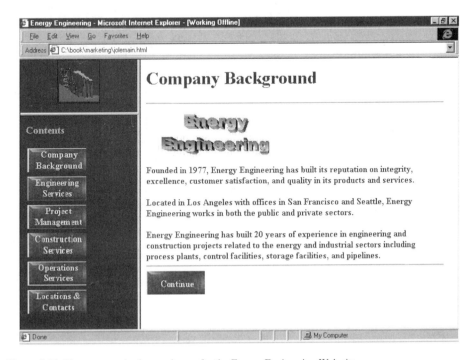

Figure 9-12 The company background page for the Energy Engineering Web site.

the differences. In the age of electronic marketing, initial answers to these questions must be given electronically. Figure 9-12 illustrates the company background page for Energy Engineering.

Objectives

The third element in the material package is the page containing the organization objectives. As discussed in Chapter 3, vision, mission, and objectives are used to convey a message to both internal and external constituents. Electronic marketing is one avenue through which this message gets transferred to the external world. Placing a strong set of strategic statements on the Web page gives the customer an initial understanding of the objectives that the organization is aspiring to achieve. This is not the time to hesitate about sharing a vision or objective with external constituents. These statements contribute to an aggressive, competitive attack on market leaders and challengers. Placing a statement that the organization has a vision to become a nationally recognized leader in its field not only has an impact on potential customers, but it can also cause hesitation in the competition. Although they may not believe that the organization has the capacity to achieve the goal, the fact that it is published in the marketing materials requires competitors to take notice of the aggressive intentions. Once again, the business battlefield has a strong psychological component. The objectives page can serve as a force in manipulating psychological perceptions. Figure 9-13 illustrates the objectives page for Energy Engineering.

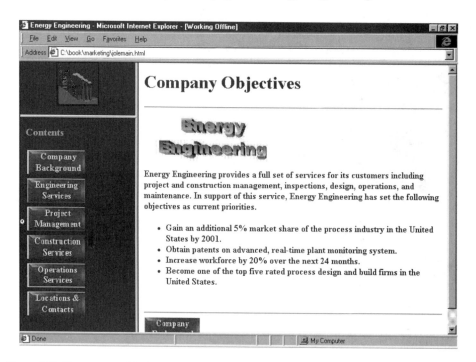

Figure 9-13 The company objectives page for the Energy Engineering Web site.

Project History

Once the stage has been set for the customer by the introduction, background, and objectives, the organization can present its project history. The portfolio of projects selected for the history is just as important in an electronic medium as it is in a paper medium. Customers want to obtain the essence of the organization experience without being overwhelmed by every project that the organization has ever developed. As such, the project history pages should include a representative sampling of the types of work in which the organization has participated. Where the electronic medium does provide a difference is in the ability to showcase these projects. Rather than restricting the presentation to a single picture, the organization has the opportunity to enhance the projects through multimedia effects. Examples of these effects include a short audio message from the user stating their satisfaction, a video clip showing a tour of the completed project, an animation sequence showing the evolution of the project from design through completion, or perhaps a link to a real-time camera taking pictures of the project on a constant basis. These options are only representative of the abilities provided by electronic technology. The project page is a primary selling point to a customer. The organization that is able to make the greatest use of these pages obtains a competitive advantage. Figure 9-14 illustrates one part of Energy Engineering's project history pages.

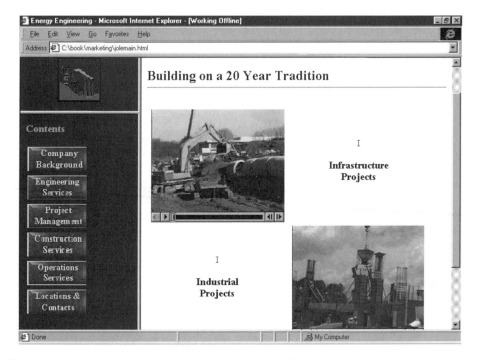

Figure 9-14 The project history page for the Energy Engineering Web site.

The Personnel

The greatest asset of any organization is its people. From the entry-level designer to the senior executive, the members of a civil engineering organization are its primary assets. Assembly lines do not deliver civil engineering products; civil engineers deliver these products. As such, civil engineering organizations rely on developing the strongest collection of people as primary elements in their competitive force. The effort to develop this force should be reflected in the personnel pages. Electronic marketing materials provide an organization with the flexibility to give a broader organization portrait than is available in the paper medium. The time and expense associated with paper brochures often limit organizations to introducing only the senior executive staff to potential customers. With electronic media, an organization has greater opportunities to introduce customers to the range of talent comprising the organization force. Although a portrayal of the senior staff is important, the organization should emphasize strength throughout the company. A link to staff members of all levels portrays to the customer that the organization believes its members can compete with any other organization in the market. Given the ability to update the pages, the organization does not have to be concerned with the information being out of date due to individuals being promoted or leaving the organization. The bottom line is that the organization will battle for customers based on its people, and thus it should not hesitate to project the capabilities of these individuals to both customers and competitors. Figure 9-15 illustrates the personnel page for Energy Engineering.

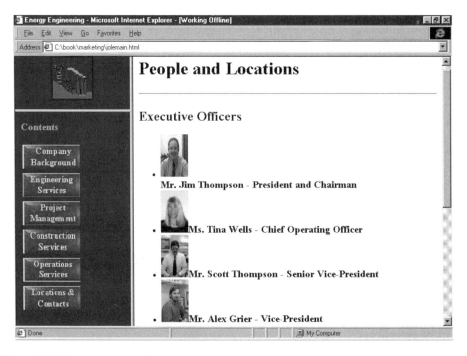

Figure 9-15 The personnel page for the Energy Engineering Web site.

Real-Time Communication

The final component of the electronic marketing materials is the communication page. One of the greatest opportunities that the Web provides for organizations is the ability to include a direct communications link to any member of the organization directly from the marketing materials. The customer does not have to leave a message with a secretary or go through endless rounds of phone tag. Rather, an electronic link can be provided that allows a message to be delivered directly to a specific individual. Although these messages do not replace direct telephone conversations, they do provide an initial communication link that tells the customer that the organization is serious about speaking with them. Although it is a simple addition, psychologically it conveys a significant meaning. Figure 9-16 illustrates this last component for Energy Engineering as it completes the electronic marketing materials.

■ Summary

In summary, competition in the civil engineering domain will increasingly reflect the business challenges that have long been associated with other industries. The age of congenial competition and equitable sharing of customers will begin to change as new markets, new challengers, and new forces combine to influence the industry. As these

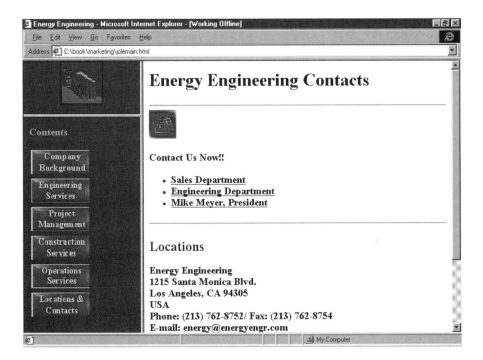

Figure 9-16 The communication page for the Energy Engineering Web site.

changes occur, the traditional market leaders and challengers will be forced to reevaluate their positions. During this reevaluation, competitive forces will result in organizations changing their relative positions within the industry hierarchy. The question for organization leaders is where do they want their organization to emerge after this change occurs. For those who want to remain at the top or challenge for a leadership position, an aggressive, competitive stance is required. Business competition is like battle. Winners and losers emerge from the business battlefield. Developing a strategic management philosophy that includes elements from the preceding six chapters is a great start, but failing to close the transformation with an aggressive competitor perspective may put all the previous work to waste.

DISCUSSION QUESTIONS

1. Analyzing the competition is a fundamental component of competitive positioning. This chapter introduces several options for obtaining this knowledge. Given a limited budget, how should a civil engineering organization prioritize their approach to competitive analysis?

2. A market leader spends considerable effort achieving a leadership position. However, once this position is achieved, the risk of organization complacency becomes a tangible issue. How can civil engineering organizations prevent complacency from setting in to organizations in a market leadership position?

3. A market challenger needs to identify weaknesses in a market leader's position. How can a civil engineering organization with a limited marketing budget develop a strategy for identifying these weaknesses?

4. Returning to the historic lessons from military leaders such as Napoleon that state a challenger should turn a defender's strengths into weaknesses, how can these lessons be applied by a new market entrant against an established civil engineering organization?

5. Technological marketing is the future of civil engineering organizations. However, the changing face of technology makes it difficult to predict where the technology future will reside. Given these factors, what advice would you give a growing civil engineering organization concerning a technological marketing plan?

| **Case Study** | **Battling the Home Field Advantage** |

Home field advantage. The phrase is used to denote advantages in everything from sporting events to international diplomacy. Is there such an element as home field advantage? Does it have a place in the civil engineering marketplace? The answers to these questions do not have a definite yes or no response. In some areas of business, being a local organization provides infinite advantages. In other areas, the playing field is level, provided that an organization is familiar with the local conditions under which it must play. However, for civil engineering organizations, the home field advantage becomes significant when organizations attempt to enter new markets that place an emphasis on local connections. Whether it is a multi-billion-dollar organization or a sole proprietorship, the home field advantage can be just as real and just as powerful a barrier to market expansion. Compounding this issue is the ability for significantly smaller organizations to invoke the home field advantage over nationally and internationally recognized firms. With discussions focusing on the emerging twenty-first-century business environment, will these advantages continue to exist? Does an organization have to succumb to playing the rules as defined by the local players?

One organization facing these questions is Turner Construction Company. Founded in 1902, Turner has expanded from its home office in New York City to over 40 regional offices throughout the United States. With a staff of 3,000 salaried employees and revenue exceeding $3.5 billion per year, Turner is consistently ranked as one of the largest commercial contractors in the United States. Focusing on a philosophy that emphasizes clients as "respected friends," Turner has developed experience in building almost every kind of structure in the commercial field. Areas of specific expertise include government facilities, healthcare, sports arenas, and pharmaceutical campuses. Offering a range of services from preconstruction consulting through building maintenance, Turner has developed a reputation for quality that is unquestioned in the building industry.

With this as a background, Turner continues aggressively to enter new markets where its services can be marketed as competitive to existing options. Focusing on a strategy of local service backed by a national corporation, Turner places responsibility for local operations in the hands of regional offices. Each office is responsible for suggesting market targets, getting projects, and running operations. In this manner, the New York emphasis is downplayed in favor of a local connection to potential customers.

The Focus

A specific area of interest for Turner is the Southeast United States. As one of the fastest-growing regions in the country, the Southeast presents significant market and expansion opportunities. Given the national Turner reputation, the expansion in this market is a logical extension of current market objectives. At the center of this expansion effort is Peter Leslie. Leslie is a long-time Turner employee with almost 20 years of service to the organization. With this background, Leslie has progressed through the traditional career ladder, holding positions from field engineer through project manager and up to his current position as a vice president in charge of an office overseeing two southeastern states. This is the third such assignment for Leslie. Previously, he has been given the responsibility to build up offices in both the Rocky Mountain and Sunbelt regions. Leslie's success in these ventures provided an excellent foundation to enter the economically strong southeastern office.

Starting slowly 5 years ago, Leslie has built the local office from $50 million in revenue to over $120 million in revenue and a staff of almost 100 people. Even with the strong economy in the region, it is difficult to find fault with this successful expansion. However, the office is now approaching a critical juncture. Up to this point in time, much of Turner's work in the area has been a direct result of personal contacts built by Leslie. Originally from the South, he has been able to project a local feeling to customers while emphasizing the quality of the Turner product. Downplaying the New York connection due to regional biases, this personal connection has successfully built the office to where it is today. However, Leslie's success has not gone unnoticed by corporate headquarters. Given his success, Turner executives have determined that the economic and business scenarios exist for Turner to make a significant gain in the region. Specifically, the goal has been set for Leslie to increase his revenue to $250 million in the rapidly expanding market.

The Challenge

The instruction to double revenue would be a daunting challenge for any organization. However, for Peter Leslie this challenge brings with it an extra component. Specifically, to achieve this goal, Leslie must attract customers away from the three local construction companies that dominate the market. Although very small in comparison to the national Turner organization, the competitors each have revenues of 2–3 times that of Leslie's current revenue level. Although he believes, and his current clients agree, that the quality of Turner's work is unsurpassed in the region, the local organizations pose a significant challenge to market expansion. Specifically, the competitors bring three advantages to the marketing battle:

1. Each of the organizations has long family ties in the region with the family names being associated with civic, education, and corporate activities.
2. Each of the organizations has strong name recognition to potential customers, with current projects being constructed throughout the region.
3. Although Peter focuses on the local office connection, a wariness of New York influences persists among many potential customers.

With these challenges, the local competitors have established an entry barrier into the local marketplace. Although the Turner services are equal or better than the local competition, the competition is effectively using local ties and regional bias against Leslie's organization. For him, the difficulty is magnified when placed in the context of corporate objectives. Given Leslie's previous success, Turner executives have placed significant confidence in his ability to achieve the goals. Additionally, Turner recently made it clear to area managers that lifelong employment is no longer a standard at the corporation. Economic competition has forced Turner to adopt tighter business controls and reduce overhead. In terms of senior managers, this translates to potential difficulties for individuals who are unable to reach their specified objectives.

Thus Leslie is not only facing local barriers in his region, but is being forced to attack those barriers by a corporate office that is expecting results.

A final difficulty for Leslie is the rapid growth within his own office. Whereas current clients were predominantly brought on board through his personal contact, the size of the operation now precludes Peter from being personally involved in every interaction. As such, Peter is being forced to rely on business development staff to absorb some of the corporate development activities. For an individual who has been successful in developing a business, this transfer of responsibility is a difficult undertaking when placed on a backdrop of increased expectations.

The Proposed Solution

Given a mandate to increase revenues, Peter Leslie has been forced to develop a plan of attack against the local competition. Understanding the local emphasis on doing business with local organizations, Leslie is using the approach of emphasizing the local office as a complete service provider for the customer. Downplaying the New York connection, he emphasizes to customers that Turner has the capabilities within its local office to undertake any project that the customer wishes to construct. Using his personal southern background, he reassures customers that he understands the local conditions and methods of operations. However, to differentiate Turner from its rivals, Leslie emphasizes to the customers that Turner also brings to the table the economic and experience advantages that exists with a 3,000-member organization. Although local service is the focus, he has the ability to draw from an internationally recognized group of individuals to solve unique customer issues. Holding this as an advantage that none of the local competitors can match, he is challenging the core market of the competition.

To support this challenge, Leslie is additionally making it a priority personally to create a connection to the public and private decision makers in the region. Using the Turner marketing capabilities as a resource, he spends considerable time making the connections that competitors have held for generations. Relying on his personal marketing strengths, Leslie is bringing the Turner name to the attention of individuals who make project decisions that could increase Turner's revenue. As stated by several public officials when asked about Turner's product, the Turner product is of superior quality and the organization can be relied upon to deliver it professionally. Given this endorsement, Leslie has positioned the office to achieve significant growth. However, even with this increase in local recognition, Leslie recognizes that a challenge remains. Increasing revenues from $50 million to $100 million is a notable achievement, but increasing those revenues to over $250 million in a compressed period of time is another level of difficulty when the largest contractor in the area has revenues of just over $300 million and has been in existence in the local market since the family founded the business over 50 years ago.

QUESTIONS

1. Put in Peter Leslie's position, explain your initial strategy for addressing the market expansion challenge.

2. Has Peter Leslie adopted the right approach to challenging the local market?

3. Do the local organizations have a real home field advantage, or is it a perception that is used to erect barriers against outside competitors?

4. Placed in the position of the local market leader, how would you continue to defend against the national resources represented by Peter Leslie and the Turner organization?

PUTTING STRATEGIC MANAGEMENT INTO ACTION

10

The completion of the final question in the question–answer methodology signals the return to the beginning of the strategic management process. Once again, the organization should answer the question of a 5-year strategy. However, the completion of the seven questions should result in a different response than that provided at the beginning of this process. Specifically, the answers to the questions in the preceding chapters provide a foundation for a strategic plan with a greater understanding of the emerging marketplace. Combining insights of internal organization functions and external marketplace developments, the answers place the organization in a position to examine the final strategic management component, putting strategic management into place within the organization.

Although the answers given to the strategic management questions provide a basis for strategic management, these answers alone do not ensure that an organization is going to succeed in either obtaining these goals or moving from project-based management to organization-based management. Rather, to succeed, the organization must change its focus from the exploratory mode of answering questions to an action mode of implementing responses. While it is true that the effort expended in answering the questions is extensive, this effort can be wasted if the organization does not choose actively to follow this process with a concentrated implementation effort. Similarly, the act of answering the strategic management questions the first time does not imply that the strategic management process is complete. Rather, the responses to the questions represent the state of the organization at a particular snapshot in time. To build on the results of the process, the organization must return to these questions on an annual or semiannual basis to ensure that constant adjustments are conducted in the face of a rapidly changing economy and marketplace.

It is this need to reevaluate the strategic situation of the organization and develop implementation plans that leads to the final summary point within this book, the need to translate analysis into action. In the following sections, four steps are presented as cornerstones to the translation of the strategic answers into strategic actions. Similar to the seven strategic management questions, the following steps are intended as guidelines for the organization. The steps must be reevaluated on a regular basis to ensure that the organization continues to move along a path toward a developed vision. However, these steps raise one final challenge to the organization—*is the organization prepared to put the effort into making changes that result in strategic rather than project actions?*

■ Step 1: Where Are You Now?

The first step in the process of moving to strategic action is to reexamine the complete list of answers that were generated for the strategic management questions. Rather than treating each question as an independent item, the organization must place each answer into an overall organization context. Specifically, the organization must determine where current strengths exist, where gaps exist, and where the priorities will be set to build upon these answers. With this reevaluation step as a first priority, the strategic management questions are revisited as shown in Table 10-1.

The topics, questions, and challenges in Table 10-1 summarize the strategic questions in the preceding chapters. However, unlike the single question per chapter format used in the main body, the table provides the organization with the challenge of examining the answers to the questions as a whole. In this process, an organization should not be discouraged if it finds one or more areas have significant gaps at the present time. Every organization has room to improve. The difference between the organization that is destined to succeed and the one that is destined to ride the waves of the marketplace is the desire to fill these gaps. At the same time, the organization needs to be realistic about its efforts to fill these gaps. In some instances, significant investment is required to move forward toward strategic objectives and vision. In these instances, the organization must set priorities and balance available resources. For example, if an organization finds itself with gaps in education and competitive analysis, then a decision must be made as to which of these gaps requires the greater attention at the current time. Since each of these gaps will require an investment of time, planning, and monetary resources, the organization must determine where the resources will be allocated. However, this process does not have to be an either–or situation. Rather, advancement in each area is an appropriate response. In this example, that response may be to establish a lunchtime seminar series with in-house personnel (a low-cost action to address education concerns), while at the same time investing in a team to analyze the current competitive situation in an identified market expansion area (a larger resource commitment, but one that is considered critical to long-term success by the organization).

Examining the Gaps

The balancing of organization resources such as in the previous example should become the normal operating process in all civil engineering organizations. Similar to

TABLE 10-1
The Seven Strategic Management Questions Revisited As the First Step in the Move To Strategic Management Implementation.

Topic	Question	Challenge
Long-term goal	Where do you want your company to be in 5 years? Do you know how to get there?	Is the organization prepared to start the strategic management process?
Vision, mission, and goals	What are the vision and the associated mission and goals that the organization has set as the foundation for long-term achievement?	Examine the fundamental purpose of the organization and set the forward-looking, almost out of reach goals that are the hallmark of innovative organizations.
Core competencies	What are the core business and strength of the organization?	Examine the resources and operations of the organization to determine if the greatest strengths are being identified as the emphasis for emerging opportunities.
Knowledge resources	What processes have been put in place to enhance knowledge and information resources within the organization?	Examine the integration of information technologies and human resources to determine if the work force is maximizing the knowledge-based opportunities presented by emerging technologies.
Education	What formal and informal procedures are in place for lifelong learning within the organization?	Examine the educational needs and opportunities within the organization and set the education objectives required to create a civil engineering learning organization.
Finance	What financial risk analysis is in place to forecast and protect the organization from economic swings?	Examine the strategic economic opportunities within the organization and set the economic objectives required to create long-term financial stability.
Markets	What market analysis procedures are in place proactively to identify new and expanded organization opportunities?	Examine evolving market opportunities and establish an expanded organization vision that encourages exploration and innovation.
Competition	What positions are established to protect against new competitors and leverage the organization into new competitive positions?	Identify existing barriers to strategic positioning and establish a vision to aggressively pursue and defend market opportunities.

the balancing of resources on a project, the balancing of resources at an organization level is required to keep the organization on a continued path of advancement. Rather than advancing an individual area of the organization, organization leaders must retain an overall perspective acknowledging that each of the seven strategic areas are equally important to achieving a long-term vision. At the core of this overall perspective is a realistic evaluation of gaps in the strategic management areas.

Human nature often leads individuals to focus on positive rather than negative circumstances. Obviously, people prefer to emphasize the good things that are occurring around them rather than the bad things. While this is a positive approach to going through daily life, it can lead to unrealistic business evaluations in an organization. The desire to emphasize the positive, and downplay the negative, provides organizations with a positive attitude toward achieving long-term goals, but it hinders the development of a strategic implementation plan. In the development process, the highest priority for civil engineering organizations is the identification of strategic management gaps and the acceptance that these gaps exist. Without this recognition, the organization will not take the aggressive steps required to correct the identified situation. As such, it is recommended that organizations examine each of the questions and attempt to place the current state of the organization in one of the following seven categories:

1. Never heard of the concept
2. Aware of the concept, but no action taken
3. A conceptual plan has been developed, but no formal actions
4. A plan has been developed, but not initiated
5. A plan has been initiated within the organization
6. A plan has been fully implemented in the organization
7. The actions are being evaluated on a regular basis for a given area

Developed through multiple surveys by the authors, the preceding scale provides an organization with an evaluative scale on which to place their current state of strategic management. At the lower end of the scale, the organization is indicating that resources have to be placed toward developing a set of actions that can be implemented in response to the strategic questions. Although the rankings at this end of the scale are the lowest, they should not be confused with unalterable. Rather, the rankings should be viewed as an impetus to formalize what previously may have been acknowledged as something to achieve in the future. The middle rankings in the scale indicate progress by the organization, but additional effort is required. This status is often observed by organizations as the stage in which actions are acknowledged as needing to be taken, but for one reason or another, none is taken. However, in these situations, the rankings need to serve as a push to advance the organization to full implementation of strategic actions. Finally, the upper end of the rankings provides the organization with an acknowledgement that appropriate actions are being taken, but complacency must be avoided. Based on the constant state of change, both internally and externally, reevaluation of the strategic management process is required at regular intervals. In this response mode, the organization is pushed not only to reach the highest levels of action implementation, but additionally to evaluate actions that may lead to the appearance of gaps where none previously existed.

■ Step 2: Action Perspectives

The identification of strengths and gaps is completed; it is time to put the need for action in a larger perspective for success. While the need for action is apparent to move from analysis to implementation, the need to place actions in a context of success may be less apparent. Specifically, implementation cannot be undertaken without a methodology to ensure that actions will result in ultimate success. In the context of strategic management, the authors propose a three-part concept for organizations to consider. As illustrated in Table 10-1, the strategic management questions serve as a foundation to support a larger organization structure. The pillars of this structure are composed of three elements: a plan of action, leadership, and action evaluation. Each of these pillars must be given equal strength successfully to support the top of the structure, the organization vision. As outlined in this book, the development of a strong foundation requires extensive effort by every member of the organization. However, this effort may return minimal benefits if implementation and an accompanying emphasis on leadership and evaluation do not follow it.

Leadership

In terms of leadership, Chapter 3 outlined the need for organization leaders to serve as caretakers for the mission, vision, and objectives. In the final context of placing the strategic management concepts into action, this concept of leadership is expanded to include the caretaking of the overall strategic management focus. As outlined in the earlier context, the caretaker role is designed to give leaders a context in which to conduct organization affairs. Specifically, by assuming the role of personal responsibility to organization success, the organization leader is committing to the pursuit of the agreed-upon vision. This role is reemphasized at the overall strategic management level. Simply, the organization cannot achieve strategic management success without an individual or group of individuals leading the effort to achieve this success.

As represented in Figure 10-1, the leadership pillar receives equal weight to the action pillar. There is no possibility of understating the importance of this role. It is inevitable that every organization will encounter some degree of rough times in its existence as a business concern. Leadership is required to see an organization through these rough times. Similarly, every organization will experience a time where the vision appears impossible to achieve or the existence of the organization is questioned. Strong leadership is once again required to keep the organization members focused on the long-term goal. In the absence of this leadership, negativity will quickly spread as organization members begin to believe any and all negative statements as statements of inevitable fact. Taken to an extreme, the negativity will result in poor morale and defections to other organizations. Leadership is required to preempt this occurrence. There are no stronger terms that can be used to emphasize the role of leadership in the strategic management process. A great plan may be developed to address the gaps found during the question–answer process, but the absence of a strong leader will result in this plan turning into one more futile attempt to achieve a goal that is unattainable.

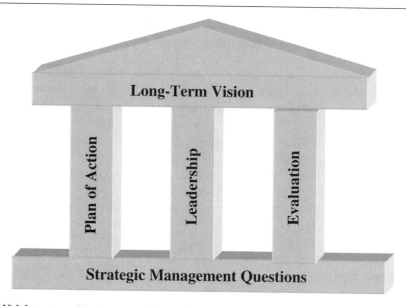

Figure 10-1 Long-term vision is supported by the pillars of action, leadership, and evaluation supported by the strategic management questions.

Evaluation

Often overlooked by organizations except in the context of market share or revenue projections, the final pillar of the action perspective is the evaluation of the action steps. As the third pillar in the action perspective, the evaluation pillar emphasizes the need to evaluate the progress of each strategic management component on a regular basis. Similar to the current emphasis on business development evaluation, strategic management evaluation is required to determine progress toward achieving strategic objectives. However, in contrast to the business development evaluation, strategic management evaluation may not be quite so clear and well defined. Items such as education and core competencies cannot be measured in terms of dollars and market share. Rather, these strategic management issues need to be evaluated in terms of organization progress and movement toward an ultimate goal. For example, an organization that sets the goal of increasing the focus on core competency operations may not be able to place a dollar return on that specific objective. However, that does not translate into a failure of ability to evaluate progress toward that goal. Rather, the evaluation criteria must be altered to fit the criteria. In this case, the criterion may be the increase in clients requesting services related to the core competency, or the number of new hires related to the core strengths. In either case, the evaluation criteria focus less on dollars and more on building the foundation for long-term success.

The difficulty associated with this third pillar is illustrated in the above example, the difficulty in identifying appropriate evaluation criteria. As a new topic in the management literature, clear guidelines cannot be extracted from other disciplines. While the need for evaluation is made clear, the civil engineering industry faces an independent trail in this pursuit. As such, the development of evaluation criteria must become as much a part of strategic management development as implementation plans. Specifically, organizations must continually pose the questions of how a specific objective or pursuit can be evaluated for progress at the same time that the objectives and pursuits are being identified. Additionally, the evaluation process must become an integral part of the implementation process. Whether the organization determines that monthly, semiannual, annual, or some other evaluation interval is appropriate, the essential factor is that the organization remains committed to the evaluation process.

A final thought for organizations entering this evaluation phase is the matter of what to do when the evaluation returns a negative response. First and foremost, the response cannot be panic. The strategic management process is not an overnight action item. Every organization will undergo positive and negative moments. The item for organizations to remember is that the evaluation returned from the process is an additional input to the implementation plan. As illustrated in Figure 10-2, feedback is critical to the success of strategic management implementation. If the evaluation process determines that a successful step has been taken, then the organization should use this success as a validation of its process and redouble its efforts to move forward. However, if the evaluation returns a negative response, then the organization should consider this a warning to change course, and subtle adjustments should be made accordingly. In either case, the feedback must be treated with no greater or less importance than what it appropriately deserves. Feedback is a cyclical process. Incremental steps must be taken as part of the implementation plan and overseen with effective leadership as a means to support and achieve the overall vision.

■ Step 3: Putting a Strategic Management Team in Place

Every organization member has received an edict from top management at some time in his or her career. With little or no justification or explanation, the edict from the top instructs organization members that a change in policy or procedure is going to

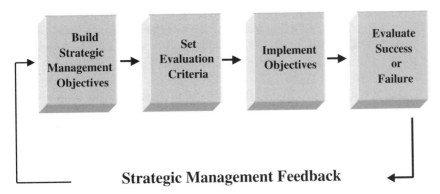

Figure 10-2 Strategic management implementation is dependent on long-term and regular feedback to indicate success and failure of current implementation initiatives.

occur. While this edict may have very good grounds on which to stand, the recipients may have little or no understanding of this foundation. Changes in policy such as purchasing or technology may result in grumbling from the organization members subsequent to such an edict, but it usually will not expand to widespread discontent unless it is part of a broader practice. In contrast, edicts from the top relating to strategic management carry a far stronger potential for negative repercussions. Specifically, the decision to alter an organization's market focus or long-term goals goes to the heart of why many organization members elected to join the organization. As such, changes in these core matters directly affect the perception of the organization by the members.

Given this potential for discontent, organization leaders are faced with the dilemma of instituting progress and potential change without instigating widespread negative response. In the past, this issue has been addressed by management trends such as TQM by advocating the inclusion of members from many organization levels. While laudable on the surface, the overselling of these management techniques in the civil engineering industry has left many individuals skeptical of inclusionary techniques. While a challenge to the implementation of strategic management techniques, this skepticism must not be allowed to stop the implementation process. Rather, organization leaders must reach out to all areas and levels of the organization to emphasize the need for input from all organization sectors and levels. This inclusion lies at the heart of the implementation process for strategic management since inclusion equates to ownership. As outlined earlier in the text, ownership of objectives and goals is essential to ensuring that each member strives to achieve the long-term objectives. Without this ownership, the organization exists as a conglomeration of individuals who are striving for little more than the receipt of a paycheck.

In achieving this ownership and inclusion, organizations should focus on three elements as minimum factors in developing the strategic management implementation team: representation, forward thinking, and challenge. The first of these elements, representation, addresses the often-stated need to obtain broad participation in the organization management process. Avoiding the perception of an edict from the top while at the same time developing ownership in long-term plans can be greatly assisted if the organization puts a team together that includes members from all areas of the organization. Yes, this inclusion means that individuals who have a primary responsibility to projects or customers may be removed from their assignments for a day or two from time to time. However, while this removal may have a temporary effect on a project, it provides long-term benefit to the organization. Specifically, these individuals bring fresh perspectives to the strategic management process as well as customer perspectives from a daily interaction position. Without these perspectives, the organization may fall into the trap of redefining existing practices rather than making a true push towards new objectives.

The second element that is required in the strategic management team is forward-thinking individuals. Specifically, any individual who insists on focusing on tradition as a barrier to change should not be allowed in the process. While this statement may appear harsh, the organization must realize that it has expended considerable resources analyzing the seven strategic areas. Discounting these efforts due to a lack of forward thinking is detrimental to the organization as a whole and to the individuals who participated in the analysis process from the beginning. In contrast, encouraging individuals to look to the future as a context for implementing strategic management concepts signals to the organization that management is committed to moving the organization forward. Therefore, the need to identify individuals who are looking to the future is a critical component of the team-building process. This is not to say that the team must be composed of dreamers who enjoy nothing better than to create grand schemes, but rather to say that the team must include a mix of futurists and pragmatists to ensure that all perspectives of future development are appropriately addressed.

The final element for the strategic management team is challenge. In the current context, challenge is related to individuals who are not afraid to challenge the status quo and the establishment of an environment that encourages such challenges. Far too often, management teams are assembled only to witness a senior individual assert a personal position of power and intimidation that results in the remaining individuals not having the opportunity freely to challenge existing directions. The ramifications of this type of team are significant for an organization endeavoring to implement strategic change. The absence of challenges prevents the opportunity to analyze fully appropriate steps to achieve strategic objectives. Without this analysis, valid options may be discarded without consideration, resulting in a lost opportunity for the organization. As such, the commitment to assemble a team with the responsibility to move the organization forward requires a second commitment encouraging that team to challenge the status quo. When combined with a focus on the future and broad representation, this dynamic environment will produce the innovative suggestions required to turn strategic management analysis into strategic management actions.

■ Step 4: First Steps

The team is set, the implementation requirements are identified, and the strategic management gaps have been prioritized; now it is time to take the first steps toward achieving the strategic management goal. Unfortunately, "pulling the trigger" may be difficult to achieve. Why? Very simply, up to this point, the move to a strategic management focus was primarily a paper exercise. Although existing processes were analyzed and challenged, change has not been instituted. It is not until this point that real changes may occur, and starting this change is frightening to many individuals. Although status quo may not be the preferred manner of operation, it nonetheless is a known manner of operation, whereas new processes introduce the unknown. As such, organizations must endeavor to take a first step and move past this stage of action paralysis.

In achieving this first step, an organization can start by doing two things: setting a small goal and taking a first step. In terms of the former, the organization should take a high-priority item identified during the question–answer process such as knowledge resources and set a short-term, achievable implementation goal such as the development of a project web site. In this small goal, the organization institutes multiple positive steps. First, the organization sets itself up to succeed. Organization members will be waiting to see a success from the implementation process. Providing a small, but visible, success story sets the stage for larger endeavors in the next stage. Second, the organization sets its sights on a goal with demonstrable results. Rather than implementing a first step that requires substantial explanation by management in terms of its success, a demonstrable result allows each organization member to extend his or her ownership to the process. Finally, the organization gives the strategic management team a foundation on which to build. The concept of the three pillars is essential for long-term success. However, if these pillars become unstable, then the entire structure is in danger of collapse. As such, strengthening the pillars at every opportunity is a desirable result. A first goal and success provides such an opportunity. By demonstrating that the process can succeed, the implementation team is given the opportunity to move forward to bolder implementation steps.

However, taking a first step is arguably the most important component of the implementation process. In the same manner as deciding to undertake the question–answer process at the beginning of the text was the first step toward achieving a strategic management perspective, taking a first action is the first step in achieving strategic management implementation. The hurdle at this point is just taking this first step. However, actions speak very loudly to organization members. Many of these members may be tired of hearing promises by organization leaders that are never realized. The work has been done at this point in the process to support a first action; take it.

■ Closing Thoughts

As closing thoughts to this text on strategic management in civil engineering, the authors put forth both a warning and an encouragement. First, a warning: Strategic management is not a magic cure for organization problems. Rather, strategic management sets the path to address organization problems on a long-term basis rather than a project-by-project basis. The path will not be completed overnight, but it will be put in place to address issues from a new perspective. Second, an encouragement: Undertaking a move to strategic management may be painful for some organizations, but this discomfort should be tempered by the thought that the organization is setting in place a roadmap for the future. In contrast to organizations that ride the waves of the marketplace, the organization that institutes a strategic management perspective will be setting its own direction and path through the changing waters of the market. It is through this independence, aggressiveness, leadership, and vision that organizations will move to the forefront of the civil engineering industry and ensure themselves an opportunity to respond to the constant changes in the global marketplace.

REFERENCES

ACM (1968). *IEEE Design Automation Workshop*, New York: Institute of Electrical and Electronics Engineers.

Anumba, Ch. J., and Duke, A. (1997) "Internet and Intranet Usage in a Communication Infrastructure for Virtual Construction Project Team," *Proceedings in IEEE Journal of Engineering and Applied Science*, 56–61.

Arditi, David, and Davis, Larry (1988), "Marketing of Construction Services," *Journal of Management in Engineering* **4**(4), 297–315.

"Arthur D. Little," Arthur D. Little School of Management, available at http://www.arthurdlittle.com/som/som.html.

Bady, Susan (1995), "Builders Find Benefits in Creating a Brand Name," *Professional Builder* **60,** 30.

Ballard, Glen, and Howell, Greg (1998). "Shielding Production: Essential Step in Production Control," *Journal of Construction Engineering and Management* **124**(1), 11–17.

Barnes, B. Kim (1993), "Intelligent Risk-Taking," *Executive Excellence* **10**(9), 11–12.

Barnevik, Percy (1994). *Global Strategies: Insights from the World's Leading Thinkers*, Boston: Harvard Business School Press.

Bartlett, Christopher A., and Ghoshal, Sumantra (1997). "The Myth of the Generic Manager," *California Management Review* **40**(1), 92–116.

Bernstein, Leopold (1993), *Analysis of Financial Statements,* 4th Ed., Homewood, Illinois: Business One Irwin.

Betts, M., and T. Wood-Harper (1994). "Reengineering Construction: A New Management Research Agenda," *Construction Management and Economics* **12,** 551–556.

Billingsby, J. Lee (1992), "Reaching for the Top: How One Manager Turned a Branch Around," *Managers Magazine* **67**(2), 18–21.

Binder, John D. (1997). "CATIA Assembles Next Generation Aircraft," *Aerospace America* **35**(3), 18–20.

Birkett, W. P. (1995). "Management Accounting and Knowledge Management," *Management Accounting* **77**(5), 44–48.

Boston Consulting Group (1975). *Strategy Alternatives for the British Motorcycle Industry*, London: Her Majesty's Stationery Office.

Boudreaux, Gregory (1984). "On Peter Drucker and Managing in Turbulent Times, Part II," *Management Quarterly* **25**(4), 23–28.

Boynton, A. C., Victor, B., and Pine, B. J., II (1993). "New Competitive Strategies: Challenges to Organizations and Information Technology," *IBM Systems Journal* **32**(1), 40–64.

Bradley, Stephen P., Hausman, Jerry A., and Nolan, Richard L. (1993). *Globalization, Technology, and Competition*, Boston: Harvard Business School Press.

Brisley, Chester L., and Fielder, William F., Jr. (1983). " 'Unmeasurable' Output of Knowledge/Office Workers Can and Must be Measured," *Industrial Engineering* **15**(7), 42–47.

Brown, Steven P., Cron, William L., Slocum, John W., Jr. (1997), "Effects of Goal-Directed Emotions on Salesperson Volitions, Behavior, and Performance: A Longitudinal Study," *Journal of Marketing* **61**(1), 39–50.

Calem, Robert E. (1995), "Black & Veatech Power," *Forbes*, Supplement 62–64.

Carr, Stuart C., Pearson, Sallie-Anne, and Provost, Stephen C. (1996), "Learning to Manage Motivational Gravity: An Application of Group Polarization," *Journal of Social Psychology* **136**(2), 251–54.

Cayes, Kim (1998). "Need to Learn, and Why Engineers May Be Poor Students," *Journal of Management in Engineering* **14**(2), 31–33.

Chinowsky, Paul S. (1997). "Introducing Multimedia Cases into Construction Education," *Proceedings of the Fourth Congress on Computing in Civil Engineering*, T. Adams (ed.), 122–28.

Chinowsky, Paul S. (1998). "A Benchmark Study of World Wide Web Communications," *Proceedings of the Fifth Congress on Computing in Civil Engineering*, Kelvin Wang (ed.), 822–31.

Chinowsky, Paul S., and Guensler, Randall (1998). "Multidisciplinary Civil Engineering Education through Environmental Impact Analysis," *Proceedings of the 1998 ASEE Southeastern Section Conference*, J. P. Mohsen (ed.), 79–84.

Christ, Carl F. (1993). "Assessing Applied Econometric Results," *Federal Reserve Bank of St. Louis Review* **75**(2), 71–94.

Clark, Jim, and Koonce, Richard (1995). "The Secrets of Organizational Success: Aligning Employees behind New Corporate Goals and Objectives," *Training & Development* **49**(8), 28–29.

Cleland, David I., and King, William R. (eds.) (1983). *Project Management Handbook*, New York: Van Nostrand Reinhold.

Clough, Richard H., and Sears, Glenn A. (1991). *Construction Project Management*, New York: John Wiley & Sons, Inc.

Cocco, Anthony F. (1995). "Using Performance Goals to Motivate Workers: A Practical Guide for Project Managers," *Project Management Journal* **26**(2), 53–56.

Collins, James C., and Porras, Jerry I. (1991). "Organizational Vision and Visionary Organizations," *California Management Review* **34**(1), 30–52.

Collins, James C., and Porras, Jerry I. (1996). "Building Your Company's Vision," *Harvard Business Review* **74**(5), 65–77.

Collis, David, and Montgomery, Cynthia (1991). *Corporate Strategy: A Conceptual Framework*, Boston: Harvard Business School Press.

Cookson, Gillian (1997). "Family Firms and Business Networks: Textile Engineering in Yorkshire 1780–1830," *Business History* **39**, 1–20.

Coppula, Deborah (1998). *Continuing Education: Hyperlink to Growth*, American Society for Engineering Education, Washington, D.C.

"Corps of Engineers," U.S. Army Corps of Engineers, available at http://www.usace.army.mil.

Davis, Robert T. (1987). *Strategic Planning Revisited*, Stanford, CA: Graduate School of Business, Stanford University.

Davy, Jo Ann (1998). "Education and Training Alternatives," *Managing Office Technology* **43**(3), 14–15.

Dearden, John (1987). "Measuring Profit Center Managers," *Harvard Business Review* **65**(5).

Donkin, Richard (1997). "Luring a Visionary Leader," *The Financial Times*, November 19, 1997, 32.

Dornbusch, Rudiger, and Fischer, Stanley (1987). *MacroEconomics*, 4th Ed., New York: McGraw-Hill Book Co.

Douglas, Paul H. (1921). *American Apprenticeship and Industrial Education*, New York: Columbia University.

Dreyer, R. S. (1996). "Make Them Want to Be Motivated," *Supervision* **57**(11), 19–20.

Dun & Bradstreet (1996). *Corporate Management Reference Book*, Murray Hill, New Jersey: Dun & Bradstreet.

Drucker, Peter F. (1967). "Effective Decision," *Harvard Business Review* **44**(1).

Drucker, Peter F. (1974). "New Templates for Today's Organizations," *Harvard Business Review* **51**(1).

Drucker, Peter F. (1988). "Coming of the New Organization," *Harvard Business Review* **65**(1).

Drucker, Peter F. (1993). *Post-Capitalist Society*, New York: HarperCollins Publishers.

Dutton, Gail (1997). "What Business Are We in?" *Management Review* **86**(8), 54–57.

Edwards, John S., Yakemovic, K. C., Cowan, Denis P., Gaiser, Ted J., Gancz, John, Levin, Ed, Vezina, Jim, Wynn, Eleanor (1996). "E-Team: Forming a Viable Group on the Internet," *Proceedings of the 1996 ACM SIGCPR/SIGMIS Conference*, Association of Computing Machinery, 161–72.

ENR (1996a). "The Top 400 Contractors," *Engineering News Record*, May 20, 1996, 37–82.

ENR (1996b). "The Top 100," *Engineering News Record*, June 10, 1996, 32–42.

ENR (1996c). "The Top 500 Design Firms," *Engineering News Record*, April 1, 42–76.

ENR (1998). "Big Tests Ahead for Design–Build," *Engineering News Record*, June 15, 47–62.

Evans, Philip B., and Wurster, Thomas S. (1997). "Strategy and the New Economics of Information," *Harvard Business Review* **75**(5), 70–82.

Fairhurst, Gail T., Jordan, Jerry Monroe, and Neuwirth, Kurt (1997). "Why Are We Here? Managing the Meaning of an Organizational Mission Statement," *Journal of Applied Communication Research* **25**(4), 243–63.

Farkas, Charles M., and Wetlaufer, Suzy (1996). "The Ways Chief Executives Lead," *Harvard Business Review* **74**(3), 110–22.

Farr, John V. (1997). "Are Changes Needed in Engineering Education?," *Journal of Management in Engineering* **13**(6), 34–36.

Farrell, Christopher (1993). "A Wellspring of Innovation. (From 18th Century Mills to Worker-Empowered Auto Plants)," *Business Week*, Enterprise Special Issue, 56–62.

Federal Register (1997). *Two-Phase Design–Build Selection Procedures*, Federal Register **61**(1), Subpart 36.3.

"Federal Reserve Board" (1998). *The Beige Book*, Federal Reserve Bank of Chicago, Chicago, IL, available at http://www.bog.frb.fed.us/fomc/BeigeBook.

Fink, Michael (1998). "Resources for Ongoing Professional Development," *Journal of Management in Engineering* **14**(2), 29–30.

Fruchter, Renate (1996). "Conceptual, Collaborative Building Design through Shared Graphics," *IEEE Expert* **11**(3), 33–41.

Fruchter, Renate (1997). "A/E/C Virtual Atelier: Experience and Future Directions," *Proceedings in ASCE Computing in Civil Engineering*, 395–402.

Gallon, Mark R., Stillman, Harold M., and Coates, David (1995). "Putting Core Competency Thinking into Practice," *Research-Technology Management* **38**(3), 20–28.

Garrison, William L., and Souleyrette, Reginald R., II (1996), "Transportation Innovation, and Development: The Companion Innovation Hypothesis," *Logistics & Transportation Review* **32**(1), 5–38.

Gelertner, David (1998). "Fatal Beauty." *Across the Board* **35**(2), 38–43.

"Georgia." Georgia House of Representatives Comparative Budget Summary of HB 204, available at http://www2.state.ga.us/Legis/1997_98/budget/sfy1998/agency.htm.

Goddard, Jules (1997). "The Architecture of Core Competence," *Business Strategy Review* **8**(1), 43–52.

Goodman, Robin E. (1998). *Taxonomy of Knowledge Requirements for Executives of General Contracting and Construction Management Enterprises*, Ph.D. Dissertation, School of Civil and Environmental Engineering, Georgia Institute of Technology, P. Chinowsky, Advisor, September 1998.

Goodman, Robin E., and Chinowsky, Paul S. (1997). "Preparing Construction Professionals for *Executive* Decision Making," *Journal of Management in Engineering* **13**(6), 55–61.

Goodson, Jane R., and McGee, Gail W. (1991). "Enhancing Individual Perceptions of Objectivity in Performance Appraisal," *Journal of Business Research* **22**(4), 293–303.

Goold, Michael (1992). "Design, Learning and Planning: A Further Observation on the Design School Debate," *Strategic Management Journal* **13**, 169–70.

Gould, Frederick E. (1997). *Managing the Construction Process*, Upper Saddle River, New York: Prentice Hall.

Grant, Robert M. (1991). "The Resource-Based Theory of Competitive Advantage," *California Management Review* **33**(3), 114–34.

Grant, R. M. (1996). "Toward a Knowledge-Based Theory of the Firm," *Strategic Management Journal* **17**(Winter), 109–22.

"Green" (1998). *The Green Report—Engineering Education for a Changing World*, American Society for Engineering Educators, New York: ASEE.

"Grinter" (1955). *The Grinter Report*, ASEE Committee on Evaluation of Engineering Education, Washington, D.C.: ASEE.

Halpin, Daniel W., and Woodhead, Ronald W. (1998). *Construction Management,* 2nd Ed., New York: John Wiley & Sons, Inc.

Hamel, Gary, and Prahalad, C. K. (1989). "Strategic Intent," *Harvard Business Review*, **67**(3).

Hammond, J. L., and Hammond, Barbara (1937). *The Rise of Modern Industry*, New York: Harcourt, Brace, and Co.

Hancher, Donn E. (1985). "Construction Education in Civil Engineering," *Proceedings of ASCE Construction Education in Civil Engineering*, George K. Wadlin (ed.), 232–40.

Handy, Charles (1995). "Trust and the Virtual Organization," *Harvard Business Review* **73**(3).

Hanna, Awad S., and Brusoe, Jacqueline K. (1998). "Study of Performance Evaluations in Electrical Construction Industry," *Journal of Management in Engineering* **13**(6), 66–74.

Hansen, Karen, and Tatum, C. Bob (1996). "How Strategies Happen: A Decision-Making Framework," *Journal of Management in Engineering* **12**(1), 40–48.

Harrigan, John E., and Neel, Paul R. (1996), *The Executive Architect*, New York: John Wiley & Sons, Inc.

Hart, B. H. Liddell (1954). *Strategy: The Indirect Approach*, New York: Frederick A. Praeger.

"Harvard" (1997). "75th Anniversary Timeline," *Harvard Business Review*, 75(5).

Hausman, Eric (1997). "Wang Looks to Get the Word Out," *Computer Reseller News* **764,** 51–52.

Hecker, P. A. (1997). "Successful Consulting Engineering: A Lifetime of Learning," *Journal of Management in Engineering* **13**(6), 62–65.

Hekman, John S., and Strong, John S. (1981). "The Evolution of New England Industry," *New England Economic Review*, Mar–Apr, 35–46.

Heller, Laura (1997). "CompUSA Targets Diverse Markets," *Discount Store News* **36**(21), 6.

Hilton, Marvin H. (1995). "The Importance of Civil Engineering Leadership in the Government Sector," *Journal of Management in Engineering* **11**(5), 12–16.

"Intel," "Processor Hall of Fame," available at http://www.intel.com/intel/museum/25anniv/Hof/hof_main.htm.

Jansen, David (1998). "Speak Out: The Engineer as Communicator," *Journal of Management in Engineering* **14**(2), 19–22.

Jeffords, Raymond, Scheidt, Marsha, and Thibadoux, Greg M. (1997). "Getting the Best from Staff," *Journal of Accountancy* **184**(3), 101–5.

Jennings, Alan, and Ferguson, J. D. (1995). "Focusing on Communication Skills in Engineering Education," *Studies in Higher Education* **20**(3), 305–14.

Jin, Yan, and Levitt, Raymond E. (1996). "Approaches to Simulating Organizational Behavior of Concurrent Design Teams," *Computing in Civil Engineering*, 281–87.

Jones, Thomas O., and Sasser, Jr., W. Earl (1995). "Why Satisfied Customers Defect," *Harvard Business Review* **73**(6).

Kacapyr, Elia (1996). *Economic Forecasting—The State of the Art*, Armonk, New York: M. E. Sharpe.

Kangari, Roozbeh, and Lucas, Chester L. (1997). *Managing International Organizations*, New York: ASCE Press.

Kaplan, Robert S., and Norton, David P. (1996). "Using the Balanced Scorecard as a Strategic Management System," *Harvard Business Review* **74**(1), 75–85.

Katzenbach, Jon R. (1997). "The Myth of the Top Management Team," *Harvard Business Review* **75**(6), 82–91.

Katzenbach, Jon R., and Smith, Douglas K. (1993). "Discipline of Teams," *Harvard Business Review* **71**(2).

Kerzner, Harold (1998). *Project Management,* 6th Ed., New York: John Wiley & Sons, Inc.

Khedro, Taha, Case, Michael P., Flemming, Ulrich, Genesereth, Michael R., Logcher, Robert, Pedersen, Curtis, Snyder, James, Sriram, Ram D., and Teicholz, Paul M. (1995). "Development of Multi-Institutional Testbed for Collaborative Facility Engineering Infrastructure," *Proceedings of the ASCE 2nd* Congress on Computing in Civil Engineering, J. P. Mohsen (ed.), 1308–15.

Klein, Howard J., and Kim, Jay S. (1998), "A Field Study of the Influence of Situational Constraints, Leader–Member Exchange, and Goal Commitment on Performance," *Academy of Management Journal* **41**(1), 88–95.

Koehn, Enno (1995). "Practitioner and Student Recommendation for an Engineering Curriculum," *Journal of Engineering Education* **84**(3), 241–48.

Kostem, Celal N., and Shephard, Mark S. (1984). *Computer Aided Design in Civil Engineering,* New York: ASCE Press.

Kotter, John P. (1990). "What Leaders Really Do," *Harvard Business Review* **68**(3), 103–11.

Kotter, John P., and Schlesinger, Leonard A. (1979). "Choosing Strategies for Change," *Harvard Business Review* **57**(2).

Kozin, Marc D., and Young, Kevin C. (1994). "Using Acquisitions to Buy and Hone Core Competencies," *Mergers & Acquisitions* **29**(2), 21–26.

Lammers, Teri (1992). "The Effective and Indispensable Mission Statement," *Inc.* **14**(8), 75–77.

Learned, Edmund P., Christiansen, C. Roland, Andrews, Kenneth R., and Guth, William D. (1969). *Business Policy: Text and Cases,* Homewood, Illinois: Richard D. Irwin.

Leavitt, Harold J., and Lipman-Blumen, Jean (1995). "Hot Groups," *Harvard Business Review* **73**(4), 109–16.

Lemmon, David L., and Early, Stewart (1996). "Strategy & Management at Amoco Pipeline Company," *Planning Review* **24**(1), 12–14.

Levenbach, Hans, and Cleary, James (1984). *The Modern Forecaster*, Belmont, California: Lifetime Learning Publications.

Levitt, Theodore (1975). "Marketing Myopia," *Harvard Business Review* **53**(5), 26–48.

Levitt, Theodore (1983). "Globalization of Markets," *Harvard Business Review* **61**(3).

Liedtka, Jeanne, and Rosenblum, John W. (1996). "Shaping Conversations," *California Management Review* **39**(1), 141–57.

Lih, Marshall (1997). "Educating Future Executives," *Prism*, American Society for Engineering Educators, January, 30–34.

Locke, Edwin A., Latham, Gary P., and Erez, Miriam (1988). "The Determinants of Goal Commitment," *Academy of Management Review* **13**(1), 23–39.

Long, R. P. (1997). "Preparing Engineers for Management," *Journal of Management in Engineering* **13**(6), 50–54.

Lowe, John G. (1991). "Interdisciplinary Postgraduate Education for Construction Managers," *Journal of Professional Issues in Engineering Education and Practice* **117**(2), 168–75.

Magnan, Christopher J. (1997). "How the Vermont Agency of Transportation Uses Visualization for Public Communication," *Proceedings in ASCE Computing in Civil Engineering*, 136–42.

Mandt, Edward (1978). "Managing the Knowledge Worker of the Future," *Personnel Journal*, March, 138–43.

Mansfield, Edwin (1985). *Microeconomics: Theory and Applications*, New York: W. W. Norton & Co.

Markides, Constantinos (1997). "To Diversify or Not to Diversify," *Harvard Business Review* **75**(6), 93–99.

Mauboussin, Michael, and Johnson, Paul (1997). "Competitive Advantage Period: The Neglected Value Driver," *Financial Management* **26**(2), 67–74.

Maxwell, Steve (1998). "A Merger Manual," *Civil Engineering* **68**(2), 73–75.

McCabe, Donald L., and Narayanan, V. K. (1991). "The Life Cycle of the PIMS and BCG Models," *Industrial Marketing Management* **20**, 347–52.

McClenahen, John S. (1997). "Closer to Boom than Bust," *Industry Week* **246**(11), 173–79.

McKinney, Kathleen, and Fischer, Martin (1997). "4D Analysis of Temporary Support," *Proceedings in ASCE Computing in Civil Engineering*, 470–76.

McSulskis, Elaine (1998). "Why Do Employees Leave?" *HR Magazine* **43**(4), 24.

Meeker, J. Edward (1922). *The Work of the Stock Exchange*, New York: The Ronald Press Company.

Miles, Raymond E., and Snow, Charles (1986). "Organizations: New Concepts for New Forms," *Harvard Business Review* **64**(2).

Miller, Roger, and Lessard, Donald R. (1998). *High Risk, Uncertain Returns: The Game of Project Venturing*, Report of the IMEC Research Program, Universite du Quebec, Montreal, Canada.

Mintzberg, Henry (1994). "The Fall and Rise of Strategic Planning," *Harvard Business Review* **72**(1), 107–14.

Mintzberg, Henry (1996a). "The 'Honda Effect' Revisited," *California Management Review* **38**(4), 78–79.

Mintzberg, Henry (1996b). "Musings on Management," *Harvard Business Review* **74**(4), 5–11.

Mitchell, Jan (1995). "Small Builders, Big Ideas. (Innovative Advertising Concepts in the Home Building Industry)," *Builder* **18**, 153–57.

Mitchell, William A. (1931). *Outlines of the World's Military History*, Harrisburg, PA: Military Services Publishing.

Moore, James F. (1993). "Predators and Prey: A New Ecology of Competition," *Harvard Business Review* **71**(3).

Moore, Sarah (1995). "Making Sense of Strategic Management: Towards a Constructive Guide," *Management Decision* **33**(1), 19–23.

Moore, Thomas E. (1997). "The Corporate University: Transforming Management Education," *Accounting Horizons* **11**(1), 77–85.

Mowshowitz, Abbe (1997). "Virtual Organization," *Communications of the ACM* **40**(9), 30–37.

Nadler, David A., and Tushman, Michael L. (1990). "Beyond the Charismatic Leader: Leadership and Organizational Change," *California Management Review* **32**(2), 77–97.

National Academy of Sciences (1995). *Reshaping the Graduate Education of Scientists and Engineers,* Washington, D.C.: National Academy Press.

National Science Foundation (1996). *Shaping the Future: Strategies for Revitalizing Undergraduate Education,* Arlington, VA: Naional Science Foundation.

Nelson, Daniel (1980). *Frederick W. Taylor and the Rise of Scientific Management*, Madison, Wisconsin: The University of Wisconsin Press.

Nickerson, Clarence (1986). *Accounting Handbook for Nonaccountants*, New York: Van Nostrand Reinhold.

Nohria, Nitin, and Eccles, Robert G. (eds.) (1994). *Networks and Organizations: Structure, Form, and Action*, Boston: Harvard Business School Press.

Nonaka, Ikujiro (1991). "The Knowledge-Creating Company," *Harvard Business Review* **69**(6), 96–104.

Nord, Jeretta H., and Nord, G. Daryl (1995). "Why Managers Use Executive Support Systems," *Industrial Management & Data Systems* **95**(9), 24–28.

Novick, David (1990). "Life-Cycle Considerations in Urban Infrastructure Engineering," *Journal of Management in Engineering* **6**(2), 186–96.

Oberlender, Garold D. (1993). *Project Management for Engineering and Construction*, New York: McGraw-Hill.

Oglesby, C. H. (1990). "Dilemmas Facing Construction Education and Research in the 1990s," *Journal of Construction Engineering and Management* **117**(2), 4–17.

Oliva, Terence A., Day, Diana L., and DeSarbo, Wayne S. (1987). "Selecting Competitive Tactics: Try a Strategy Map," *Sloan Management Review* **28**(1), 5–16.

O'Sullivan, Tom (1997). "Chainstore Massacre," *Marketing Week* **20**(18), 30–31.

"Outlook" (1997). "Outlook Is Mixed for Orange County," *The New York Times*, December 27, 1997.

Panitz, Beth (1998). *The Integrated Curricula*, American Society for Engineering Educators.

Pascale, Richard T. (1996). "Perspectives on Strategy: The Real Story Behind Honda's Success," *California Management Review* **38**(3).

Paulson, Boyd C. J. (1995). *Computer Applications in Construction*, New York: McGraw-Hill.

Pawar, Kulwant S., and Sharifi, Sudi (1997). "Physical or Virtual Team Colocation: Does It Matter?" *International Journal of Production Economics* **52**(3), 283–90.

Peak, Martha H. (1997). "Go Corporate U!" *Management Review* **86**(2), 33–37.

Pearce, John A., II (1982). "The Company Mission as a Strategic Tool," *Sloan Management Review* **23**(3), 15–24.

Peurifoy, Robert L., and Oberlender, Garold D. (1989). *Estimating Construction Costs,* 4th Ed., New York: McGraw-Hill.

Pfeffer, Jeffrey, Hatano, Toru, and Santalainen, Timo (1995). "Producing Sustainable Competitive Advantage through the Effective Management of People," *Academy of Management Executive* **9**(1), 55–72.

Pfeiffer, J. William (ed.) (1986). *Strategic Planning—Selected Readings*, San Diego: University Associates.

Porbahaie, M. Ali (1994). "Engineers and Managerial Challenges: A Transitional Perspective," *Civil Engineering* **64**(6), 73–74.

Porter, Michael E. (1975). *Note on the Structural Analysis of Industries*, Boston: Harvard Business School Press.

Porter, Michael E. (1979). "How Competitive Forces Shape Strategy," *Harvard Business Review* **57**(2), 137–45.

Porter, Michael E. (1980). *Competitive Strategy*, New York: Free Press.

Porter, Michael E. (1987). "From Competitive Advantage to Corporate Strategy," *Harvard Business Review,* **65**(4).

Porter, Michael (1996). "What Is Strategy?" *Harvard Business Review* **74**(6).

Pospisil, Vivian (1997). "When Good Companies Fail," *Industry Week* **246**(14), 31.

Prahalad, C. K., and Hamel, Gary (1990). "The Core Competence of the Corporation," *Harvard Business Review* **68**(3), 79–91.

Price, Robert M. (1996). "Executive Forum: Technology and Strategic Advantage," *California Management Review* **38**(3), 38–56.

Prokesch, Steven E. (1997). "Unleashing the Power of Learning: An Interview with British Petroleum's John Browne," *Harvard Business Review* **75**(5).

Quinn, James B. (1980). "Managing Strategic Change," *Sloan Management Review* **21**(4), 3–17.

Rao, Asha, Schmidt, Stuart M., and Murray, Lynda H. (1995). "Upward Impression Management: Goals, Influence, Strategies, and Consequences," *Human Relations* **48**(2), 147–67,

Reichers, Arnon E. (1986). "Conflict and Organizational Commitments," *Journal of Applied Psychology* **71**(3), 508–14.

Reisman, Arnold (1994). "Technology Management," *IEEE Transactions on Engineering Management* **41,** 342–46.

Rigby, Rhymer (1998). "Mission Statements," *Management Today*, March 1998, 56–58.

Riggs, Joy (1995). "Empowering Workers by Setting Goals," *Nation's Business* **83**(1), 6.

Ritner, Joseph A., and Taylor, Lowell J. (1997), "Economic Models of Employee Motivation," *Federal Reserve Bank of St. Louis Review* **79**(5), 3–21.

Robar, Tracy Y. (1998). "Communication and Career Advancement," *Journal of Management in Engineering* **14**(2), 26–28.

Roesener, Larry A., and Walesh, Stuart G. (1998). "Corporate University: Consulting Firm Case Study," *Journal of Management in Engineering* **14**(2), 56–63.

Rogers, Kathy A. (1993). "Go for the Goals," *Incentive* **167**(12), 67–70.

Rosener, Judy B. (1990). "Ways Women Lead," *Harvard Business Review* **68**(6), 119–25.

Ross, Steven S. (1997a). "A/E/C Systems Nets' Surprises," *Architectural Record*, August, 139–42.

Ross, Steven S. (1997b). "Collaboration by Wire," *Architectural Record*, September, 131–37.

Russell, Jeffrey S., Pfatteicher, Sarah K. A., and Meier, John R. (1997). "Improved Management of the Engineering Educational Process," *Journal of Management in Engineering* **13**(6), 37–41.

Safley, Thomas M., and Rosenband, Leonard N. (1993). *The Workplace before the Factory*, Ithaca, NY: Cornell University Press.

Schoemaker, Paul J. (1992). "How to Link Strategic Vision to Core Capabilities," *Sloan Management Review* **34**(1), 67–82.

Schoemaker, Paul J. (1995). "Scenario Planning: A Tool for Strategic Thinking," *Sloan Management Review* **36**(2), 25–40.

Schwartz, Susana (1997). "Laying the Groundwork for Mobile Computing," *Insurance & Technology* **22,** 38–40.

Selznick, Philip (1957). *Leadership in Administration*, New York: Harper Row.

Senge, Peter M. (1990). "The Leader's New Work: Building Learning Organizations," *Sloan Management Review* **12**(1), 7–24.

Shapiro, Benson P. (1988). "Close Encounters of the Four Kinds: Managing Customers in a Rapidly Changing Environment," *Harvard Business Review* **66**(4).

Sillars, David N. (1998). *Pre-Operational Attributes as Predictors of Organizational Success within a Joint Venture*, Ph.D. Dissertation, School of Civil and Environmental Engineering, Georgia Institute of Technology, R. Kangari, advisor, May, 1998.

Simons, Robert, and Davila, Antonio (1998). "How High Is Your Return on Management?" *Harvard Business Review* **76**(1), 71–80.

Singh, Amarjit (1992). "Experience-Based Issues in Construction Education," *Journal of Professional Issues in Engineering Education and Practice* **118**(4), 388–402.

Skaggs, Neil, and Carlson, J. Lon (1996). *MicroEconomics: Individual Choice and Its Consequence,* 2nd Ed., Cambridge, Massachusetts: Blackwell Publishers.

Smail, John (1997). "Demand Has Shape: Exports, Entrepreneurs, and the Eighteenth Century Economy," *Business & Economic History* **26**(2), 354–64.

Snyder, Amy V., and Ebeling, H. William, Jr. (1992). "Targeting a Company's Real Core Competencies," *Journal of Business Strategy* **13**(6), 26–32.

Sobek, Robert S. (1973). "A Manager's Primer on Forecasting," *Harvard Business Review* **51**(3), 6–28.

Sobel, Robert (1965). *The Big Board—A History of the New York Stock Market*, New York: The Free Press.

Souder, H. Ray (1983). "Implementing a Decision Support System," *Proceedings of the 1983 ACM Computer Science Conference and SIGCSE Symposium*, Orlando, FL, Association of Computing Machinery.

Spindel, Paul D. (1971). *Computer Applications in Civil Engineering*, New York: Van Nostrand Reinhold.

Stalk, George, Evans, Philip, and Shulman, Lawrence E. (1992). "Competing on Capabilities: The New Rules of Corporate Strategy," *Harvard Business Review* **70**(2), 57–69.

Stanton, William J. (1984). *Fundamentals of Marketing,* 7th Ed., New York: McGraw-Hill.

Stata, Ray (1989). "Organizational Learning—The Key to Management Innovation," *Sloan Management Review* **10**(3), 63–74.

Stewart, Thomas A. (1990). "How to Manage in the New Era," *Fortune* **121,** 58–62.

Stiltner, Kenneth R., and Barton, David R. (1990). *Econometric Models and Construction Forecasting*, U.S. Department of Commerce.

"Strategic" (1996). "Strategic Planning," *Business Week*, August 26, 1996.

Sunoo, Brenda Paik (1996). "Weighing the Merits of Vision and Mission Statements," *Personnel Journal* **75**(4), 158.

Sunoo, Brenda P. (1998). "Corporate Universities—More and Better," *Workforce* **77**(5), 16–17.

Sviokla, John J. (1996). "Knowledge Workers and Radically New Technology," *Sloan Management Review* **37**(4), 25–40.

Szakonyi, Robert (1990). "How do you Measure R&D Productivity?," *R&D* **32**(7), 83–84.

Szostak, Rick (1989). "The Organization of Work: The Emergence of the Factory Revisited," *Journal of Economic Behavior & Organization* **11**(3), 343–58.

Szostak, Rick (1996). "Economic Impacts of Road and Waterway Improvements," *Transportation Quarterly* **50**(4), 127–41.

Tannenbaum, Robert, and Schmidt, Warren H. (1973). "How to Choose a Leadership Pattern," *Harvard Business Review* **51**(3), 3–11.

Taylor, Alex, III (1998). "Consultants Have a Big People Problem," *Fortune* **137**(7), 162–66.

Taylor, Bernard (1995). "The New Strategic Leadership—Driving Change, Getting Results," *Long Range Planning* **28**(5), 71–81.

Teicholz, Eric (1985). "The Current State of Design Automation," *Computer Graphics World*, February, 81–83.

Tersine, Richard J., and Hummingbird, Edward A. (1995). "Lead-Time Reduction: The Search for Competitive Advantage," *International Journal of Operations & Production Management* **15**(2), 8–18.

Thompson, Patti, and Brooks, Kathy (1997). "A Creative Approach to Strategic Planning," *CMA Magazine* **71**(6), 20–22.

Tubbs, Mark E. (1993). "Commitment as a Moderator of the Goal–Performance Relation: A Case for Clearer Construct Definition," *Journal of Applied Psychology* **78**(1), 86–97.

Tzu, Sun (1963). *The Art of War*, as translated by Samuel B. Griffith, London: Oxford University Press.

Unyimadu, Stephenson O. (1989). "Management and Industrial Revolution in Europe, United States of America and Japan," *Engineering Management International* **5**(3), 209–18.

U.S. Census Bureau (1992). *1992 Geographic Area Series*, U.S. Department of Commerce, Bureau of the Census, SC(92)-A-1-52, Washington, D.C.

U.S. Census Bureau (1996). *Service Annual Survey: 1996*, U.S. Department of Commerce, Bureau of the Census, Washington, D.C.

U.S. Census Bureau (1997). *Statistical Abstract of the United States: 1996*, U.S. Department of Commerce, Bureau of the Census, Washington, D.C.

Volti, Rudi (1996). "A Century of Automobility," *Technology and Culture* **37**(4).

Wadlin, George K. (Ed.) (1985). *Challenges to Civil Engineering Educators and Practitioners— Where Should We Be Going?*, New York: ASCE

Warren, Carl, Fess, Philip, and Reeve, James (1996). *Accounting,* 18th Ed., Cincinnati: South-Western College Publishing.

Watson, Bibi S. (1995). "The New Training Edge," *Management Review* **84**(5), 49–51.

Webber, Alan M. (1993). "What's So New about the New Economy?" *Harvard Business Review* **71**(1), 24–42.

Weise, Elizabeth (1998). "Net Use Doubling Every 100 Days," *USA Today,* April 16, 1998, 1A.

Williamson, Mickey (1993). "Models in the Making," *CIO* **6**(18), 48–53.

Wilson, John L., and Shi, Chenggang (1996). "Computational Support for Distributed and Concurrent Design Team," *Computing in Civil Engineering*, 544–50.

Winter, Drew (1996). "The Mass-Production Revolution," *Ward's Auto World* **32**(5), 101–2.

Wysocki, Robert K., and DeMichiell, Robert L. (1997). *Managing Information across the Enterprise*, New York: J. Wiley.

Yates, Janet K. (1994). "Construction Competition and Competitive Strategies," *Journal of Management in Engineering* **11**(1), 58–69.

Zipf, Robert (1995). *How Municipal Bonds Work*, New York, New York Institute of Finance.

INDEX